Understanding Scientific Reasoning

Third Edition

Ronald N. Giere
Minnesota Center for Philosophy of Science
University of Minnesota—Twin Cities

Holt, Rinehart and Winston, Inc.

Fort Worth Chicago San Francisco Philadelphia Montreal Toronto London Sydney Tokyo

Publisher Ted Buchholz
Acquisitions Editor Jo-Anne Weaver
Project Editor Steve Welch
Production Manager Kathy Ferguson
Art & Design Supervisor Vicki McAlindon Horton
Text Designer Bill Maize, Duo Design Group
Cover Designer Vicki McAlindon Horton/Bill Maize, Duo Design Group

Library of Congress Cataloging-in-Publication Data

Giere, Ronald N.
Understanding scientific reasoning / Ronald N. Giere. — 3rd ed.
p. cm.
Includes bibliographical references and index.
1. Science—Philosophy. I. Title.
Q175.G49 1991
501—dc20 90-49877
CIP

ISBN: 0-03-026419-7

Address for Editorial Correspondence: Holt, Rinehart and Winston, Inc., 301 Commerce Street, Suite 3700, Fort Worth, TX 76102.

Address for Orders: Holt, Rinehart and Winston, Inc., 6277 Sea Harbor Drive, Orlando, FL 32887. 1-800-782-4479, or 1-800-433-0001 (in Florida).

Printed in the United States of America

1 2 3 4 090 9 8 7 6 5 4 3 2 1

Holt, Rinehart and Winston, Inc.
The Dryden Press
Saunders College Publishing

For permission to use copyrighted materials the author is grateful to the following:

Chapter 2: Section 2.10, "Gene Analysis Upsets Turtle Theory," *New York Times,* March 14, 1989. © 1989 by The New York Times Company. Reprinted by permission; "New View of the Mind Gives Unconscious An Expanded Role," *New York Times,* February 7, 1984. © 1984 by The New York Times Company. Reprinted by permission; "Was That a Greenhouse Effect? It Depends on Your Theory," *New York Times,* September 4, 1988. © 1988 by The New York Times Company. Reprinted by permission; Exercise 2.1, "Einstein's Impossible Ring: Found," by M. Mitchell Waldrop, *Science,* June 24, 1988, Vol. 240, p. 1733. © AAAS. Reprinted by permission; Exercise 2.2, "Why is the World Full of Large Females," by Roger Lewin, *Science,* May 13, 1988, Vol. 240, p. 884. © AAAS. Reprinted by permission; Exercise 2.4, "Green Antarctica," by Bill Lawren, *Omni Magazine,* August 1987, p. 30; Exercise 2.5, "Hot Extinction Theories," *Sky and Telescope,* July 1988. Reprinted by permission of Sky and Telescope magazine. Exercise 2.6, "A Heresy in Evolutionary Biology," by Roger Lewin,

(continued on page 314)

Preface

Understanding Scientific Reasoning was originally motivated by a desire to make some aspects of the philosophy of science relevant to the needs of all students. It now also serves the widely recognized goals of improving critical thinking skills and contributing to general scientific literacy. It is explicitly directed toward first- and second-year college students who have not yet chosen a major area of study. Its specific purpose is to help these beginning students acquire cognitive skills in *understanding* and *evaluating* scientific material as found in college textbooks and in a wide variety of both popular and professional printed sources.

The text has three parts. Part One (chapters 2, 3, and 4) develops techniques for understanding and evaluating *theoretical* hypotheses of the sort typically found in the physical sciences as well as in the more theoretical parts of biology and the cognitive sciences. Part Two (chapters 5 through 8) develops techniques for understanding and evaluating *statistical* and *causal* hypotheses typically found in the social and biomedical sciences. Part Three (chapters 9 and 10) explains how scientific knowledge may be combined with individual or social values to reach personal or public policy *decisions.*

Part One, "Theoretical Hypotheses," begins with a case study of the discovery of the structure of DNA. In my own courses, this discussion is supplemented by having everyone read Watson's *The Double Helix* and by showing the film, *The Race for the Double Helix.* This provides all students in the course with a common point of reference that appeals to almost everyone.

Chapter 2, "Understanding and Evaluating Theoretical Hypotheses," analyzes this case in the course of developing a basic format for all the evaluations that follow. The analysis begins with a discussion of ordinary maps as models of spatial relationships among things like streets, buildings, rivers, and mountains. This introduces the crucial distinction between a model and the things modeled—a distinction not often made in everyday life, but absolutely crucial for understanding science. *Understanding* a bit of science, then, is presented as being mainly a matter of understanding the relevant models.

Evaluating scientific hypotheses is presented as a process of deciding whether or not given data provide evidence for regarding a particular model as a tolerably good representation of some real world objects or processes. Understanding the decision process requires distinguishing between *data* resulting from a causal interaction with the world (observation or experimentation) and a *prediction* arrived at by reasoning about a proposed model. It is the agreement or disagreement between data and predictions that provides the basis for a decision as to how well a proposed model "fits" the real world.

Chapter 2 concludes with a simple six-point program for evaluating theoretical hypotheses. There is also a corresponding flow chart. The program is easy to apply. Most students can internalize the program to the extent that they can analyze short, unedited articles with no prompting questions. The instructions to my examinations simply say: "Analyze these reports following the program we have developed in class." So the students are learning a useful skill that will stay with them for a long while.

Chapter 3, "Historical Episodes," collects improved versions of the historical cases discussed in previous editions: Newton's mechanics and Halley's comet, Lavoisier's oxygen theory, Mendel's genetics, and the 1960s revolution in geology. These are preceded by a new example: Galileo's telescopic observations of the phases of Venus. It is expected that instructors will pick and choose among these cases, perhaps even substituting some of their own favorites.

Chapter 4 focuses on "marginal sciences." Rather than emphasizing simple fallacies, it shows how marginal claims, like those of astrology, fare poorly when subjected to the same program for evaluation developed for understanding and evaluating more standard scientific cases. Moreover, it is made clear why they fare poorly: the data do not provide a sufficient basis for a clear decision between the proposed models and other plausible alternatives. This chapter is important because it brings home, in a dramatic fashion, the distinctive features of modern, experimental, science.

Part Two, "Statistical and Causal Hypotheses," begins by introducing statistical and probabilistic models as special cases of theoretical models. Chapter 5, "Statistical Models and Probability" is restricted to simple statistical models, including models for statistical correlation. Causal models are introduced later.

Chapter 6, "Evaluating Statistical Hypotheses" presents judgments of support for statistical hypotheses as following a similar pattern to that used for theoretical hypotheses, but complicated by the need to deal explicitly with probability. There is a corresponding six-step program for evaluating statistical hypotheses. The emphasis is on survey sampling (as used, for example, in public opinion polls) and the estimation of statistical parameters. Overall the presentation is much simpler, and more useful for non-specialists, than that of previous editions.

In general I have tried to provide techniques for evaluation that for the most part can be used without resorting to written calculations. Students should be able to apply the programs as they read reports of scientific findings, doing needed calculations in their heads. My experience has been that what people cannot immediately do in their heads, they are unlikely to do at all. It is only for homework exercises and examinations that things must be written down.

Chapter 7, "Causal Models," develops a model for simple *causal* relationships. The model is presented not as an "analysis" or "definition" of causality, but merely as the least complex model in a whole family of related causal models. Most examples, however, are treated in terms of the simpler model. The difference between causation and correlation is emphasized throughout.

Chapter 8, "Evaluating Causal Models," takes up the three experimental designs presented in previous editions. There is an appropriate program for evaluating causal hypotheses. Apart from a new example and simplifications necessary to bring this chapter into line with previous chapters, there are few other changes in what many have regarded as the most successful chapter in the whole book.

Part Three, "Knowledge, Values, and Decisions," presents some standard models of decision making that can be used to understand how scientific knowledge may be brought to bear on both personal and public policy decisions. It provides some simple tools for evaluating such decisions.

This edition contains many new exercises, most of which are reproduced with only minor editing from actual sources. These will be found mainly in Chapters 2, 6, and 8. The text itself contains no answers to the exercises, although it does contain many examples worked out in detail. Instructors can, therefore, use some of the exercises for homework and others for examinations. It is our plan to issue an instructor's manual that will contain answers to all exercises.

As originally conceived, and refined in the second edition, the underlying framework of *Understanding Scientific Reasoning* was that of justification by logical argument. This has been a standard framework for logic texts, both formal and informal, for many years. My original strategy was to introduce some simple deductive forms, particularly *modus tollens,* and then to develop inductive adaptations for scientific contexts.

I had reasons for dissatisfaction with this approach long before I conceived a workable alternative. From the students' side, representing scientific reasoning in terms of simple argument forms seemed to me not to contribute much to their understanding of the scientific context. Reconstructing the reasoning as an explicit argument was often a burdensome, and frequently mechanical, exercise. Faculty, particularly those with some expertise in the area of probability and induction, were not always happy with my simple inductive forms, no matter how pedagogically expedient they were, compared with available alternatives. But being dissatisfied is one thing; having a comprehensive, coherent alternative is quite another.

This third edition of *Understanding Scientific Reasoning* develops one such alternative. To that end it has been substantially rewritten, with the old introductory chapters on deductive argument forms completely eliminated. This has made possible considerable simplification and streamlining of the whole text.

The earlier editions of this text have been used by a number of people outside of philosophy, in general science programs, for example, or in programs in science, technology, and society. The present edition is even more congenial to such uses. Its overall approach is much closer to the way scientists themselves think about science.

The majority of users, however, have been philosophers. Among philosophers, philosophers of science who emphasize historical development over logical form will find the changes quite congenial. So will those philosophers who have been impressed by recent developments in the cognitive sciences. Those interested in critical thinking should also be quite comfortable with this approach. The book's emphasis on useful critical techniques fits in well with the ethos of the critical thinking movement.

Those who are apt to be most suspicious of the changes are philosophers who have always worked in the framework of justification by arguments, whether formal or informal. I understand full well that others need not share my dissatisfaction with traditional logical reconstructions of scientific reasoning. I would ask those instructors who have simply assumed a parallel with deductive reasoning at least to examine the new approach and to consider whether the gains do not outweigh the losses. The fact is that, although this new edition does not employ the traditional framework, neither does it preclude it. Indeed, the new approach provides greater freedom for instructors to provide their own preferred logical reconstructions of the reasoning if they so desire. Given the divergence of opinion as to the best reconstruction, this freedom may be very desirable.

Ronald N. Giere

Minneapolis, Minnesota
October, 1990

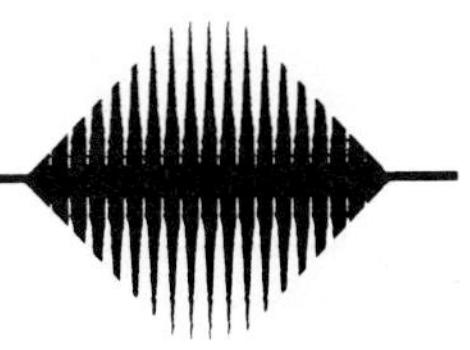

Acknowledgements

Many people have contributed to the development of this text over the past twenty years. Here I will acknowledge only those who have contributed to the preparation of this edition.

First of all, I wish to thank those who reviewed the manuscript for the press. These included Steve Carey, Portland Community College; George Gale, University of Missouri-Kansas City; Philip Gasper, Cornell University; Ronald Hough, Wright State University; Joe Pitt, Virginia Polytechnic Institute; Merrilee Salmon, University of Pittsburgh; Yaakov Schechter, Lehman College of the City of New York; Eric Stietzel, Foothill College; and Richard Van Iten, Iowa State University. I am particularly indebted to Deborah Mayo of Virginia Polytechnic Institute and State University for her advice regarding my treatment of interval estimation and statistical significance.

This edition is several years late in appearing. Mainly that is because of two moves. About the time this edition should have been in preparation, Holt, Rinehart and Winston moved its offices from New York to Fort Worth, and I moved to the University of Minnesota. Since these moves were completed, I have enjoyed having the attention of a continuing editor, Jo-Anne Weaver. She has provided the sort of institutional support that this project needed. The final production process was overseen by Steve Welch, who worked long and hard to meet a very tight schedule.

My student, Gerry Smerchanski, has for several years assisted with my own course in scientific reasoning, which has been taught along the lines of the new edition. In that capacity he has provided much useful feedback from students. He also helped in locating and preparing exercises. Finally, Gerry read over the text as it was being prepared and suggested many useful changes.

My main debt is to the principal secretary of The Minnesota Center for Philosophy of Science, Steve Lelchuk. Steve prepared the original renderings of all but a few of the diagrams, a taxing job proving that even playing with a Macintosh can be hard work. He also handled all the correspondence regarding the numerous required permissions and assisted with proofreading. He even produced a complete preliminary version of this edition for trial use in the classroom.

Finally, I must thank my wife, Barbara Hanawalt, not, as is customary, for her "understanding and encouragement," but simply for her forbearance. I hereby publicly promise never again to allow any such project to so intrude upon our life together.

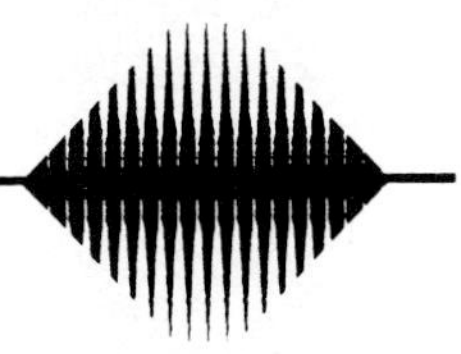

CONTENTS

PART TWO • STATISTICAL AND CAUSAL HYPOTHESES

Chapter 5

CHAPTER 1

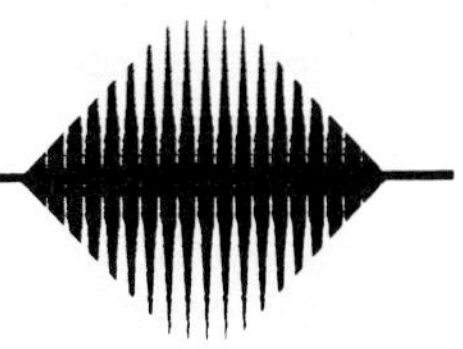

Why Understand Scientific Reasoning?

Before undertaking any course of study, one should pause to ask, "Why am I doing this?" "What benefits can I expect to receive for my efforts?" This chapter will provide you with some answers to these questions.

We will begin by examining some general reasons why it is important for anyone to understand scientific reasoning. These general reasons will then be illustrated with several examples. The examples will reappear in greater detail in the body of the text.

Having considered some reasons why it is important to understand scientific reasoning, we will examine the general strategy of the text. The chapter concludes with some tips as to how you can most effectively master this subject.

1.1 WHY STUDY SCIENTIFIC REASONING?

We all live in a world that is increasingly influenced by developments in science and technology. Even within the lifetime of most readers of this text, whole new fields of science and whole new technologies have come of age. Think, for example, of molecular biology and bio-engineering, space technology, and computers. At the middle of this century, shortly after World War II, there were no jet airliners, and most of the drugs now used in the treatment of disease had not yet been invented. As recently as 1925, the automobile had just become a prominent part of modern life, and much of what is now taught in departments of physics, chemistry, biology, and geology was unknown. Even the Gallup poll did not exist.

Of course, many things have become more and more influential during the past half century. Professional sports, television programs, and recorded music are prominent examples. But most things, such as spectator sports, you can take or leave as you wish. With science and technology, you do not have that luxury. The impact of scientific and technological developments on our day-to-day lives is so great that no one can afford to be ignorant of these developments.

In one respect, there is little danger of anyone having to remain uninformed about new trends in science or technology. The amount of publicly available information about science and technology is overwhelming. Recent doings in science and technology are regularly reported on television news programs. Most major newspapers now carry at least a weekly science section. And there are several news magazines devoted exclusively to science and technology. If, as has been proclaimed, we are in the "information age" (like the "iron age" or the "automobile age"), much of the available information concerns science or technology. But therein lies a difficulty.

To take advantage of any sort of information, you need to know something about the activity involved. Baseball statistics mean little to someone who does not understand the game of baseball. Information about new recordings of classical music means little to someone who is unfamiliar with classical composers or modern performers. The same is true of science. Assimilating scientific information requires some conception of what science is all about and some special skills in evaluating the information one receives.

Here, then, is a general reason why anyone should develop some understanding of scientific reasoning. This skill is necessary if you are to take full advantage of the scientific information that is increasingly important for functioning effectively in both your professional and your personal life. Correspondingly, the aim of this text is to provide you with the required understanding of the scientific process and the necessary reasoning skills.

Scientific reasoning is best taught by carefully mixing general principles with concrete examples. Thus, having given you a general reason for studying this subject, I will now present a few examples.

1.2 SOME PRELIMINARY EXAMPLES

Scientific subjects lie on a spectrum from those of primarily intellectual interest to those with immediate practical implications. Thus, research on the structure of the universe or the evolution of mammals may have profound implications for how you *think* about the world in general. But it will probably not have much impact on anything you do tomorrow or next year. On the other hand, research into the causes of cancer or heart disease may not do much to change your view of the universe, but it may change your choice of what to have for breakfast tomorrow morning. The following examples range along the spectrum from intellectual to practical.

Expansion of the Universe

One of the most interesting scientific findings of the twentieth century was the discovery that the universe is expanding; that is, the galaxies that make up the universe (of which our Milky Way is only one) are all moving away from each other. Most astronomers now think that the present phase of expansion began with all the matter in the universe exploding in one "big bang." The question remains, however, whether the present expansion will continue forever, or whether it will eventually stop, with all the galaxies then coming back together for another "big bang."

Scientific results now often appear in the popular press even before they are published in professional scientific journals. Thus it is not surprising to find a newspaper article with the headline "Scientists Expect New Clue to Origin of Universe." According to this article, whether the universe will continue to expand or undergo a never-ending series of expansions and contractions depends on whether the number of atoms in the universe is greater or less than one atom for every 88 gallons of space. If it is greater, gravitational attraction will be strong enough to pull everything back together again. If it is less, gravitational attraction will not be strong enough to pull things back together, and the expansion will go on forever. The reason this article appeared when it did was that some scientists have claimed that new measurements of the strength of extragalactic x-rays by earth-orbiting satellites may soon give us the answer.

Now, this type of scientific discovery has few immediate, practical implications for anyone. By the time the next contraction would occur, our sun would have long since died or exploded. But such findings may excite your curiosity. They may even influence your views about the universe as a whole and about your place in it. Moreover, because these findings are being written up for the general public, one ought to be able to understand the reported results without being an astrophysicist.

Can you identify the theories in question? Can you distinguish the theories from the facts? Do you know the difference between a theory and a fact? Can you tell which facts are relevant to which theories? Which results of the proposed experiments would support which theory? When you have finished Part One of this text, "Theoretical Hypotheses," you should be able to answer such questions. Moreover, you should be able to understand and evaluate similar scientific episodes in a variety of sciences all the way from astronomy to zoology.

Global Warming

As the decade of the 1980s came to an end, there was much talk about global warming and the "greenhouse effect." Indeed, the four hottest years of the twentieth century occurred during that decade. Prominent scientists argued that the climate of the earth is indeed warming and, moreover, that the cause of the warming is primarily the burning of hydrocarbons, such as oil and coal, by humans. The increased density of carbon dioxide and other "greenhouse gasses" in the atmosphere, they argue, has caused the warming trend we have experienced.

Other scientists disagree. They argue that our recent experience of abnormally warm weather may well just be a typical fluctuation in the earth's overall climate. So there may not even be a genuine warming trend at all. Moreover, there are other mechanisms involving ocean currents and atmospheric winds that could produce the warm years we have experienced recently.

Meanwhile, agencies of the United Nations and other international organizations have publicly urged the major industrial nations, which are responsible for most of the greenhouse gasses, to take stringent measures to reduce the production of these gasses. But many governments, including that of the United States, have argued that such measures would be very costly. They cite disagreements within the scientific community over the existence of global warming and its supposed causes as evidence that more study is needed. Meanwhile, not much has been done to reduce hydrocarbon emissions.

Does the recent unusually warm weather provide evidence of global warming? Does it provide evidence of a greenhouse effect? If not, why not? Is the lack of action by major industrial nations understandable in light of the existing scientific controversy? These are questions that can be answered by a careful reader who has no technical training in atmospheric science or other relevant disciplines. One need only know what to look for.

Cigarette Smoking and Coronary Heart Disease

It has been roughly 30 years since the surgeon general of the United States issued an official warning that cigarette smoking can be dangerous to your health. Since that time, cigarette advertising has been banned from television and warning labels have been placed on cigarette packs and printed advertisements. Airlines have banned smoking on all domestic flights. The surgeon general issues yearly reports focusing on such topics as smoking among teenagers and the effects on newborn babies of smoking by pregnant women. Yet the Tobacco Institute, the research arm of the tobacco industry, maintains that the connection between smoking and coronary heart disease, or any other disease, is "merely statistical." No causal connection, they say, has been proved. Moreover, millions of people continue to smoke, and the federal government itself continues to subsidize both the growing and the exporting of tobacco.

If you are not a statistician or a medical researcher, what are you to conclude? Whom are you to believe? Is it true that all the data are merely statistical? What is the difference between a statistical correlation and a genuine causal connection? Could any statistical data "prove" the existence of a causal connection? If not, why not? What kind of evidence could establish the existence of a causal connection? You should be able to answer these questions after studying Part Two, "Statistical and Causal Hypotheses."

Suppose the surgeon general is correct: Smoking does cause lung cancer. Does it follow that you should not smoke or that you should give it up if you already do? How do you determine the risks? Can you balance off the risks against the pleasure of smoking or the discomfort of trying to give it up? Are the concerns of public health officials necessarily relevant to individuals who may be concerned only about their own chances of contracting smoking-related diseases? Strategies for answering these questions are developed in Part Three, "Knowledge, Values, and Decisions."

Summary

The primary answer to the question, "why study scientific reasoning?" is that it will help you to be better able to understand and evaluate scientific information in both your personal life and your work. In an increasingly scientific and technological society, these abilities can literally be a matter of life and death for you, your family, and, if you should achieve a position of power in business or government, for many others as well.

I have stressed the practical value of studying scientific reasoning because this is most readily appreciated by most people. However, other motives are also important. One is simply the ability to understand and appreciate scientific findings as they are reported in the popular media. This can be a valuable ability even if the findings have no practical implications. Science is an increasingly important part of modern culture. Having some ability to understand and evaluate the latest findings makes you a more literate and cultured person. So does merely being curious about the world around you.

Finally, learning something about scientific reasoning provides some insight not only into particular scientific findings but also into the general nature of science as a human activity. It is an activity that engages an increasing proportion of our population, requires an increasing fraction of our resources, and impinges on an increasing number of our other activities. The better we all understand it, the better off we all will be.

1.3 HOW TO STUDY SCIENTIFIC REASONING

I hope you are now convinced that studying scientific reasoning is a worthwhile undertaking. But you may well be wondering how this is possible without having to study a great deal of science. In answer to this question, I will propose both a general strategy and some particular tactics for learning how to reason scientifically.

How Shall We Understand Scientific Reasoning?

Most people think of scientific reasoning as the kind of reasoning that scientists use in the process of making new scientific discoveries. The paradigmatic examples of scientific reasoning would then be the reasoning that led to the great discoveries in science: Newton's discovery of the law of gravitation, Darwin's discovery of natural selection, Mendel's discovery of the laws of inheritance, Einstein's discovery of relativity theory, the discovery of the DNA double helix by Watson and Crick, and so on.

If this is what one means by scientific reasoning, it is difficult to see how anyone could learn how to do it without a great deal of training in science and perhaps a touch of genius as well. Those of us who are not specialists in a scientific subject could not hope to reason scientifically about that subject. Even specialists in one area would find it difficult to reason scientifically about subjects other than those in their own specialty. Our understanding of what constitutes scientific reasoning must be somewhat different.

Figure 1.1 pictures some relationships between the actual practice of science by scientists in their laboratories and the rest of us. Science gets done in laboratories. The results are communicated to other scientists informally and then published in technical scientific journals, such as the *Physical Review* or *Science.* Some of these results are picked up by more popular journals, such as *Scientific American* or *Discovery.* Some are written up in news magazines, such as *Time* or *Newsweek,* and some appear in local or national newspapers, such as *The New York Times* and *USA Today.* Some appear on television newscasts or special programs, such as "Nova." Some results eventually find their way into science textbooks for high school and college students.

For the purposes of this text, then, learning to understand scientific reasoning is a matter of learning how to understand and evaluate reports of scientific findings of the type one would find in a popular magazine, a national newspaper, or a news magazine. This requires only a very general idea of what goes on in laboratories. And it does not require the skills that are necessary to do laboratory research.

Figure 1.1

How information gets from scientific laboratories to the average citizen.

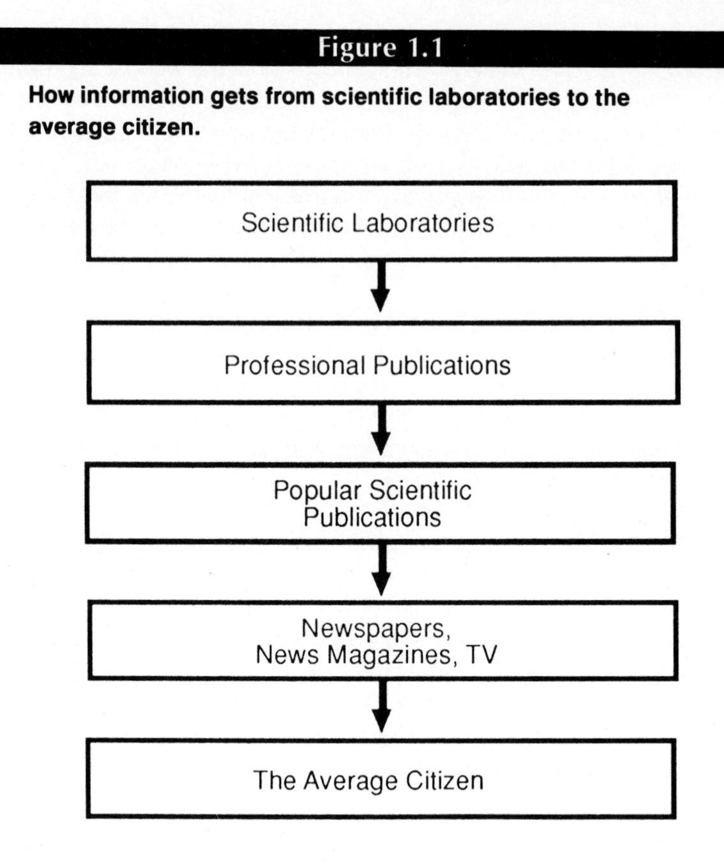

General Strategy

Now that we have a better idea of what we shall mean by scientific reasoning, we can begin to ask how one should approach learning about such reasoning. The obvious way to learn about scientific reasoning is by learning to be a scientist. But this is not necessary for our purposes. Nor is it even the best way. One could not hope to learn more than one or two scientific subjects, and, although there is a general pattern to all scientific reasoning, different subject matters lead to great differences in emphasis. So studying physics would not prepare you to deal with biomedical issues, and studying sociology would not prepare you to deal with new findings in physics. We want to be able to deal with all of these subjects, and more.

The tools we shall use have been borrowed from a number of subjects that deal with reasoning in general. These include: studies into the nature of science by philosophers of science; the study of logic by philosophers; the study of human reasoning by cognitive scientists; the study of probability and statistical inference by mathematicians and statisticians; and the study of decision making by all of the above. But we shall not ourselves be examining any of these areas in their own right. The borrowed pieces have been knit together for one purpose: to help the student learn to understand and evaluate reports of scientific findings that one meets in everyday life and work.

Computer Analogy

A computer analogy may help you better to grasp the general strategy of this text. In thinking about the operation of a computer, we can distinguish three basic components. First, there is the hardware, which includes the circuit boards making up the central processing unit and memory, the disk drives, keyboard, and so on. Then, there is software, the programs that tell the machine what to do with various inputs. Finally, there is the information that is fed into the machine for processing. These components are pictured in Figure 1.2.

Now, thinking of yourself (briefly!) as a computer, you can be quite sure that the purpose of this text is not to change your hardware. I assume that people differ somewhat in their "native intelligence," which is a function of the "hardware" one has inherited. This text, however, is designed not to require any special hardware. Any average "off the shelf" hardware will do.

More surprising, perhaps, it is not my purpose to feed you new information. At least, not in the usual sense. You will not be required to learn new facts, for example, about chemistry or biology. There are no formulas or parts of animals to memorize.

The information presented in this text is really part of a new program that is intended to improve the way you process ordinary scientific information. My objective, then, is to help you to reprogram the way you assimilate, organize, and evaluate everyday scientific information.

Tactics

Now that you have a better idea of the general aims and strategy of the text, we can turn to the tactics of how the subject will be presented and how it should be studied.

The text itself proceeds on three levels. First, I will be presenting some general ideas about the nature of science and such things as causality and probability. But you should not regard this as material to be memorized.

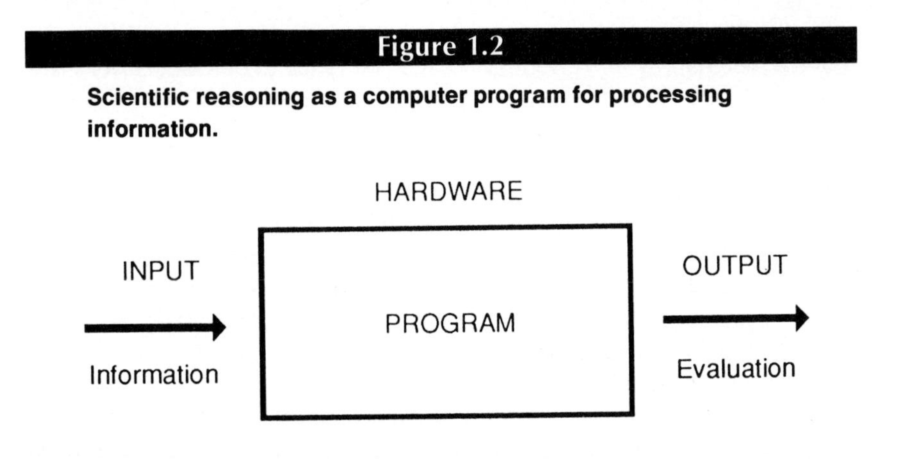

Figure 1.2

Scientific reasoning as a computer program for processing information.

Rather, it forms the background for learning how to process scientific information. To assimilate these general ideas you must read actively and critically, even skeptically. If you do not understand something, do not just reread it, although that may be necessary. Try to formulate explicit questions about what you do not understand. Then look for answers to your questions in the text—or ask your instructor.

Accompanying all the general ideas are examples, some of which are discussed in great detail. The purpose of the examples is to help you grasp the general ideas in such a way that you can use them in dealing with new cases. What you are trying to acquire is the ability to deal with new cases.

Finally, there are exercises for you to work out on your own. Most of the exercises consist of reports of some scientific finding that you are to analyze using the techniques explained and illustrated in the text. Very few questions focus on the background ideas themselves. Your goal is to learn to use these ideas in real life.

Reasoning as a Skill

I suggest that you think of learning scientific reasoning as being like learning a new skill, such as playing the piano or playing tennis. Learning a new skill is something that cannot be done simply by reading about it. You have to practice actually doing it. The point of the exercises, then, is to provide you with an opportunity to practice your reasoning skills.

The correct way to approach the exercises is as follows: First, study the relevant sections of the text and any related lecture notes until you think you understand the general principles. Pay particular attention to the examples that are worked out in the text itself. Then try working the assigned exercises without looking back at the text or your notes. If you get stuck while doing a problem, spend five to ten minutes thinking about it, but not longer. Go on to the next problem. Maybe it will give you a clue to doing the previous one. If, after a while, there are still some problems you cannot figure out, go back to the text and your notes. See if you can discover what you are doing wrong from the text and your notes. If, after an honest effort—that is, another ten minutes of hard thinking—you still do not see what is wrong, it is time to consult with someone, perhaps a classmate or your instructor. Remember that the material within a given part of the text tends to be cumulative. If you miss something important at the beginning, you will probably be confused all the way through.

Keep in mind that the object of doing the exercises is not simply to get the right answer down on a sheet of paper. The object is to develop the skill that will enable you to do the exercises without having the text at hand. You will most likely not have access to the text during quizzes and examinations. More important, you will not have it when you need to evaluate scientific reasoning in day-to-day situations. That is where the real payoff from developing reasoning skills should be.

Another implication of thinking of reasoning as a skill is that you must practice regularly. Do the exercises in each chapter as they are assigned. Do not let things go and then try to cram the night before a test. No one would dream of staying up all night to practice before a tennis match or a piano recital. Staying up all night studying for a test of reasoning skills is almost as silly. Believe me, it will not work.

Finally, as in music or sports, becoming very good at scientific reasoning requires both practice and talent. Becoming tolerably good, however, requires mainly practice and only a little talent. For most people, tolerably good is good enough. So work at developing your skills little by little. By the time you have to face a quiz, you will just be doing what comes naturally. And you will be able to do it naturally in real life too.

PART ONE

Theoretical Hypotheses

CHAPTER 2

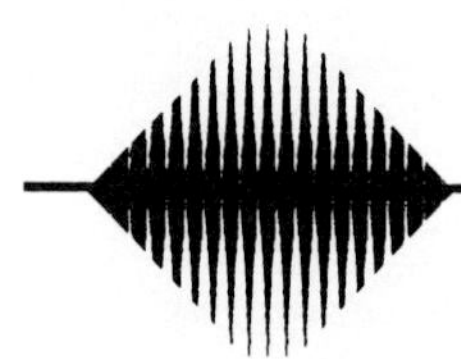

Understanding and Evaluating Theoretical Hypotheses

In this chapter, we will develop a framework for understanding and evaluating a wide range of scientific cases. The chapter begins with a case study, the discovery in 1953 of the structure of DNA. The case study will provide everyone with enough background on this particular episode so that it can be used as an example for most of the chapter. After the framework is in place, it will be applied to a variety of other examples.

2.1 THE DOUBLE HELIX: A CASE STUDY

In the fall of 1951, a 23-year-old American named Jim Watson arrived at The Cavendish Laboratory of Cambridge University in Cambridge, England. He had come in pursuit of his personal scientific quest to discover the physical structure of DNA—deoxyribonucleic acid—a discovery he was sure would bring him fame. Why DNA?

The idea that human inheritance is transmitted from parents to offspring by identifiable bits of matter in germ cells (sperm and eggs) has a long history. Since around 1900, there had been slow, but steady, progress in determining the chemical structure of these particles, now called genes. By 1950, it was well known that germ cells contained both DNA and proteins, which are large chains made up of units called amino acids. In spite of Oswald Avery's 1944 experiments, which strongly suggested that genes are made of DNA, in 1950 most biologists and chemists still thought that genes are made of proteins rather than DNA. One of the relatively few people who took Avery's work seriously was Salvador Luria, an Italian-born geneticist teaching at Indiana University. Watson, after completing his B.A. in zoology at the University of Chicago in 1947, went on to study for his Ph.D. with Luria. When Watson finished his Ph.D. in 1950, he and Luria decided that the best route to further progress in genetics would be through detailed knowledge of the structure of DNA. Watson got a grant for further study in biochemistry with an expert in Copenhagen.

The state of knowledge in 1951 concerning the makeup of DNA is summarized in Figures 2.1 and 2.2. A DNA molecule was thought to consist of one or more chains of nucleotides. Each nucleotide consists of a sugar molecule (deoxyribose), a phosphate molecule, and a base. There are four different possible bases, two each of two kinds: purines (adenine and guanine) and pyrimidines (cytosine and thymine). Such chains of nucleotides,

Figure 2.1

A short section of DNA as represented by organic chemists in 1951. Note that this representation omits any reference to the three-dimensional arrangement of the atoms.

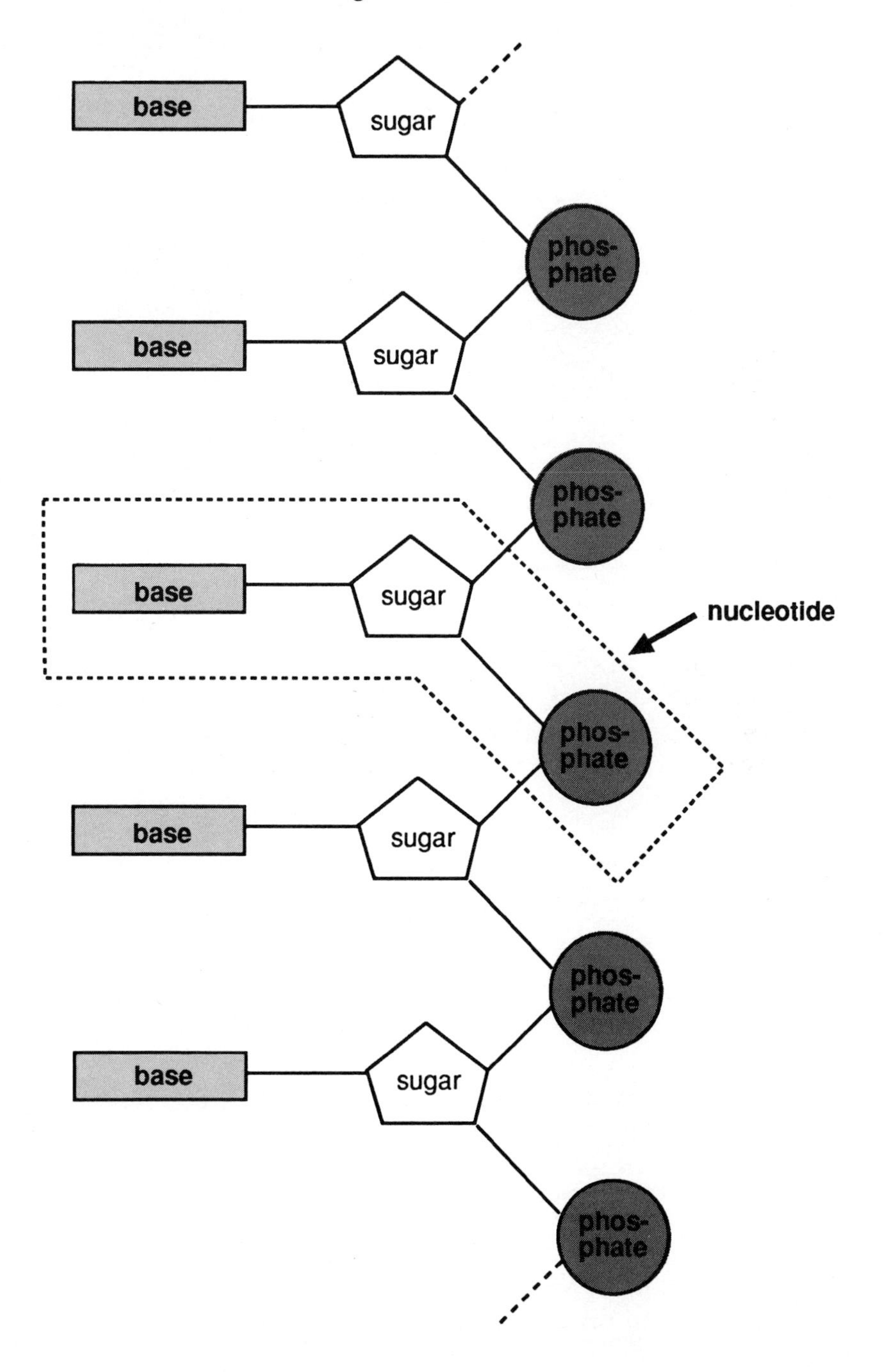

Figure 2.2

Representations of the four bases known in 1951 to be present chemically in DNA.

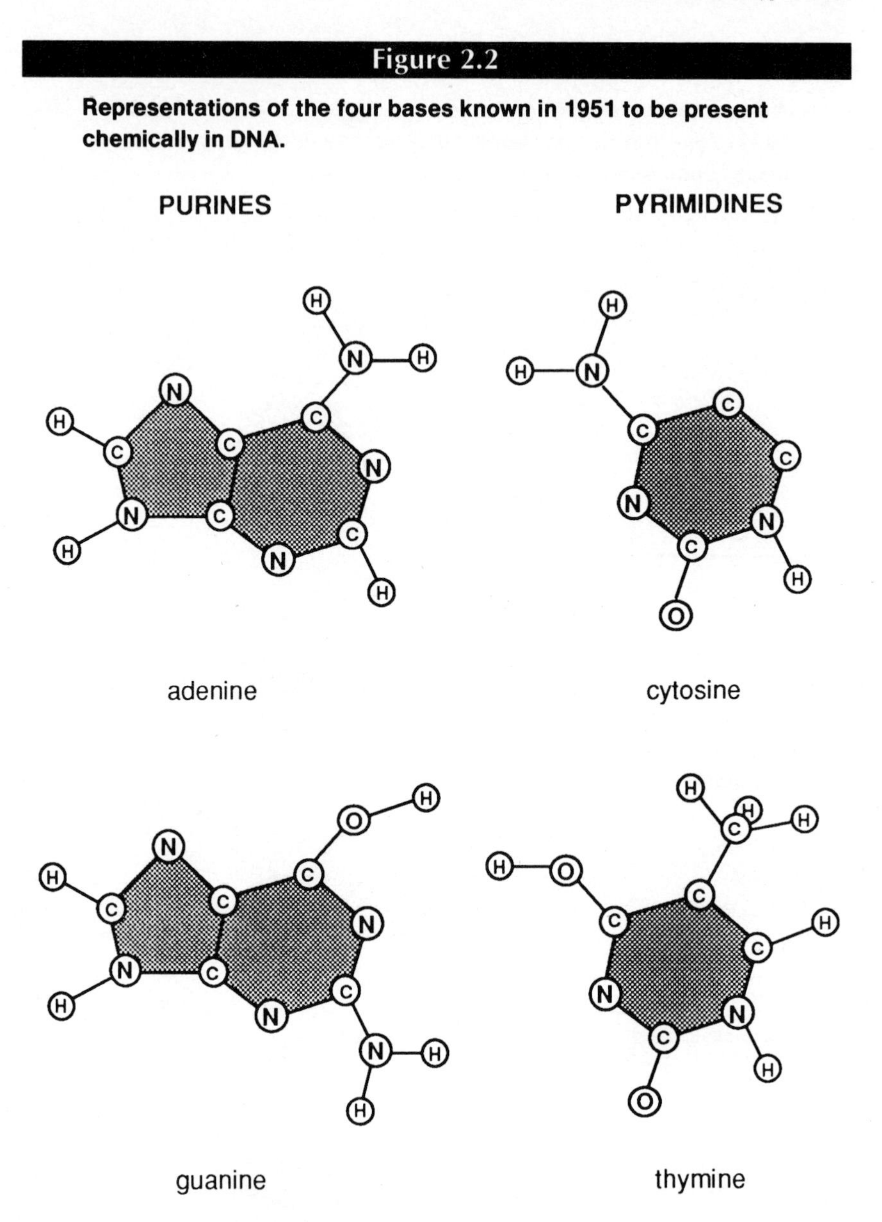

called "polynucleotides," were thought of as consisting of a "backbone" composed of sugar and phosphate supporting a sequence of bases. What Luria and Watson wanted to know was how all these pieces physically fit together. They believed that such structural knowledge would make clear how genes function in the process of inheritance.

Watson found the work in Copenhagen a waste of time. The following spring he went off to Naples, Italy, for two months where he spent his time reading articles from the early days of genetics and daydreaming about discovering the secret of the gene. While in Naples he attended a small, scientific meeting concerning the structure of large molecules in living organisms. One talk excited him. Maurice Wilkins, an Englishman working at Kings College at the University of London, showed a photograph taken by focusing x-rays on a small amount of chemical DNA. The picture indicated that DNA had a fairly regular crystal-like structure. That meant that there was a reasonable chance of actually figuring out some details of the structure. Watson was delighted, and began to think that x-ray methods were a better route to the structure of DNA than the biochemistry he had vainly been trying to master. He tried to make friends with Wilkins but without success.

While visiting a friend in Geneva on the way back to Copenhagen, Watson learned that the world's greatest living physical chemist, Linus Pauling, had just discovered the structure of a significant protein molecule, α-kerotin. The structure was a helix, and Pauling had discovered it by building a physical model of the molecule, using information obtained from x-ray photographs. Now Watson was sure this was the way to go. But where could he go to pursue his quest? He had been put off by Wilkins and he was sure Pauling would pay little attention to someone with so poor a knowledge of physical chemistry. The only place he could think of was The Cavendish, where he knew that some people were using x-ray techniques to study large molecules. He wrote to Luria for help. By good fortune, shortly after receiving Watson's letter, Luria met one of The Cavendish scientists at a meeting in the United States and arranged for Watson to begin work there the following fall. In science, as in other endeavors, it helps to have influential friends.

Three-Chain Model

It was at The Cavendish that Watson first met Francis Crick. Although a dozen years older than Watson, Crick was still working on his Ph.D. He kept getting distracted by other interesting theoretical problems. Crick, however, shared Watson's belief in the importance of DNA and his hunch that the best strategy for discovering its structure was to build models as Pauling had done in discovering the α-helix. They speculated that DNA also possessed a helical structure.

One possible embarrassment was that Wilkins was already working on the problem, and, unlike Americans, English scientists tended to respect such territories. However, Wilkins's own work was going very slowly because he did not get along with the person in his laboratory engaged in x-ray studies of DNA, Rosalind Franklin. Partly because of his conflicts with Franklin, Wilkins voiced no objections to Watson and Crick's fiddling with models of DNA.

Crick soon provided a major contribution to the project by developing a theoretical account of how x-rays are diffracted by helically-shaped molecules. If one is going to use x-ray pictures in building models of helically-shaped molecules, one needs to know what an x-ray picture of such molecules should look like.

They still needed more information about existing x-ray photographs of DNA. Luckily, Franklin was scheduled to give a talk on her recent work in London the middle of November. Watson was dispatched to the talk to learn what he could. The day following the talk found Watson and Crick on the train to Oxford for a weekend visit with a friend. Crick was excited. His theory of x-ray scattering, together with the data Watson related from Franklin's talk, indicated that there could be only a small number of possible helical structures for DNA molecules. It should consist of at least two, but not more than four, polynucleotide chains. They decided to try a model with three chains.

The next big question was the position of the sugar phosphate backbone, as it was called, relative to the bases. There were only two major alternatives. Either put the intertwined backbone in the center and let the bases hang out on the outside, or put the backbone on the outside and try to fit the bases into the inside. Fitting the bases into the inside seemed too complicated, so they decided to try building a model with the bases on the outside.

Upon returning to Cambridge, they set about building a model using pieces of wire and specially fabricated metal plates to represent the various components of the polynucleotide chains. In this task, their major reference work was Pauling's book *The Nature of the Chemical Bond.* This book provided the best available information about the distances and angles between the various groupings of atoms held together by chemical bonds. A good model had to reflect these basic features of atoms.

In less than a month they had completed what they regarded as a quite satisfactory model. They invited Wilkins and Franklin up from London to inspect their handiwork. It took Franklin only a few minutes to discover a major flaw in the model. Natural DNA is surrounded by water, which is loosely bound to the molecule. Watson and Crick's three-chain model left far too few places for water molecules to hook on to the DNA molecule. In fact, real DNA accommodates ten times the amount of water permitted by the model. Indeed, Franklin had given the correct information in her talk the previous month. Watson had misremembered what she had said!

In the aftermath of their humiliation by the group from London, the director of The Cavendish, Sir Lawrence Bragg, forbade Watson and Crick to engage in any more DNA model building. Watson went off to spend Christmas with the family of a friend at their manor house in Scotland. His dreams of fame and glory seemed far from being realized.

Two-Chain Model

Returning from vacation, Watson took up learning how to take x-ray pictures of the tobacco mosaic virus (TMV). He was not wasting his time because TMV should have a helical structure. And, indeed, several months and a new x-ray tube later, he obtained good pictures clearly indicating a helical structure. But neither Watson nor Crick stopped thinking about DNA, even though they were officially forbidden to work on it.

Meanwhile, two scientists at Cold Spring Harbor reported an experiment that strongly supported the idea that the primary genetic material is DNA, not proteins. In eight years, the scientific climate had changed. Unlike Avery's work, these new experiments were being taken very seriously by many other geneticists. For Watson, news of these new results was both good and bad.

It confirmed that he was right to focus on DNA. But now many other people would start working on DNA. His advantage was slipping away.

Another new result aroused Watson and Crick's interest. An Austrian-born biochemist at Columbia University, Erwin Chargaff, had carefully measured the base contents of DNA from several different biological species. The relative amounts of the pyrimidines, adenine and cytosine, varied from species to species. But, remarkably, the amounts of adenine and thymine were the same in all samples, as were the amounts of cytosine and guanine. Like Chargaff, Watson was sure these results were highly significant, but no one seemed to have any idea just what the significance might be. Crick, too, became increasingly preoccupied with the Chargaff results.

September of 1952 found Watson turning his attention to the idea that bacteria come in male and female pairs. If true, this meant that the genetics of bacteria are much more like that of higher organisms than had earlier been thought. Crick was once more back at work on his still unfinished Ph.D. dissertation. One new aspect of their lives was that Linus Pauling's son, Peter, had joined their group at The Cavendish. Through Peter they were able to keep abreast of the news from Pasadena.

The first ominous word was that Pauling was working on α-coils. A little later came word that he was working on DNA, but no details. Then, in the middle of January, came a draft of a paper in which Pauling outlined a model of DNA. To their great relief, Watson and Crick found that Pauling had come up with a model superficially resembling their own ill-fated, three-chain model. In addition, it had several other features that they felt sure had to be mistaken. They figured they had at most six weeks before Pauling discovered his mistake and turned the full power of his genius to rectifying the blunder. Nevertheless, they were determined to turn all their energies to the problem once again. The serious prospect that Pauling, an American, might beat his British group to the solution was enough to convince Bragg to let them try again.

Watson journeyed down to London to show Pauling's paper to Wilkins and to enlist his support for their new effort. Not finding Wilkins immediately, he went around to Franklin's lab. Citing Pauling's paper, he tried to convince her of the urgency of the situation and to enlist her, and her carefully acquired x-ray data, in the effort. She chased Watson out of her lab, reportedly insisting that she would have no part of their "little boys' games."

Wilkins consoled Watson by showing him a picture Franklin had taken of what she called the "B form" of DNA, which contained much more water than the then standard "A form." The pattern, as shown in Figure 2.3, was incredibly simpler than any Watson had seen before. The strong black crosses could only come from a helical structure. That followed immediately from Crick's theory about how x-rays are diffracted by helixes. Watson returned to Cambridge more excited than ever.

They still did not have enough information to answer the big questions that had faced them the year before. How many chains are there? Are the bases outside or inside? If the bases are inside, how are they arranged? This time Watson decided to try two-chain models, appealing to the general idea that important biological entities come in pairs.

Figure 2.3

A schematic rendering of the pattern revealed by Rosalind Franklin's 1952 x-ray photograph of the B form of DNA.

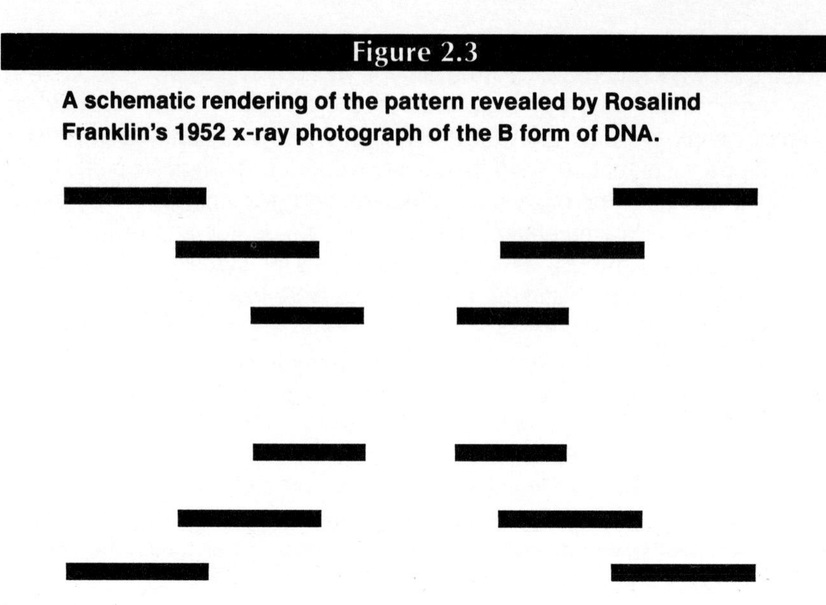

The decision whether to put the bases outside or inside was more difficult. They had always worried that there would be too many possible ways of arranging the bases on the inside, thus making it difficult to decide which is correct. But seeing that they were getting nowhere with base-outside models, they decided to have a go at models with the bases inside.

The idea was that a base attached to one sugar phosphate backbone should bond with a base on the opposing sugar phosphate backbone, thus forming a kind of miniature spiral staircase. The immediate problem was that while the distance between the two strands, the diameter of the spiral, should be constant, the bases are all of different sizes. Hooking up any two bases across the inside could either deform the bases or make bulges in the spiral backbone. Nevertheless, in spite of deformations and bulges, Watson proceeded to build a model with bases bonded like-with-like, that is, adenine with adenine, and so on.

He was shortly set straight by an American crystallographer, Jerry Donohue, who had worked with Pauling. Watson had taken information about the hydrogen bonds on guanine and thymine out of a standard textbook. Donohue informed Watson that the standard texts were wrong. The bases could not possibly bond the way Watson's like-with-like model required. Crick voiced still other objections. Watson reluctantly gave up on the like-with-like scheme.

The machine shop at The Cavendish was late in producing the little tin plates they needed to represent the bases. Being impatient to get on, Watson cut his own set of bases out of stiff cardboard. Playing around with his cardboard models Watson discovered that the combination adenine-thymine has a very similar shape to the combination guanine-cytosine. With these "steps" one could build a spiral staircase with a uniform diameter. Donohue confirmed that the required hydrogen bonding would work. Moreover, this scheme provided an immediate explanation for the Chargaff results because each pair consists of one purine and one pyrimidine, and only these particular combinations would bond in the required manner. Crick, who later claimed independently to have come to the same conclusion without benefit of Watson's cardboard models, was quick to proclaim that they had found "the secret of life."

But there was still work to do. When the shop delivered the metal cutouts, they took a plumb line and measuring stick to the model, carefully aligning all the pieces to make sure that they did fit together in a configuration consistent with knowledge of the relevant chemical bonds. It seemed to be all in order. Even Sir Lawrence Bragg was pleased. More to the point, Wilkins, and even Franklin, agreed that the proposed structure was confirmed by a detailed examination of their own x-ray data.

"We wish to suggest a structure for the salt of deoxyribose nucleic acid (D. N. A.)." So began the 900-word paper by Watson and Crick published in *Nature,* March 25, 1953. Figure 2.4 shows the schematic rendering of the structure of DNA as it appeared in this first paper. By prior arrangement, Watson and Crick's paper was followed by two papers by Wilkins and Franklin, respectively. These papers set the direction for work in molecular biology that continues as we traverse the final decade of the twentieth century.

2.2 UNDERSTANDING EPISODES IN SCIENCE

Now that we know how the double helix was discovered, let us go back over the case and begin developing a few general analytical tools for understanding this and similar episodes in science. We will first survey some general features of such episodes, and then go on to examine a few of these features in more detail.

Human Context of Science

The Double Helix, Watson's own personal account of the discovery, was first published in 1968, fifteen years after the event. Many people objected that the book was too personal. In particular, it exposed the all-too-human foibles and weaknesses of the participants. Prominent among these were the conflicts between Rosalind Franklin and Maurice Wilkins. There were also the sometimes unflattering characterizations of Francis Crick, as talking too much, Sir Lawrence Bragg, as being too stiff, and Linus Pauling, as being a showman. And of course there is also the picture of Watson himself as obsessively pursuing a Nobel Prize in active competition with Wilkins, Franklin, and Pauling.

Why did people object to the personal nature of *The Double Helix?* Some, like Crick and Wilkins, objected that Watson had distorted the story. Others, however, objected mainly because they wished to preserve an image of science as a "rational," "objective" activity. Besides being flattering, this image often serves scientists well in dealings with people outside science. Watson showed that the official image of science is simply not correct.

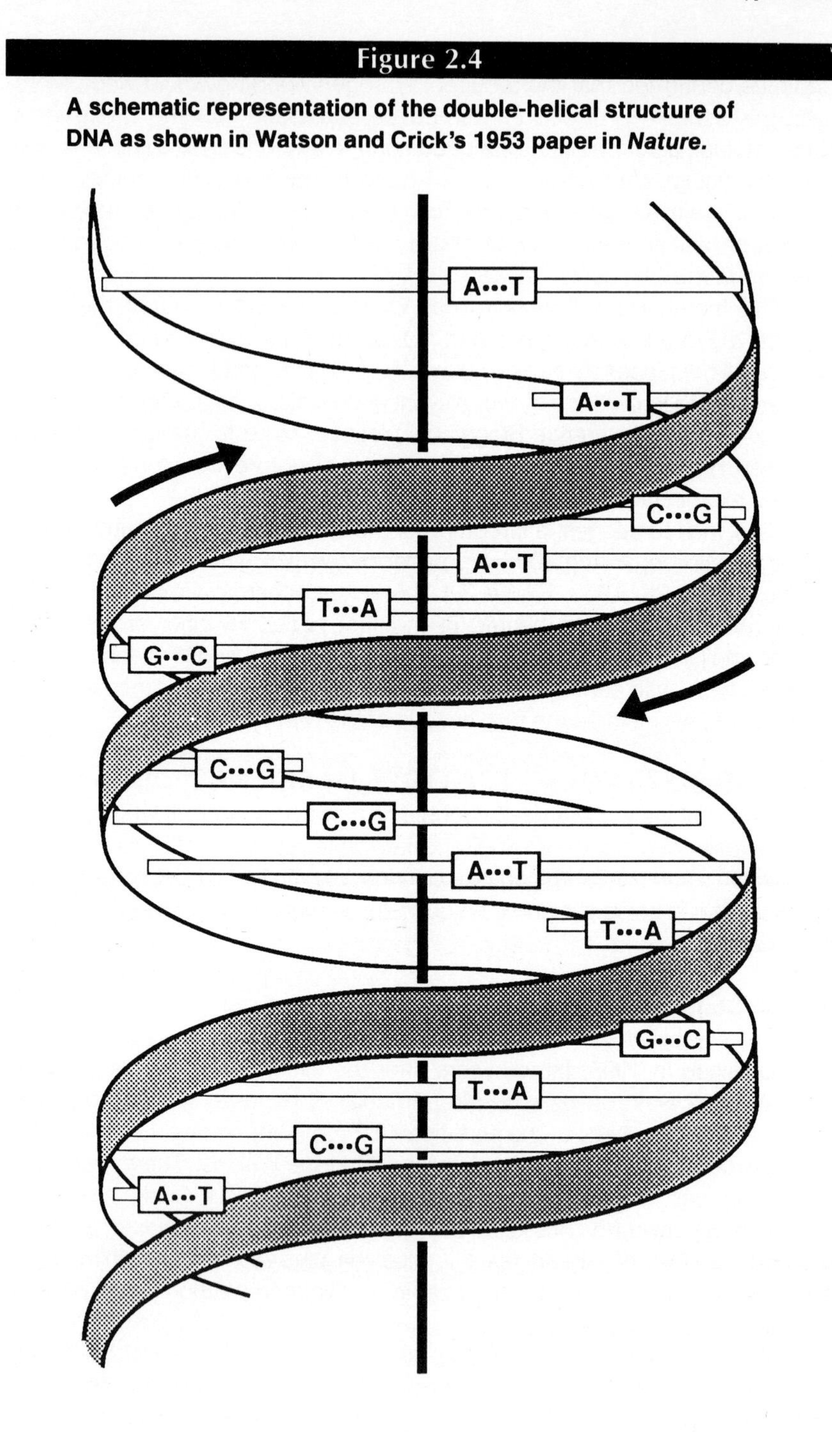

Figure 2.4

A schematic representation of the double-helical structure of DNA as shown in Watson and Crick's 1953 paper in *Nature*.

Of course, no one should ever have been taken in to start with. Science is a human activity, and scientists are ordinary human beings. Thus, among scientists, as among members of any other profession, one should expect to find a full range of human strengths and weaknesses. So one general lesson from this episode is that the successes of science are not due to scientists being unusually blessed with virtues like objectivity, modesty, or honesty. We must look for another explanation why science works as well as it does.

A more practical lesson is that behind every episode you read, or hear about, there are many typically human stories. People's reputations and careers are on the line. The typical report of new scientific findings one sees in newspapers or magazines, however, reveals only snatches of these stories, if anything at all. One can only imagine what the real flesh-and-blood scientists were doing. Nevertheless, in trying to understand an episode, it always helps to remember that real people were there behind the scenes all along.

Exploring How the World Works

Let us grant, then, that science is a human institution run by people with a wide range of interests and motivations. Still, there must be some things that distinguish science from other institutions, such as commerce, the military, the arts, politics, or religion. The most general claim one can make is this: Insofar as they participate in science as an institution, scientists are engaged in exploring how the world works. Whatever their differences in personal motivation and style, Pauling, Watson, Crick, Wilkins, and Franklin were all engaged in trying to figure out the structure of DNA. This was part of the larger project of discovering how genes replicate, and, thus, how parents pass on characteristics to their offspring.

One can also say something in general about how scientists explore the workings of the world. They engage in careful and deliberate interactions with the world. They do experiments and make observations, some of which are designed to help them decide which of several possible ways the world might work is most like the way it really does work. This activity distinguishes a scientific tradition, for example, from a religious tradition that seeks understanding of the world through the interpretation of sacred texts, or a literary tradition based on particular literary forms, such as the novel.

It is often said that science is distinguished by the use of something called "the scientific method." This is doubtful, at best, because the methods scientists actually employ are as varied as the subjects scientists study, from astronomy to zoology. What one can say in general about the methods of science, about experimentation, for example, cannot be specific enough to be of direct use to any scientist engaged in actual research. Some things, however, can be of use to an onlooker seeking to understand and, to some extent, even to evaluate the goings-on in particular scientific inquiries.

Finding a Problem

No scientist is simply engaged in the general pursuit of discovering how the world works. They all focus on some particular aspect of the world.

They have special problems they seek to solve. Watson, for example, set himself the problem of determining the three-dimensional structure of DNA.

How an individual scientist comes to focus on a particular problem is largely a matter of the accidents of personal history. Watson was, apparently, interested in biology from an early age. He had been an avid bird watcher and claimed that he first developed an interest in the nature of genes as a senior in college. His interest in DNA in particular seems to have been sparked by his graduate school adviser, Salvador Luria.

The lesson for us is that knowing how a particular scientist came to work on a problem is not likely to be of much use in understanding or in evaluating specific scientific results. It may be interesting as biography and useful for understanding why particular avenues of inquiry were pursued when they were, but that is not our concern here.

Constructing Models

Watson and Crick spent a great deal of time thinking about, talking about, and actually building, a model of DNA. Not all scientists literally build scale models like the wire and metal-plate model Watson and Crick constructed. In a more abstract sense, however, most scientists can be said to be engaged in constructing models of some aspect of the world, if only in their minds. Understanding episodes in science requires some understanding of the particular models being developed. Also, learning to analyze reports of scientific episodes requires an appreciation of the nature and role of models in science. Acquiring such an appreciation will be one of our first tasks.

Deciding Whether a Model Fits

Scientists do not construct models just for the fun of it. Constructing models is part of the process of figuring out how the world works. While working on their models, scientists are all the time trying to decide whether these models actually exhibit a reasonably good fit to the real world. In making such decisions, scientists obviously consider all sorts of facts, particularly the results of careful experiments. But their decision-making process is influenced by many other factors as well. Among these are the desire to be the first to make an important discovery and the fear of being shown wrong. Watson's desire for success clearly outweighed his fear of making mistakes, as demonstrated by his initial enthusiasm for the disastrously ill-fitting, three-helix model. To understand episodes in science, we will have to learn more about how scientists decide whether a model really does fit the world.

Convincing Others

It is not enough for an individual scientist to decide that a model fits. Other scientists must be persuaded to make the same decision. This requires data and arguments that will appeal to other scientists who are approaching the subject with a wide variety of interests, backgrounds, and skills. Franklin, for example, came to the conclusion that DNA has a helical structure long after Watson, Crick, and Wilkins were quite convinced it must. Pauling fairly quickly agreed that the double-helix model was correct after seeing both the model and the x-ray data at Kings. Other scientists in the field quickly agreed after hearing about it or reading the papers by Watson, Crick, Wilkins, and Franklin.

Spreading the Word

Once most of the scientists directly involved in a scientific area decide that a model fits, there begins a much slower process by which that conclusion spreads to the general, nonscientific public. Scientists may be involved in this process, as Watson has been, but so are many others, particularly teachers, journalists, and even filmmakers. It is at this stage of the game that the rest of us learn most of what we know about science. To understand and evaluate this information requires learning how to use what is presented to reconstruct some features of the models and the decision-making processes that went into producing the information in the first place. We turn now to this task.

2.3 MODELS AND THEORIES

Scientists often describe what they do as constructing and using models. Understanding science requires knowing something about models, and how they are used. In fact, there are at least three different types, or uses, of models to keep in mind.

Scale Models

Watson and Crick were helped greatly by actually trying to construct a physical model of DNA. This was a model in the ordinary sense in which model airplanes and doll houses are models. They are all scale models. The big difference between Watson and Crick's model and more familiar scale models was the extreme scale, which in the case of the DNA model was roughly a billion to one. That is, an inch in the model represented roughly one one-billionth of an inch in an actual DNA molecule.

Scale models are widely used in science and even more widely used in engineering. For example, one can learn a lot about the wind resistance of various automotive designs by testing scale models of automobiles in small wind tunnels. This is a lot easier, and cheaper, than building wind tunnels large enough to hold full-sized cars. Nevertheless, when scientists talk about models, they are often not talking about scale models.

Analog Models

In *The Double Helix,* Watson talks about noticing spiral staircases, and of thinking that the structure of DNA might be like a spiral staircase. He also had the example of Pauling's α-helix. Here we would say that Watson was using a spiral staircase and the α-helix as analog models for the DNA molecule. He was suggesting that the DNA molecule is analogous to the α-helix or to a spiral staircase. One might also say that he was modeling the structure of DNA on that of the α-helix or a spiral staircase.

Probably the most famous analog model in modern science is that of the solar system as an analog model for an atom. The nucleus of an atom, containing protons and neutrons, is said to be analogous to the sun. The electrons are said to be analogous to planets circling the sun. There is no doubt that this analogy between the solar system and atoms was extraordinarily fruitful during the first half of the twentieth century. It suggested all sorts of questions that formed the basis of much research. For example, "How fast are the electrons moving around in their orbits?" "Are the orbits circular or elliptical?" In investigating such questions, scientists learned much about atoms. In particular, they learned about many respects in which atoms are not like the solar system. In the end, a good analogy often leads to its own demise.

Analog models are typically most useful in the early stages of research when scientists are first trying to get a handle on the subject. At this point, almost any suggestion as to how they might construct a new model may be helpful. At later stages, when the question turns to evaluating how well the new model fits the real world, the original analog model is less useful. In trying to convince Wilkins and Franklin that they had the right structure for DNA, Watson and Crick did not appeal to features of spiral staircases. Nor did they simply appeal to Pauling's success with the α-helix. Similarly, facts about the orbits of the planets were not used as evidence that the solar system model of the atom is correct. For these evaluations, other evidence was needed.

In sum, thinking about analog models may be very useful in attempting to understand a proposed new model. Such models are much less useful in attempting to evaluate a proposed new model.

Models and Maps

The models most commonly referred to in scientific contexts are theoretical models. In attempting to understand what theoretical models are, it is helpful to invoke an analogy between theoretical models and maps. That is, we are going to use maps as analog models for theoretical models. Maps are more abstract than scale models, but still less abstract than a typical, theoretical model.

It is instructive if at this point the student can produce his or her own map. This could be a map showing a trip between home and school, between a dormitory and a classroom building, or between home and work. Figure 2.5, for example, is a map that depicts part of the University of Minnesota campus, including the main library and the building housing a particular department. Draw your own map before proceeding.

In the map, as shown in Figure 2.5, there is a solid arrow. What is the object to which the arrow is pointing? Stop and answer this question before reading any further.

The standard answer to the question is that the arrow is pointing to a building, presumably the university building in which the department office is located. What would you think if someone told you that your answer is mistaken? Not only is it mistaken, it is not even close to being right. It is totally off base. No doubt, you would begin to suspect that there is some trick being played on you. You would be right. But the trick has an important point.

The answer is that the arrow is pointing to a rectangle drawn on the page. That is, quite literally, what the arrow is pointing to. The reason one is inclined to say that the arrow points to a building is that it is pretty clear from the

Figure 2.5

A partial sketch of the Twin Cities campus of the University of Minnesota.

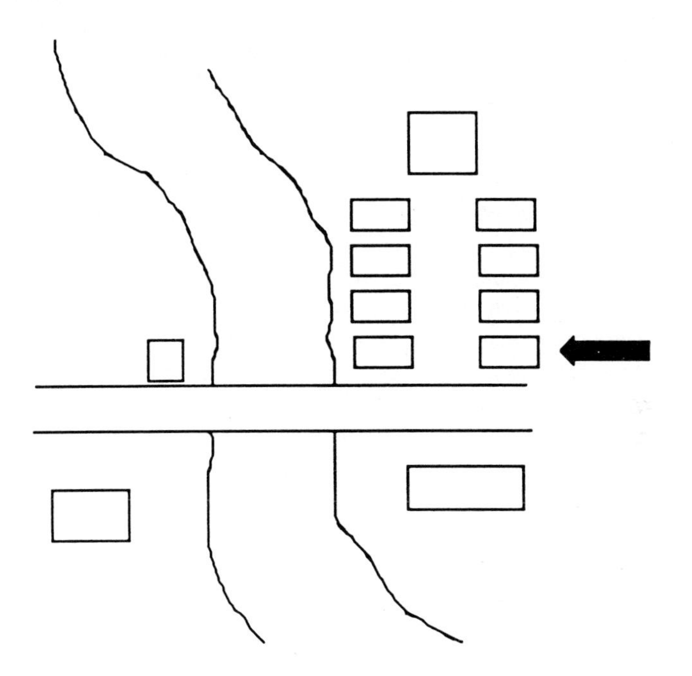

map and accompanying text that the rectangle represents a building. One, therefore, interprets the arrow as pointing to what the rectangle represents, rather than to the rectangle itself.

The point of this exercise is that a map is not the same thing as what it represents. In the case of maps, no one is likely to make this mistake. After all, one can fold up a street map and put it in one's pocket. One cannot fold up the city and put it in one's pocket. Nor is one likely to mistake a scale model for the thing modeled. Surely no one was in danger of confusing Watson and Crick's wire and tin scale model for a real molecule of DNA. Theoretical models, as we shall see, are another matter.

Granting that a map is distinct from what it maps, what is the relationship between the two? It is true, but not very informative, to say that the map maps the area mapped. It is somewhat more informative to say that the map represents the area mapped. The next question is, how does a map manage to represent a particular space?

The first part of the answer is that a map exhibits a particular similarity of structure with the space mapped. In the case of maps, the particular similarity of structure is spatial. The spatial relationships among marks on a street map, for example, correspond to the spatial relationships among the streets in the city represented by the map.

The second part of the answer as to how a map manages to represent the area mapped is that we have a whole set of fairly well understood social conventions for constructing and reading maps. Without these conventions, a map would be just a piece of paper with lines drawn on it. The conventions for street maps are so well known that most people are not even aware of them. But this is a special case. Few people know the conventions for interpreting Watson and Crick's scale model of DNA. One has to know a lot of physical chemistry to be able to interpret a particular tin plate as representing a purine base. Franklin could do it easily and quickly recognized that the three-chain model did not have enough places to attach water molecules.

The analogy between maps and models suggests further interesting questions. One is, "Could there be a perfect map, for example, a perfect map of Chicago?" The answer depends on what one means by a "perfect map." Suppose it means a map that contains a perfectly accurate representation of every feature of the city. Is that possible? Hardly. To represent every feature would mean representing every alley, house, garage, tree, bush, broken sidewalk, and abandoned car. It would mean representing not just the locations of buildings, but their height as well. That is an impossible task. So one way in which maps are not perfect is that they are incomplete, that is, they represent only selected features of their subject, such as streets, and ignore others, such as heights of buildings.

Restricting our attention to those features that are represented, there remains the question of how accurately those features are represented. For example, does the map accurately portray the relative distances between the Water Tower and the Chicago River, or between Michigan Avenue and Halsted Street? And is it accurate to the nearest ten yards? yard? foot? inch? Clearly no map is going to be perfectly accurate down to the millimeter.

In summary, a map can be used to represent a place because there exists a set of social conventions that allow one to interpret features of the map as representing features of the place. All maps are incomplete in that they do not represent all features of the place represented. And no map gives a perfectly accurate account of the features that are represented. Nevertheless, there remains a similarity of structure between the map and the place represented. They are similar in some specifiable respects and to some specifiable degree of accuracy. All these things hold, as well, for the relationship between theoretical models and the parts of the world they represent.

Theoretical Models

A theoretical model is part of an imagined world. It does not exist anywhere except in scientists' minds or as the abstract subject of verbal descriptions that scientists write down. When Watson was building the three-helix model, he could have written down a description of what a DNA molecule would be like if it were like this model. His description could have begun: "The model has three sugar phosphate backbones that twist in a helical structure with bases arranged . . . ." This description, obviously, cannot describe a real DNA molecule, because we now know that DNA has only two chains, not three. What it describes, rather, is a possible molecule that turned out not to exist at all.

Watson and Crick did build a scale model with three chains. What is the relationship between that scale model and the corresponding theoretical model? The scale model can be used in place of words to characterize the theoretical model. One simply says: "The theoretical model has three sugar phosphate chains with bases arranged like this," and, then, points at the scale model. This strategy works because there is a similarity of structure between the scale model and the theoretical model. One can understand that similarity if one knows the conventions used in building the scale model (such as that red wires stand for hydrogen bonds).

Why can we not just stick to scale models and dispense with the notion of theoretical models? Because not all theoretical models have corresponding scale models. Watson, for example, never completed a scale model of the two-chain molecule with like bases bonded together. But this model existed as a theoretical model. Watson even described it in a letter to Max Delbrück. More fundamentally, scientists construct theoretical models of a whole variety of complex processes for which it would be difficult if not impossible to build working scale models.

In any case, the idea of a theoretical model should not be mysterious. We all create something like theoretical models all the time. For example, one can imagine giving a party, including imagining who comes with whom and who says what to whom. Here one is constructing a theoretical model of a complex social event. The party may never occur or, if it does, it may be nothing like originally imagined. In the process of doing science, scientists imagine all sorts of complicated things and processes, including large molecules, such as DNA.

Theoretical Hypotheses

The most important question one can ask about a theoretical (or scale) model is whether it is, indeed, similar to the world in the intended respects and to the intended degree of accuracy. By way of shorthand, we will often simply ask, does the model fit the world as it was intended? Even more simply, does the model fit?

There is another way of talking about the fit between models and the world that is often used by scientists, journalists, and other commentators on science. To accommodate this way of talking, we will have to introduce some additional terminology.

When scientists make the claim that their model is, in fact, similar to the world in the desired respects, we can say that they have formulated a theoretical hypothesis and are claiming that this theoretical hypothesis is true. A theoretical hypothesis is a statement (claim, assertion, or conjecture) about a relationship between a theoretical model and some aspect of the world. It asserts that the model is, indeed, similar to the world in indicated respects and to an implied degree of accuracy. If the model is similar to the world, as claimed, then the theoretical hypothesis is true. If the model is not similar to the world, as claimed, then the theoretical hypothesis is false.

For example, at one point, Watson and Crick formulated the theoretical hypothesis that DNA has a helical structure with three polynucleotide chains. That hypothesis was shown to be false. They later formulated the theoretical hypothesis that DNA has a two-chain structure. That hypothesis turned out to be true. In general, asking whether a specified theoretical hypothesis is true or false is just another way of asking whether the corresponding theoretical model fits the real world. This relationship is pictured in Figure 2.6.

"Truth" is a heavy-duty concept whose meaning has been debated by philosophers and others for two thousand years or more. Theories of the nature of truth are discussed routinely in courses in logic and philosophy. The above discussion of the truth or falsity of theoretical hypotheses falls into the category of what is usually called the correspondence theory of truth. But there is no need to enter these troubled waters here. For the practical purposes of understanding and evaluating reports of scientific findings, it is sufficient to think in terms of the fit between a model and the world. Here the analogy of the fit between a street map and the streets of the corresponding city provides a sufficient guide. If that is not enough, one can always fall back on the more restricted relationship between a scale model and the real thing. If that does not work, an abstract inquiry into the nature of truth is unlikely to be of much help.

Finally, in everyday speech the word "hypothesis" often carries the connotation of a claim that is highly speculative—a conjecture without any real support. Thus, one may reply to a claim one disputes by saying, "Well, that is one hypothesis." Here the implication is that there are other, equally plausible, hypotheses that one might propose.

For the purpose of developing a systematic framework for understanding and evaluating scientific findings, it is best to ignore this way of talking. For us, the claims made in behalf of both the three-helix model and the two-helix model are both hypotheses. Indeed, in our preferred terminology, all general scientific claims are hypotheses. The difference is that some hypotheses are well supported by the evidence, and others are not. The important thing is to learn to distinguish between those hypotheses that are well supported and those that are not. That tells you which hypotheses it is reasonable to regard as true.

Theories

Everyone knows that scientists produce theories. Yet up to now I have not explicitly talked about theories, as such. The reason is that "theory" is a quite vague and often ideologically loaded term. The main reasons for calling

Figure 2.6

A picture of the relationship between a model and the real world. The hypothesis that the model fits the real world may be either true or false depending on whether or not the model actually fits.

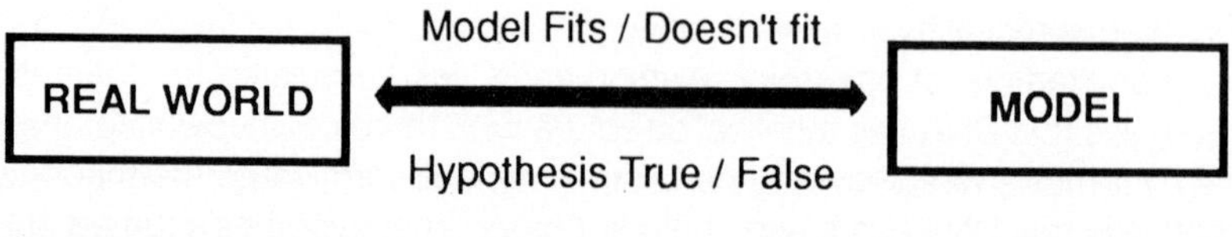

Model Fits / Doesn't fit

MODEL

Hypothesis True / False

something a theory may be to give it honorific status or, alternatively, to call it into question. Which of the two functions is served by using the word "theory" depends largely on the intentions of the speaker and the nature of the audience. Here we will attempt to employ a more neutral analysis. We already have in hand all the elements of such an analysis.

For our purposes, a scientific theory has two components. One is a family of models that may include both scale models and theoretical models. The second is a set of theoretical hypotheses that pick out things in the real world that may fit one or another of the models in the family.

In 1953, there was basically just one model, the double helix, and one chemical substance, DNA, to which this model was applied. But several years later, when people began talking more about "the theory of molecular biology," there was a whole family of similar models, and a range of other substances to which they were applied, including RNA (ribonucleic acid). Later, we will encounter many other theories that clearly include many distinct, but similar, models.

Like the word "hypothesis," the word "theory" often carries the common meaning of something speculative. It is common, for example, for those who question the theory of evolution to say that evolution is merely a theory and not a fact. Here again, we shall reject this common usage. For our purposes, the theories of molecular biology and the theory of evolution are all theories. Whether they are also facts depends on whether the corresponding models fit the world or, alternatively, whether the corresponding theoretical hypotheses are true. If so, they are facts. If not, they are not facts.

The important question is how well each theory is supported by the evidence. Here the relevant distinction is not between theory and fact but between theory and data, a distinction that is crucial for evaluating theoretical hypotheses and, thus, theories.

2.4 DATA FROM THE REAL WORLD

Everybody knows that in order to determine whether a proposed model fits the world one needs some information about the part of the world in question. But not all information is relevant. We will use the term "data" (singular, "datum") to refer to all the special information that may be relevant to deciding whether the model in question does fit. There are several, general characteristics that data must have.

The first feature that information must have in order to function as data is that it be obtained through a process of physical interaction with the part of the real world under investigation. Interaction may be active, as when one does experiments on the materials in question. Or the interaction may be passive, as when radio astronomers measure radio frequency signals from distant galaxies. In either case, the data result from a physical interaction with the relevant part of the real world.

A second general feature of data, as opposed to mere information, is that relevant differences in the data can be reliably detected. Detection may be a simple matter of looking, as when one observes a chemical solution change from blue to green. More often, detection requires elaborate instruments that produce outputs among which a scientist can discriminate just by looking. Often these outputs are computer printouts of graphs or tables of numbers. Figure 2.7 provides a schematic picture of the relationship between the real world and some data.

Among important bits of data in the story of the double helix were Chargaff's results on the one-to-one ratio of particular purines to specific pyrimidines in DNA, Franklin's results on the amount of water in DNA samples, and, of course, Franklin's x-ray pictures of DNA. In each case these data were obtained by actually working with samples of DNA.

Among information used by Watson and Crick that did not count as data in favor of their model of DNA was Pauling's discovery of the helical structure of α-kerotin. This information was influential in the decision by Watson and Crick to investigate helical models of DNA. Acquiring this information, however, required no physical interaction with samples of DNA. It could not, therefore, play a role as data for their hypothesis that DNA fits a double-helical model.

Here is one respect in which the analogy between maps and theoretical models breaks down. One can discover that a street map is deficient simply by finding a street that is not on the map. That requires no special skills or instruments. One can just look and see. Most models in modern science are not like that. Modern science typically investigates things that are very small (DNA), very large (our galaxy), very far away (distant stars), or otherwise inaccessible (the center of the earth). In all these cases, one cannot just look to see whether a proposed model fits. This fact has profound implications for how one might evaluate whether a model fits the real world.

Figure 2.7

A picture of the relationship between the real world and data generated through a physical process of observation or experimentation.

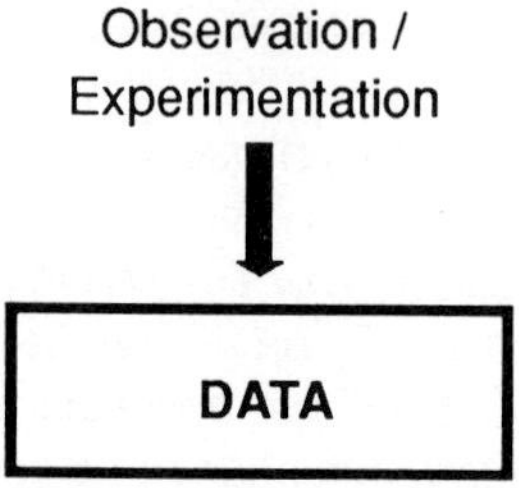

REAL WORLD

Observation / Experimentation

DATA

2.5 PREDICTIONS FROM MODELS

The fact that one can only interact indirectly with the objects of a scientific investigation means that one can only investigate the fit between parts of the objects and some limited aspects of a proposed model. It is, therefore, important to be able to figure out what kind of data an object that did fit the proposed model would produce in the circumstances of a particular experiment. Scientists often speak of using a model to make predictions about what kind of data would be produced. This requires some explanation.

Sometimes, scientists use a model to make predictions in the literal sense of trying to figure out ahead of time what the data will be. But often, predicting the data simply means using the model to determine what the data should be like, even though the experiment has already been done. For example, the double-helix model was said to "predict" the Chargaff ratios even though Chargaff's experiments were done several years earlier. Similarly, the model allowed Crick to calculate the kind of x-ray pattern a double helix would produce and thus, in this respect, "predict" the kind of picture that Franklin, at that point unbeknownst to Watson and Crick, already had in her possession.

The example of Crick's prediction of the x-ray pattern exhibits another important feature of making predictions from models. It requires more than just the model in question. It also requires that one have a well-attested model of the experimental setup. Thus, Crick had also to have a good model of how x-rays are diffracted by atoms. Otherwise, he could not calculate what the pictures should look like if the x-rays are being diffracted by a helically shaped molecule.

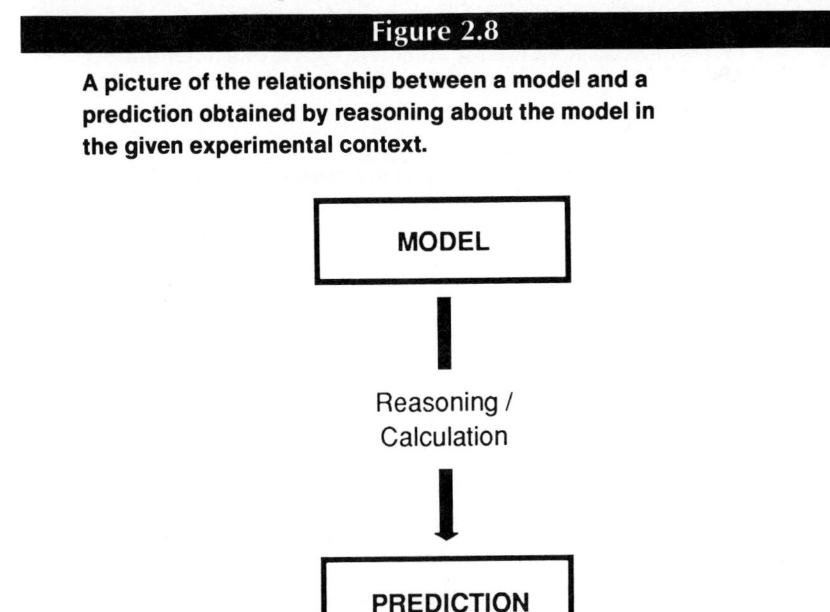

Figure 2.8

A picture of the relationship between a model and a prediction obtained by reasoning about the model in the given experimental context.

Figure 2.8 provides a schematic picture of the relationship between a model and a prediction derived from the model. Note that here the arrow represents not a physical interaction with the world but merely a process of reasoning or calculation.

2.6 THE ELEMENTS OF SCIENTIFIC EPISODES

We are now in a position to use the results of the previous three sections to construct our own simple model of a scientific episode involving a particular theoretical model (or family of similar models). There are four ingredients: (1) real world; (2) model; (3) data; (4) prediction. It is helpful to arrange these elements as shown in Figure 2.9. There are four relationships among the elements.

First, the relationship between the real world and the model is expressed by a theoretical hypothesis that asserts that the model fits the real world. It is understood that the model fits only in some respects and then only to some degree of accuracy. If the model does not fit accurately in the intended respects, then the theoretical hypothesis is false.

The real world and the data are related by a physical interaction which involves observation and, perhaps, experimentation. The model and the prediction are related by reasoning or calculation.

Finally, Figure 2.9 contains a relationship not previously noted, namely one between the data and the prediction. If what is going on in the real world, including the experimental setup, is similar in structure to the model of the world, including a model of the experimental setup, then the data and the

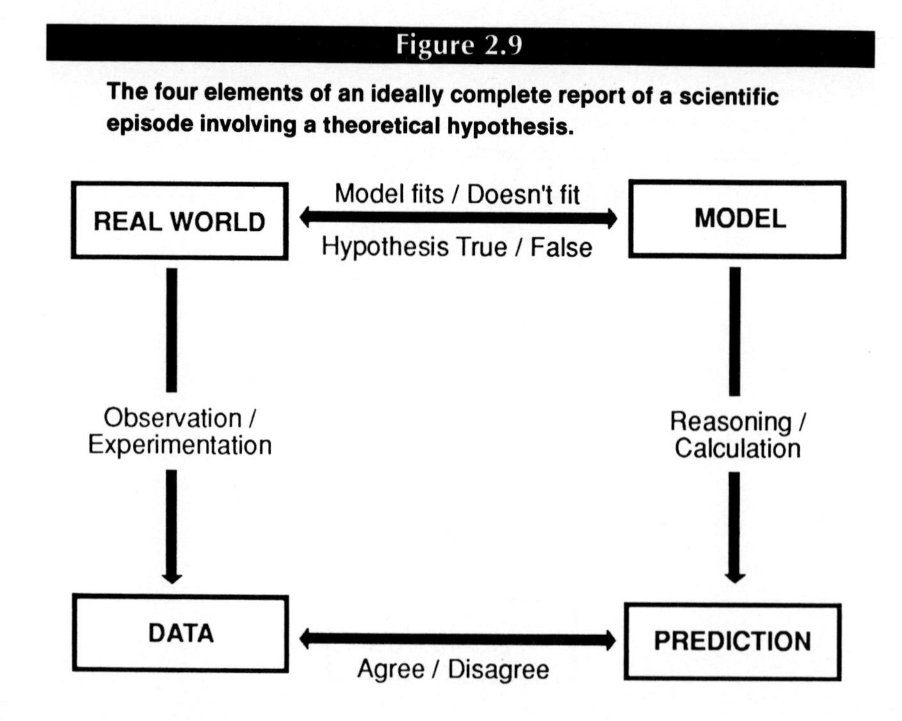

Figure 2.9

The four elements of an ideally complete report of a scientific episode involving a theoretical hypothesis.

prediction should agree. On the other hand, if the real world and the model of it are not similar in the relevant respects, then the data and the prediction may disagree.

Whether the data and the prediction agree is crucial to evaluating the truth of the hypothesis that the model fits the real world. But, as we shall soon see, there is more to evaluating the fit of a model than just agreement between data and predictions.

It should be noted that Figure 2.9 represents a picture of fully developed, scientific episodes in that it contains all four elements. Many episodes, and thus many reports of scientific findings, do not include all four elements. It is common, for example, to find reports that describe only the part of the real world under investigation together with some new data. There may be no mention of models or predictions. Similarly, one often finds discussions of new models of real world entities or processes with no mention of data or predictions. Occasionally one finds accounts of models of real world things that include predictions but no discussion of data. It is possible, but unlikely, that one would find a report including descriptions of real world things, models, and data, but no mention of predictions from the model. In such a case, the presumption would be that there is a prediction that, in fact, agrees with the data. If it did not, that would surely be noted.

One may learn a lot from incomplete episodes or reports. Unless all four elements are present, however, it is difficult to evaluate the relevance of the data to the crucial question of whether the model adequately represents the real world.

2.7 EVALUATING THEORETICAL HYPOTHESES

We are now ready to begin developing a general scheme that can be used by nonspecialists to evaluate scientific hypotheses as reported in various popular and semitechnical sources. We will continue with examples from the story of the double helix. Later in this chapter, we will work through several completely different examples.

The basic idea behind the evaluation of hypotheses is to use the agreement or disagreement between data and predictions, information that is relatively accessible, to evaluate the fit between a model and the real world, something that is not directly accessible. In the ideal case, there are only two possible cases. Either the prediction and the data agree or they do not agree. We will treat these cases separately, beginning with the case in which the prediction and data disagree. That turns out to be the simplest of the two cases.

In less than ideal cases, it may not be clear whether the prediction and data agree. Agreement may be a matter of degree. In such cases, it is difficult for a nonspecialist to make any independent evaluation of whether the model in question fits the world or not. Here the best one can do is rely on the informed judgment of specialists. Unfortunately, when agreement between data and predictions is unclear, specialists often disagree among themselves about how well the model might fit the real world. In such cases, the only safe course for the nonspecialist is to regard the data as inconclusive and to suspend judgment about the model until more decisive data become available. If use of the model in question is relevant to some practical decision that needs to be made, the problem, then, is to make that decision in a manner that takes proper account of the uncertainty as to whether the corresponding theoretical hypothesis is true. This latter sort of situation will be treated in Part Three of this text.

Evidence That a Model Does Not Fit the Real World

The story of the double helix provides a clear example of a model that was judged not to fit the real world—the three-chain model of DNA. Here the decisive data were Franklin's experimental measurements of the amount of water contained in samples of DNA. The three-chain model yielded a prediction as to how much water such a DNA molecule would accommodate. This could be determined by examining the scale model as long as one could interpret the model and knew enough physical chemistry to judge where water molecules might fit into the structure. The trouble was that the prediction from the model gave a value for the amount of water that was only one-tenth the amount Franklin had measured. So there was a clear disagreement between the experimental data from real samples of DNA and the prediction based on the three-chain model of DNA. This situation is pictured in Figure 2.10.

In this case, one is tempted to conclude without further ado that the hypothesis is false. That is, real DNA molecules do not closely resemble the proposed three-chain model. Franklin immediately drew that conclusion, although Watson and Crick took a little longer to come around. As

Figure 2.10

The elements of the episode involving Watson and Crick's three-chain model of DNA.

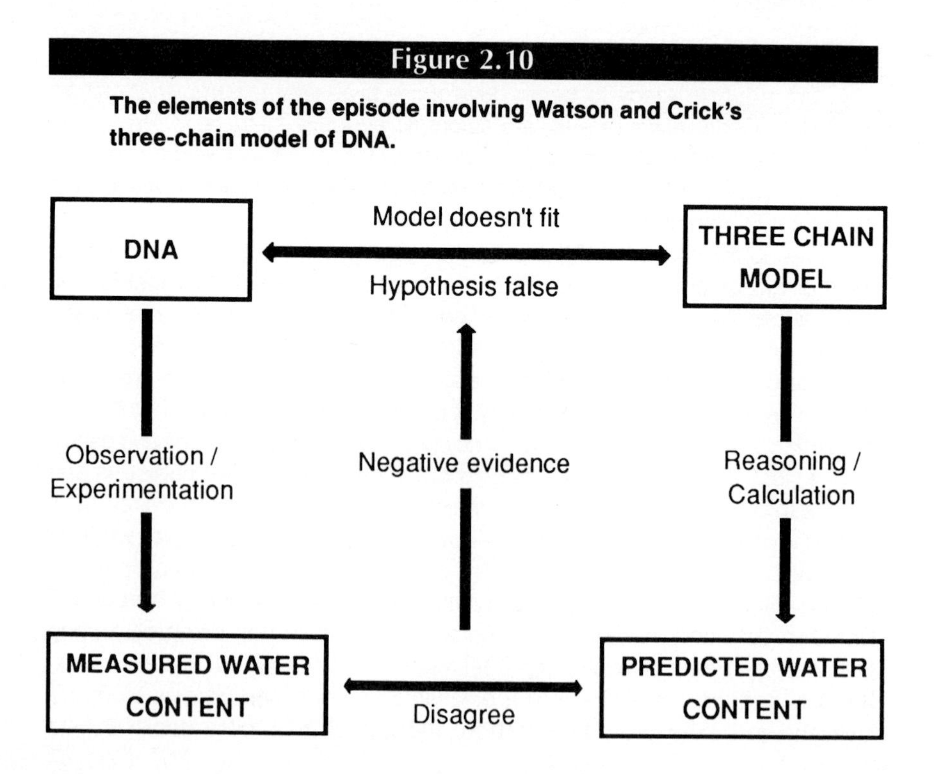

nonspecialists reading about this episode, we could just follow their lead. If we are attempting to reach an independent evaluation, however, we cannot be quite so decisive. There are two possibilities that militate against so hasty a conclusion, neither of which tend to be accessible to a nonspecialist.

One possibility is that the data were mistaken. That is, the experiment yielded a mistaken value for the amount of water. There are all kinds of things that, unbeknownst to anyone, might have gone wrong with the experiments so as to yield a value for the amount of water ten times greater than the actual amount. Only people with much experience with the actual apparatus and experimental techniques can reliably judge how likely it is that something was seriously wrong with the experiment.

A second possibility is that through misunderstanding of the model itself, or because of a mistaken model of the experimental apparatus, the prediction was mistaken. A proper understanding of the model, or the experiment, might have yielded a predicted value in agreement with the actual data. Again, this is something for which a nonspecialist must rely on the judgments of the experts.

For these reasons, we will take reports of clear disagreement between data and predictions as a basis for concluding only that there is good evidence that the hypothesis is false. That is, there is a good reason, although not necessarily a conclusive reason, for believing that the model does not adequately represent the real world.

Evidence That a Model Does Fit the Real World

One of the nice features of the double-helical model for DNA is that having the sugar-phosphate backbones on the outside left a lot of places for water molecules to attach themselves. So the double-helix model yielded a prediction for the amount of water in agreement with Franklin's data. Should one take that as a ground for thinking that the double-helix model fits?

As a matter of fact, Watson and Crick did not treat the agreement between the amount of water predicted by the double-helix model and the measured amount of water as a basis for arguing in favor of the double-helix hypothesis. Why not? Because they knew many possible ways to build models with the required places for water. It could be done with a variety of three-helix models, for example, as long as one put the backbones on the outside. Thus, predicting the measured amount of water provided no basis for distinguishing the two-helix model from a variety of three-helix models. There was, therefore, no basis for regarding this agreement between prediction and data as evidence that the two-helix hypothesis, rather than some three-helix hypothesis, was true.

This explains why the x-ray data were regarded as being so important. According to Crick's calculations, a double helix should produce a quite distinctive pattern, that is, a pattern unlikely to result from molecules with a significantly different structure. Thus, agreement between the predicted x-ray pattern and actual x-ray pictures provided a reliable basis for distinguishing between a double-helical structure and a variety of other structures. In this case, therefore, agreement between the prediction and the data did provide evidence in favor of the double-helix hypothesis. The elements of this case are pictured in Figure 2.11.

The moral of this story is that mere agreement between a prediction and some data is not enough to provide a basis for thinking that a theoretical hypothesis is true. Agreement counts only when it would not have been very likely if the hypothesis had been false, which is to say, if some significantly different model had provided a better fit to the real world. Ignoring this moral puts one in great danger of thinking that one has evidence in favor of what is, in fact, a false hypothesis.

At this point, one might have the following worry. No matter what the data happen to be, and no matter what model is being considered, is it not always possible at least to imagine some completely different model that, nevertheless, just happens to yield the same prediction as the model under consideration? Does this not mean, therefore, that agreement between a prediction from a model and observed data can never provide any basis for thinking that the model fits the world?

This sort of difficulty has been voiced by scientific skeptics since the heyday of Greek astronomy nearly two thousand years ago. The reply is that in the process of doing science it is never enough for an alternative model to be imaginable in the abstract. It must be plausible against the general background of the models being used at the time by scientists working in the same general area. This means that in principle there is almost always some possible model that would provide a better fit than any model currently

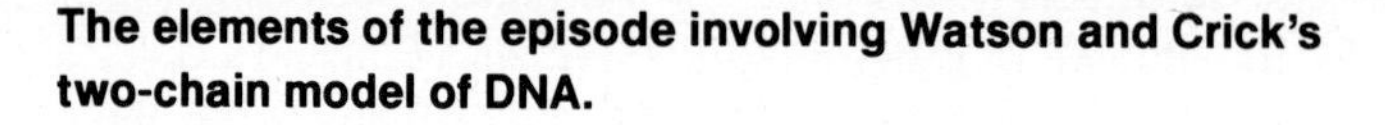

The elements of the episode involving Watson and Crick's two-chain model of DNA.

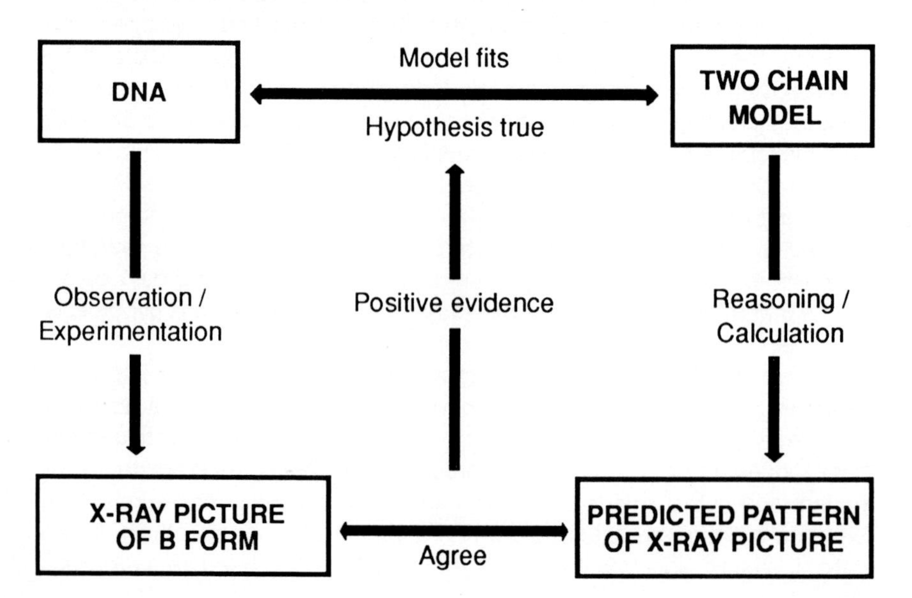

under active consideration. Nevertheless, the model currently regarded as the best fitting model will generally be the best fitting of those models that have actually been considered. It is unreasonable to expect more from any scientific investigation.

The above theoretical worry has a more practical consequence. It is often difficult for a nonspecialist to judge independently whether there are very many other plausible models that would also yield predictions in agreement with existing data. In many cases, therefore, all the nonspecialist can do is rely on the reported judgments of specialists as to whether there are any such alternative models. Worse yet, these judgments are often more implicit than explicit. One must read very carefully to determine whether there is any consensus on the availability of other plausible models yielding the same predictions. With experience, however, one can learn to recognize hints from which one can infer the existence of the relevant consensus.

Finally, as in the case of disagreement between predictions and data, we cannot take agreement between a prediction and data as a definitive basis for deciding whether the model fits. The most we can conclude is that the agreement between prediction and data provides good evidence for thinking that the model fits. The possibility of serious experimental errors, or of mistakes in determining what prediction the model yields, precludes a definitive conclusion on the basis of any single experiment. Even for cases in which it seems to the nonspecialist that scientists themselves reach a definitive conclusion on the basis of a single experiment, they may be relying on knowledge of other experiments, as well as considerable experience with the experimental and theoretical techniques involved.

2.8 A PROGRAM FOR EVALUATING THEORETICAL HYPOTHESES

In this section we will reduce the process of evaluating reports involving theoretical models to six easy steps. This does not require learning anything new. It is just a matter of organizing what we have done into a kind of program for doing an analysis. The advantage of developing such a program is that one can have in one's head a simple, uniform scheme for the evaluation of all sorts of scientific reports, a scheme that is easy to remember and to apply.

The program has two parts. The first four steps instruct one to identify the four basic elements in a complete episode. These steps provide a basis for *understanding* the case. If all four elements are reported, one can go on to the final two steps, which constitute an *evaluation* of the model. If not all of the four elements are identifiable, it may be difficult to perform any evaluation at all.

Some reports mention more than one model and prediction, and report different sets of data. The program assumes that you have identified a single model and prediction, and a single set of data, which is to be the focus of your analysis. Often, it is clear from the context which model or which set of data forms the primary focus of the report. If this is not clear, you might have to run through the program more than once, evaluating different models or considering different data sets. Later, we will consider a special case in which more than one model is evaluated in a single analysis.

The Program

Step 1. Identify the aspect of the real world that is the focus of study in the case at hand. These are things or processes in the world that you should be able to describe in your own words with, perhaps, just a bit of existing scientific terminology.

Step 2. Identify a theoretical model used to represent the real world. Describe the model, using scientific terminology as needed. A diagram may be helpful in presenting a model. Indeed, a diagram may be a version of a model.

Step 3. Identify data that have been obtained by observation or experiment involving the real world objects of study.

Step 4. Identify a prediction, based on the model, that says what data should be obtained if the model actually provides a good fit to the real world.

Step 5. Do the data agree with the prediction? If not, conclude that the data provide good evidence that the model, in its present form, does not fit the real world. If the data do agree with the prediction, go on to Step 6.

Step 6. Was the prediction likely to agree with the data even if the model under consideration does not provide a good fit to the real world? This requires considering whether there are other clearly different, but plausible, models that would yield the same prediction about the data. If there are no such alternative models, the answer to the question is "no." In this case, conclude that the data do provide good evidence that the model does fit the real world. If the answer to the above question is "yes," conclude that the data are inconclusive regarding the fit of the model to the real world.

Figure 2.12 exhibits the above program as a "flow chart." This is easier to remember than the verbal instructions, and may help you to recall the details of the written program.

Although the first four steps in the flow chart are numbered in the order given in the program, the chart itself suggests that there is nothing sacred about this particular ordering. What is important is that one identify all four elements before proceeding to the evaluation in Steps 5 and 6, where the order is important. As a general rule, it is advisable to begin with the real world objects. This tells you what the whole episode is about. But the order in which you identify the remaining three elements may vary. The structure of a particular report, particularly the order in which information is presented, may make one order seem to you more natural than another. It is fine for you to follow that order, as long as you clearly identify the elements of the case.

Finally, the program is only a guide, an outline to help you evaluate reports of scientific findings. Learning to use it effectively requires experience. The only way of acquiring that experience is to do many evaluations.

Figure 2.12

A flow chart corresponding to the program for analyzing reports of scientific episodes involving theoretical hypotheses.

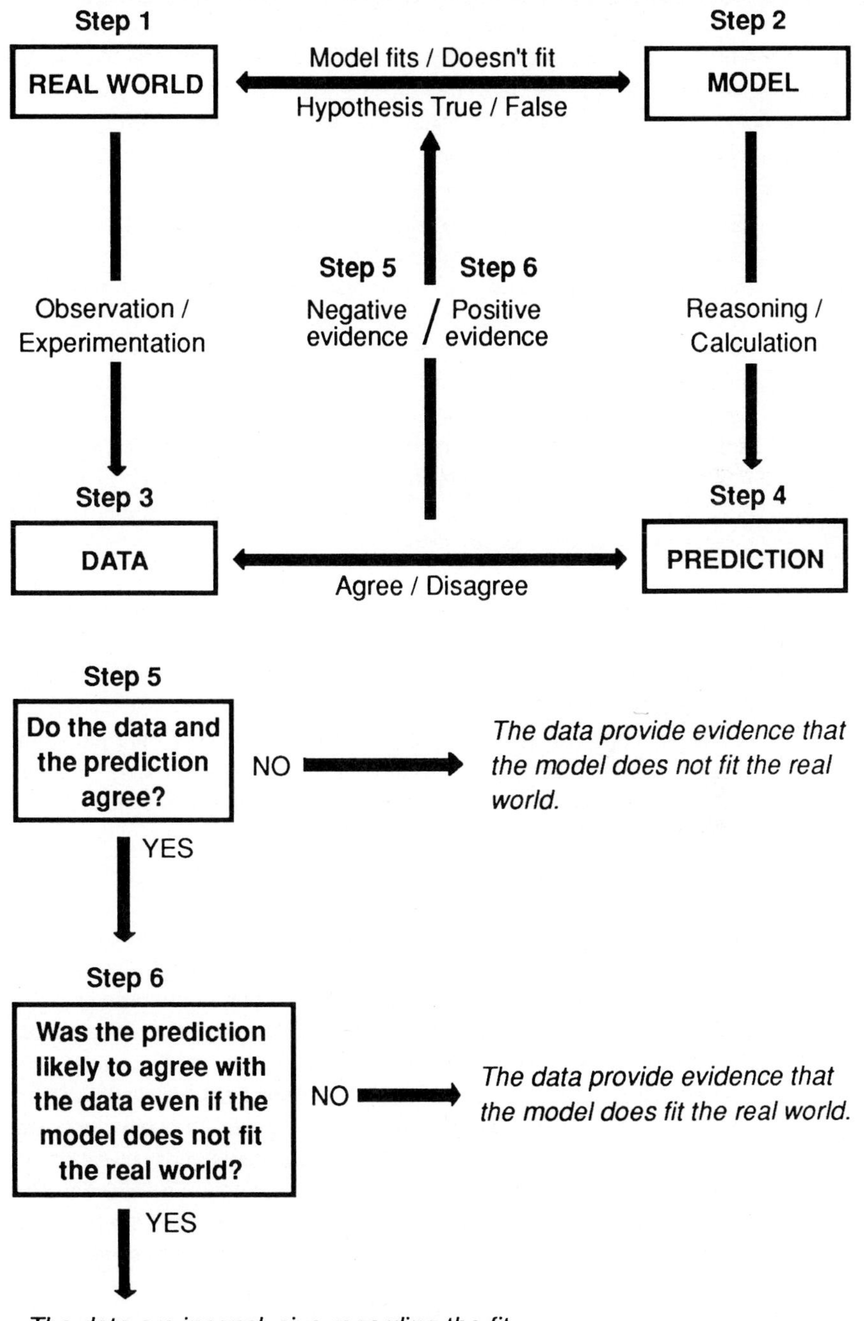

2.9 WHY THE PROGRAM WORKS

Why should you use the suggested program for evaluating theoretical hypotheses? Because, by following the program, you will have a good chance of reaching the right conclusion, whatever the right conclusion happens to be. There are three possibilities: (1) the data may provide evidence that the model does not fit the world, (2) the data may provide evidence that the model does fit the world, (3) the data may be inconclusive. In each case, the program will lead you to the appropriate conclusion.

Case 1. When the data and a prediction disagree, typically, that is because there are some respects in which the model does not adequately represent the real world. Steps 4 and 5 of the program are particularly relevant to this case. Step 4 is relevant because it requires you to consider whether there is a prediction that is clearly derived from the model. That is important because, if the prediction has little to do with the model, the relationship between the prediction and the data would not be much of an indicator of any similarity, or lack of similarity, between the model and the real world. Step 5, then, asks you to consider whether the data and the prediction agree. If they do not, that is an indication that something is wrong with the model, and the program directs you to the appropriate conclusion.

Case 2. If the data and prediction agree, the program directs you to Step 6. It asks you to consider the question: Was the prediction something that would be likely to agree with the data, even if the model under consideration does not adequately fit the real world? This question forces you to seek information in the report regarding other plausible models that could yield the same prediction. If there seem to be no other such models, then agreement between the prediction and the data is a pretty reliable indicator of adequate fit between the model under consideration and the world. The program directs you to conclude that the data provide evidence of such a fit.

Case 3. What if the agreement between the data and the prediction was quite likely even if the model in question does not fit the world? That is, what if there are other models that yield the same prediction, and whose fit to the world is, initially, as plausible as the model under investigation? In this case, the agreement between the data and the model under study provides little basis for thinking that the model under investigation fits the world better than any of the other models. On the other hand, the existence of other possible models provides no basis for thinking that the model under investigation does not adequately fit the world. So the agreement between prediction and data turns out to be inconclusive regarding any fit between the model in question and the world. The program directs you to that conclusion.

In sum, whatever the appropriate evaluation happens to be, the program will lead you to draw the right conclusion.

2.10 HOW THE PROGRAM WORKS: THREE EXAMPLES

We will now work through three examples that together illustrate the three possible results of applying the analysis program for reports involving theoretical hypotheses. Each example begins with an actual report that has been only slightly edited and shortened for the purpose of these illustrations. Readers are strongly urged to attempt their own analyses by applying the program to the reports before going on to read the analyses in the text. One always learns more from seeing how something should be done if one has already tried it for oneself.

A Case of Negative Evidence

Gene Analysis Upsets Turtle Theory

A widely accepted theory explaining why green turtles migrate 1250 miles to an island in the middle of the Atlantic Ocean to lay their eggs and then swim back to Brazil is invalid, researchers have concluded from genetic analyses.

According to the popular theory, the turtles started coming to the island more than 40 million years ago, when it was close to the shore of South America, and just kept coming as the island moved farther and farther away. But the new research concluded that the turtles had been using the island for only a few tens of thousands of years.

The now invalid explanation was advanced fifteen years ago by Archie Carr and Patrick J. Coleman. It was a bold idea, based on knowledge that the Atlantic Ocean was born 70 million years ago and began spreading. This gradually increased the isolation of Ascension Island, formed by volcanic activity on the ocean's centerline.

As long as the Atlantic was narrow, according to this hypothesis, remote ancestors of today's green turtles had no trouble reaching the island and laying their eggs in its beaches. They then returned to the shallow waters along the Brazil coast to feed on its marine grasses. After 40 million years, the ocean had grown to substantial width, but, according to the hypothesis, the turtles continued to reach the island by some mysterious form of navigation that still enables them to find it.

Now three scientists have examined the extent to which genetic material in the Ascension Island turtles has changed from that of the same species elsewhere. The difference, they believe, is far too small to have evolved over 40 million years. But it is sufficient to show that the Ascension Island turtles are a distinctive group from those nesting at other Atlantic sites, having probably used the island for not more than a few tens of thousands of years.

A comparison was made of a specific locus (the mitochondrial DNA) in the genetic material of turtles from four widely separated regions of the earth. Earlier work had indicated how fast, in a particular population, subtle changes in this material occur over centuries or millions of years. In the 1987 nesting season, eggs, or turtle hatchlings, were taken from twelve nests on French

Frigate Shoal in the Hawaiian Islands, ten on Hutchinson Island off Florida, eight on Aves Island off Venezuela, and sixteen on Ascension Island in the mid-Atlantic.

The turtles in Hawaii had presumably been isolated from the other locations since the Isthmus of Panama formed about 3 million years ago. Their DNA, as expected, was most distinctive and provided an index of how fast turtle DNA becomes modified.

Analysis

Step 1. The real world subject matter is the population of turtles that has been found to migrate between the coast of Brazil and Ascension Island 1250 miles out into the middle of the Atlantic Ocean. The question scientists have asked about these turtles is how they ever came to make such a long and difficult trip to lay their eggs.

Step 2. According to the previously standard model, the members of this species of turtles began visiting the island roughly 40 million years ago when it was quite near the coastline of Brazil. But over the next 10 to 20 million years the island moved farther out to sea as the floor of the Atlantic Ocean spread out, moving South America farther and farther from the middle of the ocean. The turtles, in this model, kept going to the island even though the trip got longer and longer.

Step 3. Examination of parts of the DNA of turtle eggs taken in 1987 from four different groups of turtles around the world provides a measure of how long a particular population of turtles has existed as a distinct population. By this method, it was determined that the Ascension Island turtles have existed as a distinct population for only a few tens of thousands of years.

Step 4. The model predicts that any measurements of the age of this population of turtles should yield an age of roughly 40 million years. According to the model, that is how long ago the turtles would have had to begin migrating to Ascension Island.

Step 5. The data and the prediction clearly disagree. The prediction is that the age of the population should be measured at roughly 40 million years. The data puts the age at more like 40 thousand years. That makes the prediction roughly a thousand times greater than the data indicate! The data, therefore, provide clear evidence that the model fails to fit the real world. As the headline says, the gene data upsets the turtle theory.

Note that the analysis has only five steps. The flow chart makes it clear that whenever the data and prediction disagree, your analysis will terminate at Step 5. Note also that, strictly speaking, the prediction is not that the turtle population is 40 million years old. Rather, the prediction must be about the data that would be expected if the model fit the world. So the prediction is that the data obtained by examining DNA samples from turtle eggs would be of the type that is known to indicate an age of 40 million years.

A Case of Positive Evidence

New View of the Mind Gives Unconscious an Expanded Role

For decades, mainstream research psychologists suppressed the notion that crucial mental activity could take place unconsciously. But now, in the wake of exciting new studies, experimental psychologists are taking the unconscious more seriously. Among the most influential of the new studies are the investigations into the role of the unconscious in the visual perception of objects and words.

One of the main researchers in this new area is Dr. Anthony Marcel of Cambridge University. He has developed a model of unconscious perception in which the unconscious mind perceives and remembers things of which the conscious mind is unaware. One of the most impressive tests of this model involves what is called "unconscious reading."

In these experiments, Dr. Marcel flashes a word on a screen for a very short time. In addition, the word of interest is masked by being surrounded with other nonsense words, such as "esnesnon." When asked directly, the subjects were unable to say what real word appeared on the screen. Dr. Marcel then asked his subjects to guess which of two words looks like the masked word. For example, the masked word might be "blood" and the two choices for look-alikes might be "flood" and "week." The subjects were correct in their guesses an astonishing 90 percent of the time.

Analysis

Step 1. The aspect of the real world being investigated is the human mind, particularly human perceptual abilities.

Step 2. The model is not described in much detail. The important thing is that, in Dr. Marcel's model, the human mind has an unconscious component that "perceives and remembers things of which the conscious mind is unaware."

Step 3. The data come from the unconscious reading experiments. In these experiments, the subjects guessed the correct word "an astonishing 90 percent of the time" even though they could not consciously report what word they had seen flashed on the screen.

Step 4. The model, as reported, does not yield any very precise predictions. It suggests, however, that the subjects might be able unconsciously to perceive and remember the masked words that they could not consciously identify. If so, then their guesses about the word should be right more often than not.

Step 5. The prediction and the data clearly agree. The subjects guessed the correct word far more often than not.

Step 6. That the data are described as "astonishing" is a good indication that these results are thought not to be very likely in the absence of something like unconscious perception. The only obvious alternative model is that the subjects were just guessing randomly. But it is very unlikely that random guessing between two alternatives would produce 90 percent correct answers. So the data do support the hypothesis that a model that includes unconscious perception does fit the human mind.

In this example, the model is only vaguely described and it is difficult to see a clear connection between the model and the prediction. Such cases often result in an inconclusive evaluation. What saves this case is that the data are quite dramatic and it is difficult to imagine another plausible model that could explain the data.

A Case of Inconclusive Data

Was That a Greenhouse Effect? It Depends on Your Theory.

The memorably uncomfortable summer of 1988 has left many Americans with a suspicion that nature is at last getting even for mankind's wanton pollution of the atmosphere. From California to the Carolinas, the summer's heat wave and drought took a sobering toll. Electric power faltered, vast forests went up in flames, river navigation was throttled, crops failed.

The greenhouse effect—the trapping of solar heat by pollutant gases in the atmosphere—became a household phrase. Some climatologists warned that unless we quickly mend our ways, the world's grain belts will turn to dust bowls, coastal regions will be flooded, forests will die, and countless species will become permanently extinct. On June 23, Dr. James E. Hansen of the National Aeronautics and Space Administration caught the nation's attention when he told a Senate committee that the warming trend almost certainly stems from the greenhouse effect. A crisis, he warned, may not be long in coming.

But forecasting climate has never been as straightforward as scientists could wish. Many are not even sure that this summer's weather was really symptomatic of any trend at all. According to the Climate Analysis Center of the National Weather Service, July of this year, when the heat wave was at its peak, was only the 11th hottest July in 58 years of record keeping. And while China also suffered a heat wave and a drought, Ireland and parts of Western Europe were unusually chilly and wet.

A. James Wagner, an analyst at the Weather Service, acknowledges that during this decade the world has seen the four warmest years of the past century—1980, 1983, 1986, and 1987. "But I don't feel that the evidence is overpowering that this is anything more than a normal fluctuation," he said.

Climatologists have invented a number of models in an attempt to understand fluctuations in the weather. One such model, which seems to mimic the real climate quite realistically, was devised by Dr. Edward Lorenz of the Massachusetts Institute of Technology. This model, which does not take carbon dioxide into account but does reckon on the interactions of the atmosphere with the ocean, exhibits large variations.

"The Lorenz model was run backward on a computer for the equivalent of about 400 years," Mr. Wagner said, "and the large fluctuations it sometimes produced, which were not entirely random but were not cyclical either, were quite startling." The swings, he said, were as much as plus or minus 3.6 degrees Fahrenheit in global temperature from one year to the next. The model sometimes produced clusters in which several years close together were unusually hot—a pattern imitating the real climate of the 1980s.

Analysis

Step 1. The real world object of study is the climate of the earth.

Step 2. The model is the greenhouse model. According to this model, the earth's atmosphere acts like the windows in a greenhouse, trapping light and heat under the atmosphere. Carbon dioxide increases the efficiency with which the atmosphere traps heat.

Step 3. The data include the drought that covered much of the United States during the summer of 1988. Also included is information about temperatures during other years, both recent and in the more distant past.

Step 4. The prediction is that the earth's temperatures should be increasing. However, the model is not developed in sufficient detail to permit precise predictions of how much the temperature should increase by specified dates, or whether the increases would be uniform around the world.

Step 5. The data and the prediction seem to agree. The summer of 1988 was unusually warm, and for most of this decade, at least in the United States, the weather has been quite warm relative to earlier years.

Step 6. The data, however, seem to be relatively likely to have occurred even in the absence of any greenhouse effect. Some climatologists have even developed an alternative model, the Lorenz model, which predicts relatively large fluctuations in the earth's temperature. These last few years could be one of those fluctuations. The data, therefore, are inconclusive regarding the applicability of the greenhouse model to our current climate. That does not mean, however, that the greenhouse effect is not operative or that it will not show up with more dramatic force in the future.

2.11 CRUCIAL EXPERIMENTS

There is one special circumstance in the evaluation of theoretical hypotheses that has fascinated scientists since the seventeenth century. This is an experimental setup that allows one to make a clear choice between two rival models. If the experiment is well designed, one model will come out the winner, and the other the loser. The fate of both is settled in one stroke. Following the terminology of the seventeenth-century scientist, philosopher, and statesman Francis Bacon, experiments of this type are called crucial experiments. So powerful are crucial experiments that scientists, and even historians of science, often reconstruct the past so as to make it seem that a particular experiment was a crucial experiment even though, in fact, no one at the time thought of it that way.

The Structure of Crucial Experiments

Figure 2.13 provides a schematic picture of a crucial experiment. Some bit of the real world is put through the apparatus, which produces a reading on a scale. The reading is the data. There are two rival models of the material under investigation. For the moment, we will simply call these $\mathbf{M_1}$ and $\mathbf{M_2}$, respectively. The experiment is cleverly designed so that if $\mathbf{M_1}$ fits the material in the apparatus, it is very likely that the apparatus would produce a reading in region $\mathbf{R_1}$ of the scale, and very unlikely that it would produce a reading in region $\mathbf{R_2}$ of the scale. Similarly, the design ensures that if $\mathbf{M_2}$ fits the material in the apparatus it

Figure 2.13

A schematic representation of a crucial experiment.

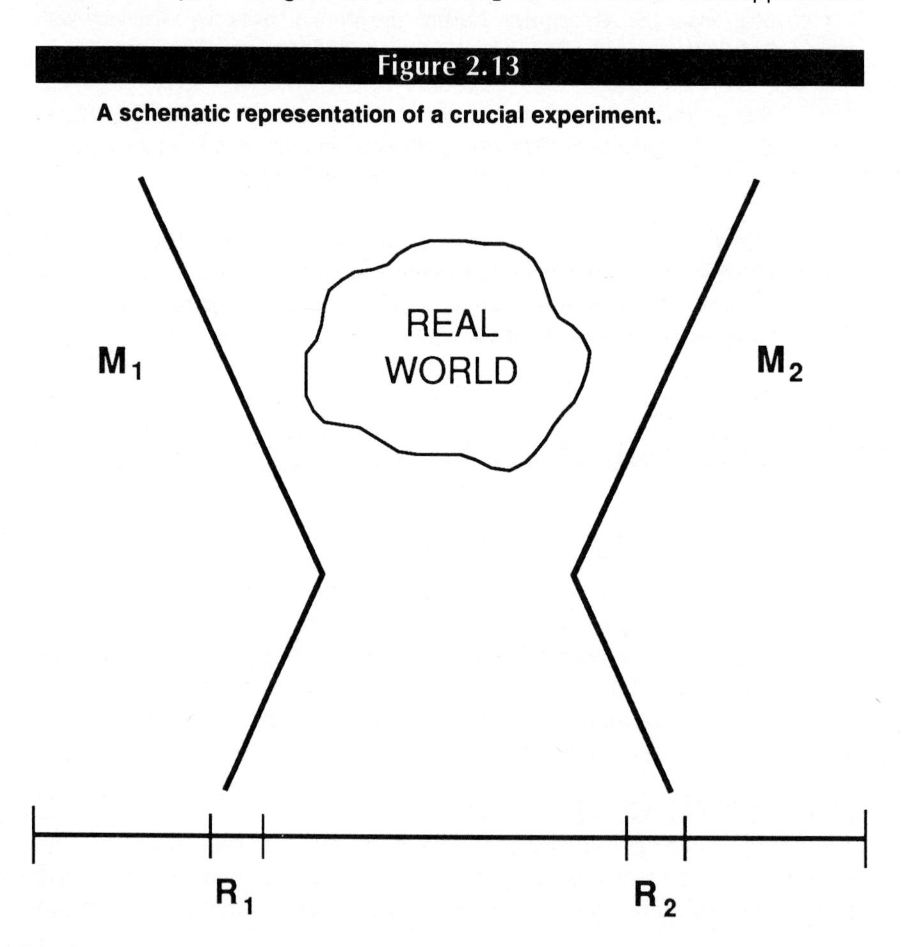

is very likely that the experiment would yield a reading in the region $\mathbf{R_2}$ of the scale, and very unlikely it would produce a reading in region $\mathbf{R_1}$. So $\mathbf{M_1}$ predicts a reading in region $\mathbf{R_1}$, and $\mathbf{M_2}$ predicts a reading in region $\mathbf{R_2}$.

The beauty of this design comes across when one considers that the result of the experiment is to be used to decide which model best fits the world. The strategy is that if the experiment produces a reading in region $\mathbf{R_1}$, then one chooses $\mathbf{M_1}$ as the best-fitting model. Likewise, if the reading is in region $\mathbf{R_2}$, one chooses $\mathbf{M_2}$ as the best-fitting model.

To see why this is an effective strategy, consider the matrix shown in Figure 2.14. Because there are two models to choose between, and two different choices that can be made, the whole experiment has four, not just two, possible final results. Two of these results are *correct* choices. These are choosing $\mathbf{M_1}$ as the best-fitting model when, in fact, $\mathbf{M_1}$ is the best-fitting model, or choosing $\mathbf{M_2}$ as the best-fitting model when, in fact, $\mathbf{M_2}$ is the best fitting model. The other two possible results are clearly *incorrect.* These are choosing $\mathbf{M_1}$ as the best-fitting model when, in fact, $\mathbf{M_2}$ fits best, or choosing $\mathbf{M_2}$ as the best-fitting model when, in fact, $\mathbf{M_1}$ fits best.

The experiment is designed with the presumption that one of the models provides an adequate fit and the other does not. So, initially, there are two possibilities. We can consider each in turn. Suppose, first, that $\mathbf{M_1}$, in fact, provides the better fit. The design ensures that the likely result of the experiment will be a reading in region $\mathbf{R_1}$. In this case, one correctly chooses $\mathbf{M_1}$ as the best-fitting model. On the other hand, suppose that $\mathbf{M_2}$, in fact, provides the better fit. The design ensures that the likely result of the experiment will be a reading in region $\mathbf{R_2}$. In this case, one correctly chooses $\mathbf{M_2}$ as the best-fitting model. Either way, one is very likely to make a correct choice and very unlikely to make an incorrect choice. The only open question is which of the two possible correct choices one will make. That depends on the data.

In actual practice, experiments that are designed to be crucial experiments often do not work out. The data obtained turn out to be readings in

Figure 2.14

A matrix representing the choice between two theoretical models in a crucial experiment.

	$\mathbf{M_1}$ BEST FITS THE REAL WORLD	$\mathbf{M_2}$ BEST FITS THE REAL WORLD
CHOOSE $\mathbf{M_1}$ AS BEST-FITTING MODEL	**CORRECT**	**INCORRECT**
CHOOSE $\mathbf{M_2}$ AS BEST-FITTING MODEL	**INCORRECT**	**CORRECT**

neither of the regions predicted by the two models. Unless there was some mistake in the execution of the experiment or in the derivation of the predicted readings, one can only conclude that neither of the two models adequately represents the real world material. The initial presumption that one of the two models fits has to be rejected.

Even though definitive crucial experiments are relatively rare, the possibility of devising such experiments provides a clear and simple, idealized model of scientific reasoning at its best. Understanding this model of scientific reasoning can help one evaluate the less than ideal cases one meets every day.

An Analysis of a Crucial Experiment

You should attempt to evaluate the following report before reading the sample analysis that follows.

Mutations

It has long been known that changes in the genes of organisms can occur. Such changes are commonly called "mutations." In the 1940s, it was not yet known how mutations occur. Part of the answer to this question was given by a famous experiment performed by Max Delbrück and Salvador Luria.

It had recently been learned that some types of viruses (bacteriophages, or phages) can attack and kill some types of bacteria. It is relatively easy to grow bacteria in covered dishes containing nourishment in which bacteria generally thrive. These are called "bacteria cultures."

Delbrück and Luria discovered that in some bacteria cultures a few of the bacteria survive attacks by phage viruses. Moreover, descendants of the surviving bacteria tend also to survive phage attacks. This shows that the genes of some of the bacteria had undergone mutations that made them resistant to the phage virus, and that these resistant bacteria passed their mutant genes on to their offspring.

The question remained as to whether the mutations that made the bacteria resistant were caused by the attacking virus itself, or whether they merely happened by chance. The experiment at issue was designed to answer this question regarding the cause of the mutations.

Delbrück and Luria considered what would happen if a number (say twenty) of bacteria cultures, each with a similar small number of bacteria, were allowed to grow for a short time, then all were injected with the same quantity of phage virus, and then were allowed to grow some more. If the phages were producing the mutations, they argued, then all the bacteria cultures should end up with roughly the same number of resistant bacteria.

On the other hand, if the mutations were arising by chance, it follows that those bacteria cultures in which the chance mutation happened to occur early in the experiment would end up with many more mutant bacteria than those cultures in which the mutation happened to occur late in the experiment. The earlier mutant bacteria would have a longer time to multiply. Those cultures in which the chance mutation happened to occur at some intermediate time would end up with an intermediate number of mutant bacteria. If it is a matter of pure chance when the mutation occurs, one would, therefore, expect that by the end of the experiment there would be a large variation in the numbers of mutant bacteria in the different bacteria cultures.

Delbrück and Luria prepared a number of bacteria cultures, then introduced the phage virus, and later found that the actual number of resistant bacteria differed widely from one bacteria culture to the next.

Analysis

One way to evaluate such reports would be to run through our standard program twice, once for each model. It is quicker, and more enlightening, however, to do just one analysis. But in this case many of the steps will have two answers. Also, the way this particular report is written makes it natural to consider the predictions before the data, thus putting Step 4 ahead of Step 3.

Step 1. The real world object of study is the process by which mutations arise in bacteria.

Step 2. There are two models: (1) according to the causal model, the mutations are caused by the action of the phage virus on the genes of the bacteria, and (2) according to the chance model, the mutations in the genes of the bacteria occur by chance, independent of any action by the phage viruses.

Step 4. There are two predictions, one for each model: (1) the causal hypothesis yields the prediction that the number of resistant bacteria will be roughly the same in all of the bacteria cultures, and (2) the chance hypothesis yields the prediction that there will be a large variation in the numbers of mutant bacteria in the different bacteria cultures.

Step 3. They found that the actual number of resistant bacteria differed widely from one bacteria culture to the next.

Step 5. The data (1) agree with the prediction derived from the chance hypothesis, and (2) disagree with the prediction derived from the causal hypothesis. The data, therefore, provide good evidence that the causal hypothesis is mistaken.

Step 6. The only rival to the chance hypothesis mentioned in the report is the causal hypothesis that yields a different prediction from the chance hypothesis. It is difficult even to imagine another hypothesis. So it does not appear likely that there would have been a large variation in the number of resistant bacteria if the chance hypothesis were false. The data, therefore, provide good evidence that the chance hypothesis is correct.

EXERCISES

Analyze these reports following the six-point program for evaluating theoretical hypotheses developed in the text. Number and label your steps. Be as clear and concise as you can, keeping in mind that you must say enough to demonstrate that you do know what you are talking about. A simple "yes" or "no" is never a sufficient answer. Many of these reports are taken directly from recent magazine or newspaper articles and are presented here with only minor editing.

Exercise 2.1

Einstein's Impossible Ring: Found

A phenomenon first predicted by Albert Einstein in 1936, and then dismissed by him as something that would be hopeless to look for, has now been found. Astronomers conducting a survey of radio sources at the Very Large Array (VLA) radio telescope near Socorro, New Mexico, have discovered an object in the constellation Leo that has been imaged into a complete ring by the gravitational-lensing effect.

"Of course, we'd all heard of Einstein rings," says team member Jacqueline Hewitt of the Haystack Observatory in Massachusetts. "But when I saw it come up on the computer screen, I thought at first it was a problem with the [VLA's image analysis] software."

It was not. Yet her skepticism was understandable. According to Einstein's general theory of relativity, gravitational lensing would happen when light or radio waves from a distant galaxy or quasar pass by a massive foreground object on the way to Earth. The object's gravity would deflect the radiation and thus produce one or more distorted images of the source. A number of such images have actually been found during the past decade. As Einstein himself pointed out, however, the image can only form a complete ring if a source and the lensing object are precisely lined up with the earth—which seems absurdly improbable.

Except that there it was on Hewitt's computer screen—radio source MG1131 + 0456, a tiny oval about 2 arc seconds across with elongated bright spots at either end. In subsequent observations, Hewitt and her colleagues were able to rule out the possibility of its being a supernova remnant, or any other such ring-like structure. Moreover, using a regular optical telescope, they obtained optical images of a candidate for the imaging mass: a 22nd magnitude object whose shape and other characteristics are those of a large elliptical galaxy.

Exercise 2.2

Why is the World Full of Large Females?

Animals come in a vast range of sizes, from the tiniest zooplankton to the largest whale. Absolute body size has a crucial influence on a species' life history, affecting such factors as metabolic rate, longevity, and territorial range. And, within a species, relative body size—females compared with males—is important in behavioral ecology terms too. In most species in the world, females are larger than males, although this rule applies more to groups, such as insects, fishes, amphibians, and reptiles, than it does to mammals and birds. Nevertheless, the largest animal that has ever lived is a female: the female blue whale.

Why females should attain a larger body size than males has long fascinated biologists. Darwin had an explanation for it, namely: "Increased size must be in some manner of more importance to the females . . . and this, perhaps, is to allow the production of a vast number of ova." This so-called, fecundity-advantage model "has achieved the status of conventional wisdom," says Richard Shine of the University of Sydney, Australia. The model appeals through its simplicity and its consistency with many empirical observations. However, it had not been formally tested, says Shine, a deficiency he has recently repaired. He finds that even though the model may apply in some species it is by no means universal.

It is no easy task, of course, to solve the question of why one sex may be bigger than the other in a particular species, not least because there are two partners in the game. Specifically, the female might be the bigger of the sexes because of the kind of selective advantage that Darwin proposed; but it is equally true that if males evolve small body size for some different, adaptive reason, then the same pattern of body size dimorphism would apply. Several biological factors are likely to be operating in any particular case, and this should always be borne in mind when looking for the *factor.*

Shine elected to test the model in something of a roundabout fashion, thereby hoping to avoid confounding variables that might affect body size in different directions. He measured male-female body size differences in a series of lizard species, some of which produce variable clutch sizes while in others the clutch size is constant. "If the main selective pressure for large female size is an associated increase in fecundity," says Shine, "the species with invariant clutch sizes would have no such advantage and females should tend to be smaller (relative to males)."

It turns out that in anoline iguanids, which produce a single egg, the proportion of species in which the female is larger than the male is about the same as in other iguanids in which clutch size is variable. "The same tends to be true for other lizards with invariant clutch sizes," says Shine. "These data, involving at least seven separate phylogenetic lineages of lizards, appear to falsify the main prediction of the fecundity-advantage model."

Exercise 2.3

Memory Transfer

In the 1950s, biologists and psychologists began speculating that memory works by changing and storing certain chemicals in the brain. That is, they began developing chemical models of memory. In the 1960s, it occurred to some psychologists at the University of Michigan that, if memory is a matter of stored chemicals, one might be able to transfer memory from one organism to another simply by transferring chemicals from the brain of one to the brain of another. Experiments were done on both worms and rats. In the rat experiments, a group of rats was taught to get a small drink of milk by pressing a lever on the opposite side of their cage from where the milk was dispensed. This is difficult for rats to learn. The average time it took a typical group of rats to learn it was about 25 hours. Then, chemicals from the brains of the trained rats were extracted and injected into the brains of other similar rats. Still other rats were injected with material from the brains of untrained rats. The injected rats, they reasoned, should learn faster if they received injections from the previously trained rats. The rats that received injections from trained rats learned the trick in an average of only 3 hours. The others averaged around the usual 25 hours.

Exercise 2.4

Green Antarctica

Most of us think of Antarctica as a Dantean wasteland, locked—past, present, and future—in the frozen, white grip of perpetual polar winter. Until recently, geologists have held much the same view. But new findings indicate that Antarctica has gone through periods during which it was warm enough to actually turn green.

Camped out for six weeks in the highlands of the Transantarctic mountain range, a team led by Peter Webb of Ohio State University found 2-million- to 3-million-year-old fossils of trees, plants, pollens, and spores.

The evidence has led Webb to paint a new, revised picture of an Antarctica that went through periodic cycles of dramatic warming. During those epochs, he says, Antarctica may have looked much like the hardy, low-lying forests currently found in southern Chile, New Zealand, and Tasmania, where dwarf shrubs and small beeches—"They look a little like bonsai trees," he says—cling to rocks and point perpetually in the direction of the strong prevailing winds. Interestingly, Webb has recently seen reports indicating that some of the same fossils have been found in the Arctic. Periodic warming, he concludes, "may be a bipolar phenomenon. It's impressive that similar species developed at both ends of the earth."

Exercise 2.5

Hot Extinction Theories

Many scientists now agree with the notion that an asteroid 10 to 20 kilometers across slammed into Earth 65 million years ago, at a time termed the Cretaceous-Tertiary (K-T) boundary. But the debate rages on about what toll it took on our planet's life forms, and even on whether the result of the impact produced a warming or cooling of the earth's atmosphere.

At a meeting of planetary scientists in Houston, Texas, John D. O'Keefe and Thomas J. Ahrens of the California Institute of Technology put forth the idea that the earth became warmer after the cataclysm. They reason that an impact occurring in shallow seas, or on dry land that had once been under water, would vaporize thick beds of carbonate sediments and thus inject huge quantities of carbon dioxide (CO_2) into the atmosphere. (By current estimates, such sediments have locked up an amount of CO_2 equivalent to between fifty and eighty times the total mass in Earth's present atmosphere.) The CO_2 would then block the escape of Earth's infrared radiation to space, creating a greenhouse effect that would persist for millennia. Calculations have shown that a global warming of only 10 degrees Kelvin would have been sufficient to trigger the K-T extinctions.

To support their model, the Caltech researchers used powerful guns to propel projectiles into calcite ($CaCo_3$) targets at 4 to 6 kilometers per second. Then they measured how much CO_2 was released as a function of impact velocity and extrapolated the results to a global scale. For example, an object 20 kilometers across, striking a 1-kilometer-thick calcite bed at 20 kilometers per second, should double the amount of CO_2 now in the atmosphere.

Admittedly, the O'Keefe-Ahrens concept requires not only that the impactor strike carbonate beds, but also that the beds be *thick enough to release an adequate mass of CO_2. Some scientists in Houston questioned the likelihood that all these conditions could be met. Others emphasized that more research is needed to determine the total response of the atmosphere to a major impact.*

Exercise 2.6

A Heresy in Evolutionary Biology

As anyone with even a passing knowledge of evolutionary biology knows, natural selection is a twofold process: the generation of genetic mutation followed by the fixation of variants that are favored by prevailing conditions. In the world of evolutionary biology, one thing has seemed certain: the generation of genetic mutations is a continuous and random process, uninfluenced by external circumstances. However, if John Cairns, Julie Overbaugh, and Stephen Miller of the Harvard School of Public Health are correct in their interpretation of certain experiments with the bacterium Escherichia coli, *that certainty may be on shaky ground.*

In a paper published in Nature, *Cairns and his colleagues aim "to show how insecure is our belief in the spontaneity (randomness) of most mutations." The Harvard researchers describe the results of a handful of experiments which, they suggest, demonstrate that "bacteria can choose which mutations they should produce." Anything more heretical can hardly be imagined. They do add, however, that "this is too important an issue to be settled by three or four rather ambiguous experiments."*

One of the experiments involves taking colonies of E. coli *that are incapable of metabolizing lactose and exposing them to the sugar. If the lactose-utilizing mutants simply arise spontaneously in the population and are then favored by prevailing conditions, then this would lead to one pattern of new colony growth. A distinctly different pattern is produced if, under the new conditions, the rate of production of lactose-utilizing mutants is enhanced. The observation is something of a mixture of patterns, indicating that directed mutation appears to be occurring. "This experiment suggests that populations of bacteria . . . have some way of producing (or selectively retaining) only the most appropriate mutations," note Cairns and his colleagues. They cite two other types of experiment that can also be interpreted in this way.*

Because the randomness of mutation has been so fundamental to evolutionary biology since the 1940s, few researchers have cared to test the notion directly. There are, therefore, no data beyond those from this handful of experiments that might indicate how general a phenomenon directed mutation might be. Nevertheless, Cairns suspects that it might well turn out to be rather widespread, at least in bacteria. Kent Holsinger, a theoretical population geneticist at the University of Connecticut, says that "if it is general and not just confined to E. coli *and other bacteria, it could have major implications for evolutionary biology. At the very least, he notes, "there is something going on here that we haven't considered."*

Exercise 2.7

Quartz Discovery Supports Theory That Meteor Caused Dinosaur Extinction

Researchers say they have strong new evidence that the age of dinosaurs ended 65 million years ago when a giant meteorite, or comet, slammed into the earth with the energy of a billion atomic bombs. Scientists with the U.S. Geological Survey office in Denver said that microscopic particles of quartz found in Europe, New Zealand, the Pacific Basin, and elsewhere contain structural cracks associated with the impact of a large body hitting the earth. The mineral debris indicates that a single catastrophic event, and not a series of volcanic eruptions suggested by other scientists, ended the 150-million-year reign of the great lizards, they said. The microscopic fracturing found in the quartz is more like that associated with the pressures of a massive impact than what would have resulted from volcanic activity, they say in a study published in the journal Science.

Bruce Bohor, Peter Modreski, and Eugene Foord said the "shocked quartz" is found in the same sediment layers that contain unusually high levels of iridium, a metal common in asteroids, meteors, and comets. The researchers said the latest findings bolster the controversial, 10-year-old theory of Nobel Prize-winning physicist Luis Alvarez and his geologist son, Walter, that a single catastrophic event led to a great extinction of life on Earth. Geological evidence appears to show a massive extinction of dinosaurs beginning about 65 million years ago, but scientists have been unsure of the reason why.

The Alvarez theory says the impact of the extraterrestrial body released energy equivalent to that of 6 billion Hiroshima atomic bombs and threw up a giant cloud of debris that encircled the globe and diminished sunlight for months, if not years. Climate cooling, resulting from the dust blocking sunlight, resulted in the death of dinosaurs and many other types of animal and plant life, according to the theory.

The theory is based upon finding up to 600 times normal levels of iridium in clay deposits from the period, and Bohor said, in a telephone interview, that the same iridium concentrations have been found at every site of the telltale quartz particles. The even distribution of the shock quartz, and certain minerals combined with it, point to a big comet or meteorite striking a continental area in the Northern Hemisphere, he said. A body six miles wide hitting the earth at 45,000 miles per hour, as calculated by the Alvarez theory, could have blasted debris high enough into the atmosphere to account for the worldwide shock quartz distribution, Bohor said.

High pressure associated with volcanic activity can fracture quartz crystals, and proponents of the volcanic theory of dinosaur extinction say this is the source of the shock crystals found in sediments from the period. However, Bohor and his colleagues said quartz fractures caused by impacts are distinct from those resulting from other pressures. "When a meteorite strikes the earth, the mass and speed of impact cause a shock," Bohor said. "This shock wave bounces around in different directions as it hits other objects and comes back into the crystal to produce multiple sets of fracture features unique in impact shocks."

The report said multiple sets of fractures were evident in quartz particles found at several sites: Stevns Klint and Nye Klov in Denmark; Petriccio and Pontedazzo in Italy; Caravaca in Spain; Woodside Creek in New Zealand; and a core taken from the north central Pacific Ocean basin. The samples matched those previously found in Montana and Wyoming, Bohor said, and Soviet scientists recently found the same type of quartz on the east side of the Caspian Sea in the Soviet republic of Turkmen.

Exercise 2.8

The Expanding Universe

One of the most interesting discoveries of the twentieth century is that the universe is expanding; that is, the galaxies are all moving away from each other. This discovery stimulated the creation of numerous models of systems in which such expansion would take place. Two of these models were widely regarded as possibly representing the structure of the real universe. One was an explosion model (the Big Bang theory) in which all matter is originally concentrated in one place and explodes outward. The other is a steady-state model, in which subatomic bits of matter are created out of nothing and eventually move outward, leaving each region of space with the same total amount of matter for all time.

If the universe is an exploding system, it follows that the density of matter (the number of galaxies per cubic light year) gets less and less the farther away from the original explosion one gets. If the universe is a steady-state system, on the other hand, the density of matter should be exactly the same everywhere. This remains true wherever in the universe one happens to be. Whichever direction one looks, the density of the most distant galaxies should be less if the explosion model is correct.

To decide which of these two models best fits the real universe, what we need to do is measure the density of matter in the most distant regions of space. In recent years, radio telescopes have made it possible to make such measurements. These measurements show a clear decrease in the density of the most distant observable galaxies.

Exercise 2.9

New Observations Reveal Cosmic Mystery

In the 1920s, it was discovered that the stars exist in large spiral- or platter-shaped clusters we now call galaxies. Each galaxy contains millions of stars with most near the center and fewer out toward the edges. Our sun and solar system are now thought to be roughly one-third the way in from the edge of the galaxy we call the Milky Way.

Since the discovery of galaxies, astronomers and astrophysicists have naturally wondered how they work, and in particular, what keeps them together. The standard idea has been that a galaxy is held together by the force of gravity. If this is right, then it is possible to calculate the motions of various stars within the galaxy. Unfortunately, it has not until now been possible to measure the motions of stars with sufficient accuracy to determine whether the calculated motions are correct.

Recently, however, computers have been used to sharpen the images of stars produced by large telescopes. This new technique has revealed that the stars on the outer edge of the Milky Way are moving much faster than they should according to the standard calculations.

The result has scientists baffled. They have not yet been able to come up with an alternative model that might explain this surprising result.

Exercise 2.10

Scientists Put a New Twist on Creation of the Universe

A Great Wall of galaxies stretching hundreds of millions of light years across the known universe has been discovered by two Harvard astronomers and threatens long-held, fundamental theories of how the universe came into existence.

The wall, some 500 million light years long, 200 million light years wide, and 15 million light years thick, consists of more than 15,000 galaxies, each with billions of stars. Described as the "largest single coherent structure seen, so far, in nature," the image of the wall emerged after about a decade of effort to map the structure of the universe in three dimensions. Correspondingly, there are large areas of relative void containing comparatively few galaxies.

The astronomers' findings were published in the journal Science. *While the area surveyed is huge, it is only a small piece of the known universe. The astronomers involved said the survey area is to the universe what Rhode Island is to the earth, a small percent. The survey was done by mapping the "red shifts," or motions, of a large group of galaxies some 200 to 300 light years from Earth.*

The survey, using the 60-inch telescope at the Whipple Observatory in Arizona, was conducted by dividing an area of the universe into thin slices, then noting the position of galaxies within the slices. As more of the slices were completed and layered together, a three-dimensional portrait of the area emerged and slowly revealed the massive wall of galaxies.

If this map of the universe proves to be a reality, it means the long-held theories of a universe evolving from a smooth, superhot plasma following a Big Bang some 15 billion years ago might be wrong. The concentration of galaxies around voids filled with either mysterious dark matter, or nothing, points to a very uneven, lumpy universe that apparently could not evolve out of the smooth beginnings described by current theories.

What, if anything, is in the voids remains unknown and could be a key to understanding the universe. Many scientists believe that 90 to 99 percent of the matter in the universe is dark and has yet to be detected. Whether the huge voids, some as large as 150 million light years in diameter, contain dark matter is not known.

But given their vast size, said Margaret Geller, one of the two Harvard astronomers to make the discovery, "it may make more physical sense to regard the voids as the fundamental, large-scale structures of the universe."

The fundamental theory most threatened by the discovery is known as the standard model. It holds that the universe originated in a Big Bang and has expanded in a fairly uniform manner for the past 15 billion years. It calls for a homogeneous universe that is very much the same in every direction. But the new findings raise questions about how such an uneven system of voids and galaxies could have developed out of a uniform, homogeneous beginning in 15 billion years.

"In my opinion, something is fundamentally wrong in how (scientists believe) the structure of the universe formed," Geller said. "Maybe the universe isn't homogeneous." She said the work is important philosophically because "mapping the universe is about finding its origins. If we can't understand where the universe came from, then we can't understand where we came from."

In their paper detailing the discovery of the Great Wall, Geller and her colleague, John Huchra, hinted that the wall may be much larger than they have measured. The "most sobering" result of their discovery, they said, is that the size of the largest structures they have found "is limited only by the extent of the survey." If the structures and the voids are larger still, then scientists may have to abandon entirely the fundamental idea of a homogeneous universe.

"There is no theory that even comes close to explaining what we are seeing," said Michael Kurtz, also an astronomer at Harvard.

Exercise 2.11

Gravity Waves

Newspapers around the world recently carried reports of a striking new experimental test of Einstein's General Theory of Relativity. According to Einstein's theory, a large rotating mass should give off gravity waves that travel at the speed of light. Several scientists have previously claimed to have detected gravity waves, but none of these claims were very well substantiated. The new experiment was made possible by the discovery of a unique pulsar—a very dense star that emits powerful radio waves in distinctive pulses. This particular pulsar appears to be rotating in orbit around another large but invisible mass (perhaps a black hole). The discovery of this unique pulsar was made using a radio telescope 1000 feet in diameter—the world's largest.

If the observed system indeed fits an Einsteinian model, it should give off gravity waves. Moreover, according to the model, because gravity waves carry off energy from the system, the rate of rotation of the pulsar should be slowing down at the rate of about one ten-thousandth of a second a year. Using the large radio telescope and clocks accurate to fifty-millionths of a second a year, the period of rotation of the pulsar was measured over a period of four years. The most recent measurements show that the pulsar has indeed been slowing down by just about one ten-thousandth of a second a year—as predicted. Even scientists who had previously doubted the existence of gravity waves had to admit that finding the system to have slowed down by the predicted amount was quite remarkable.

Exercise 2.12

Comet Source: Close to Neptune

Comets have been considered outsiders, visitors from an enveloping cloud of inactive comets on the far fringes of the solar system. But new computer simulations of how comets can be drawn into the inner solar system eliminate a far-distant, spherical cloud of comets as the source for a major class of comets. The only practicable source in these simulations for the comets that now follow small, quick orbits near the planets, is a flat disk of comets lying just outside the orbit of Neptune. That is a distance of little more than thirty times the distance between the sun and Earth (30 astronomical units) compared to the tens of thousands of astronomical units to the distant comet cloud.

Martin Duncan of Lick Observatory, and Thomas Quinn and Scott Tremaine of the University of Toronto, ran their computer model of the solar system to see how some of the peculiar characteristics of the short-period comets, those having orbital periods of 200 years or less, might have originated. Could the cloud of comets, called the Oort cloud, supply objects that behaved like these? It clearly is the source of the 589 known, long-period comets, the ones rarely seen from Earth as they loop by the sun on circuits that require 200 to millions of years to complete. Everything about their motion is consistent with having been jostled out of the Oort cloud by a passing star and into an orbit that passes near the sun.

But the short-period comets do not behave like typical Oort cloud comets gone astray. Most strikingly, the orbits of the short-period comets tend to lie within about 30 degrees of the plane of Earth's orbit, called the ecliptic. And only four of the 121 known, short-period comets have their orbits tilted more than 90 degrees so that they are orbiting in the direction opposite to that of Earth. In contrast, half the comets in the Oort cloud must have such retrograde orbits.

Duncan and his colleagues first checked to see if, as reported fifteen years ago, the four giant planets could select, from comets falling in from the Oort cloud, only those having low-inclination orbits typical of short-period comets. Using a mathematical model that included the sun, the four giant planets, and 5000 comets falling into the vicinity of massive Jupiter from the Oort cloud they ran a simulation on their own souped-up, Sun-3 microcomputer for a total of several months of computer time. They speeded up the calculation of millions of years of orbital evolution by increasing the mass of the giant planets by a factor of up to forty in the simulation.

To their surprise, these modelers found that the planets are not selective at all when they deflect comets into new, smaller orbits. The comets from the Oort cloud that achieved periods of less than 200 years formed a cloud of their own, with no preference for orbiting in a disk near the ecliptic. In addition, three-quarters of them had periods greater than fifteen years; only twenty-one of the 121 observed, short-period comets have periods greater than fifteen years.

Ruling out the Oort cloud, the modelers next tried a belt of low-inclination comets near Neptune's orbit. The idea dates back to a suggestion by Gerard Kuiper in 1951 that it would be only natural to find some debris from the formation of the solar system beyond Neptune. Comet-sized objects would not have formed planets there, but neither the giant planets nor passing stars could easily dislodge them, or even smear their disk into a cloud. In fact, an absence of comets there would imply an oddly abrupt, outer edge to the original solar system disk.

When the simulation was run with a disk of comets orbiting between a distance of 50 and 20 to 30 astronomical units, or well inside the orbit of Neptune, the comparison with reality was impressive. The mean oribital distances of both simulated and observed, short-period comets cluster around 3 astronomical units with a lesser tendency to be near 5 astronomical units, the orbital distance of gravitationally influential Jupiter. That is also where the preponderance of maximum orbital distances lies for both sets of comets. Comets in each set also have a tendency to be passing the ecliptic when they are closest to the sun.

Most crucially, about half of the simulated comets, which started out with inclinations of 0 degrees to 18 degrees, retained inclinations of less than 30 degrees. More than 80 percent of observed short-period comets are confined to such a disk. About 8 percent of the simulated comets had their orbits tilted as much as to be in retrograde motion, compared with 3 percent of observed comets and 50 percent of Oort-cloud comets. As is the case with observed comets, simulated, retrograde comets tended to have periods longer than fifteen years.

"You can't get good agreement," says Tremaine, "if you start off with inclinations far out of the ecliptic. It works if, and only if, the source is in the plane of the ecliptic." That moves the hypothesis of a comet belt lying beyond Neptune from being "very plausible to being the only plausible hypothesis." The Oort cloud as a significant source for short-period comets "is now ruled out," he says.

Exercise 2.13

Project

Find a report of the results of an experiment that is relevant to a theoretical hypothesis. You may find an example in a newspaper or newsmagazine. The Sunday supplement to your local newspaper is a good bet. The weekly science section of The New York Times *is particularly good. You might also try some popular sources that specialize in scientific findings, such as* Scientific American, Psychology Today, *or* Science. *When you have found something that you find interesting and substantial enough to work on, analyze the experiment following the standard program for theoretical hypotheses.*

This exercise may be turned into a longer project by looking for other sources of information on the same theory. You might be able to uncover a whole history of experiments relating to the theory, each of which can be analyzed. You may discover other experiments bearing on hypotheses using similar theoretical models. Alternately, you may discover cases in which such hypotheses were refuted. Perhaps more elaborate models of the same type were then developed to replace the discarded models. You can then look to see whether later evidence supported these new hypotheses, and so on.

One of the side effects of this project is that you may get an idea of the different levels of science reporting in various popular sources. Some sources tell you everything you need to know in order to evaluate the reported hypotheses. Others give so little information that you cannot tell whether the evidence supports the hypotheses or not. You are forced to take their word for it. Most sources fall somewhere in between.

CHAPTER 3

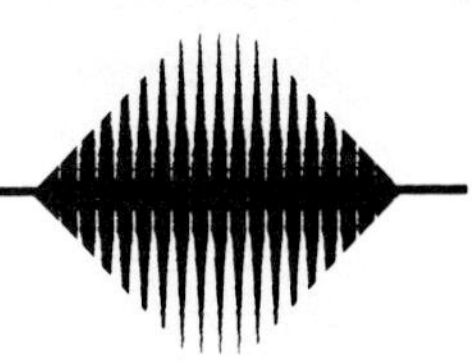

Historical Episodes

In this chapter, we will examine a number of historical episodes involving theoretical hypotheses. These examples serve several purposes. Most important, they provide a set of analog models for the analysis of reports of new cases. That is, when meeting a new case one can often find analogies between the new case and one or another of the historical cases. Exploiting such analogies can often help one quickly to understand and evaluate the new case. But this requires having the basics of the historical case carefully stored in long-term memory for easy access.

There is a second, more general, reason for studying these historical cases. The general conception of science, found among scientists and other educated members of the society at large, has to a great degree been shaped by the history of science as embodied in cases like those we shall be examining. These cases are part of our general cultural heritage. Having at least some familiarity with that heritage is part of being a member of the culture. Of course, the sketches that follow are no substitute for a genuine introduction to the history of science. But even an oversimplified understanding of these episodes is better than no understanding at all.

3.1 PHASES OF VENUS

One of the founders of modern science was Galileo Galilei, universally known simply as Galileo. He was born in Pisa, Italy, in 1564 and did most of his scientific work in the years 1590 to 1630. For some historical perspective, Shakespeare was born two months later than Galileo, Martin Luther had died in 1546, and, of course, Galileo's countryman, Christopher Columbus, first traveled to the New World in 1492. The *Mayflower* landed at Plymouth Rock in 1620.

When Galileo was born, most people thought that the earth was literally the center of the universe. As a student, Galileo was taught the astronomy of the Greek astronomer Claudius Ptolemy, as first published in A.D. 140. In Ptolemy's system, the sun, moon, planets, and stars all revolve about the earth every twenty-four hours. Figure 3.1 is a sketch of the universe according to Ptolemy.

In 1543, the Polish astronomer, Nicolaus Copernicus, had published a book, *On the Revolutions of the Heavenly Spheres,* expounding the view that the sun is the center of the universe and that the planets, including the earth, revolve in circular orbits around the sun. On Copernicus's view, the apparent motion of the rest of the universe is due to the earth itself spinning on its own axis once each day. Copernicus lived just long enough to see his book in print. Figure 3.2 is a sketch of the universe according to Copernicus. It is, of course, similar to the view we now hold.

Figure 3.1

The Ptolemaic model of the universe.

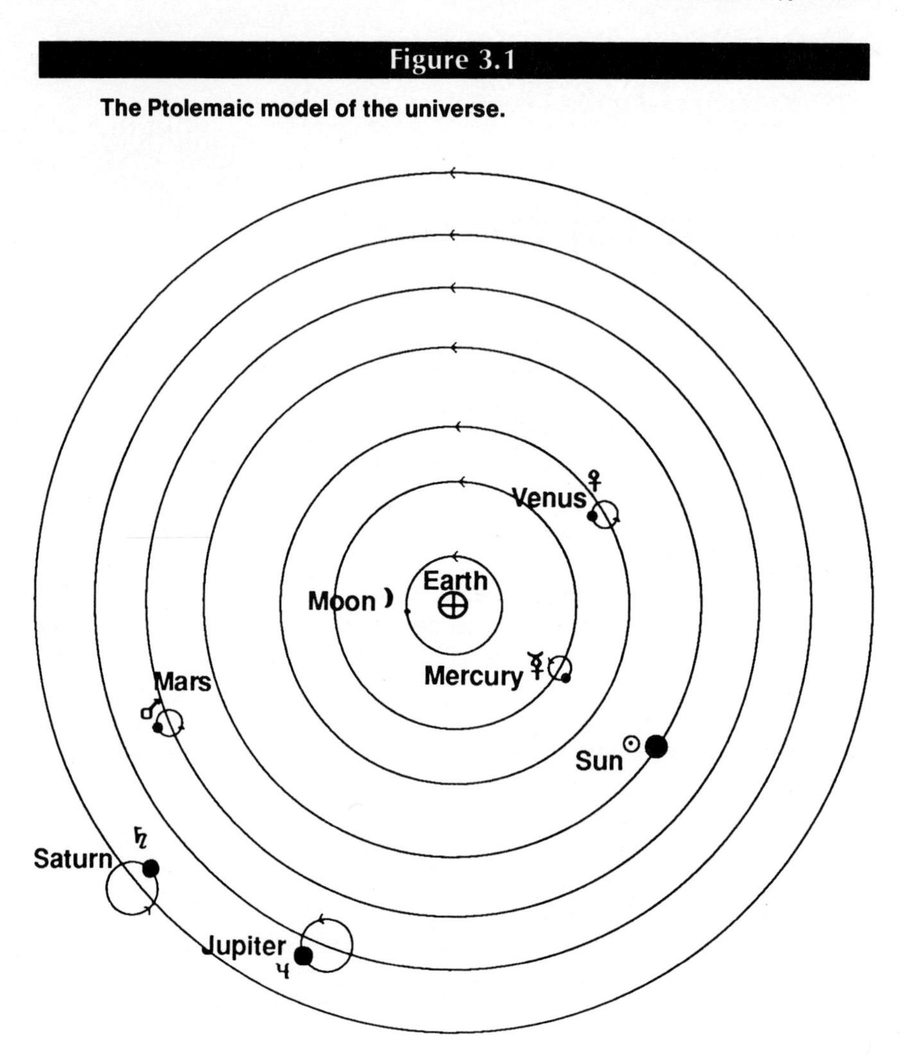

Galileo came to accept the Copernican system early in his life but for a long time kept his opinion largely to himself. In 1589, he became a lecturer in mathematics at the University of Pisa. He studied the properties both of floating bodies and falling bodies, and investigated the motions of pendulums. There is a widely believed story about his dropping two bodies of the same material but different weights from the top of the Leaning Tower of Pisa to prove that they would fall at the same rate. This story is most likely not true.

In 1602, Galileo became a professor of mathematics at the University of Padua. In 1604, a new star, which we now take to have been a supernova, appeared in the sky. Galileo studied it intensively. Sometime in 1609, he heard about the invention of a telescope and set out to build one for himself. A year and several telescopes later he had a fairly good quality instrument with a magnification of about thirty-three times. He turned it to the skies, becoming the

Figure 3.2

The Copernican model of the universe.

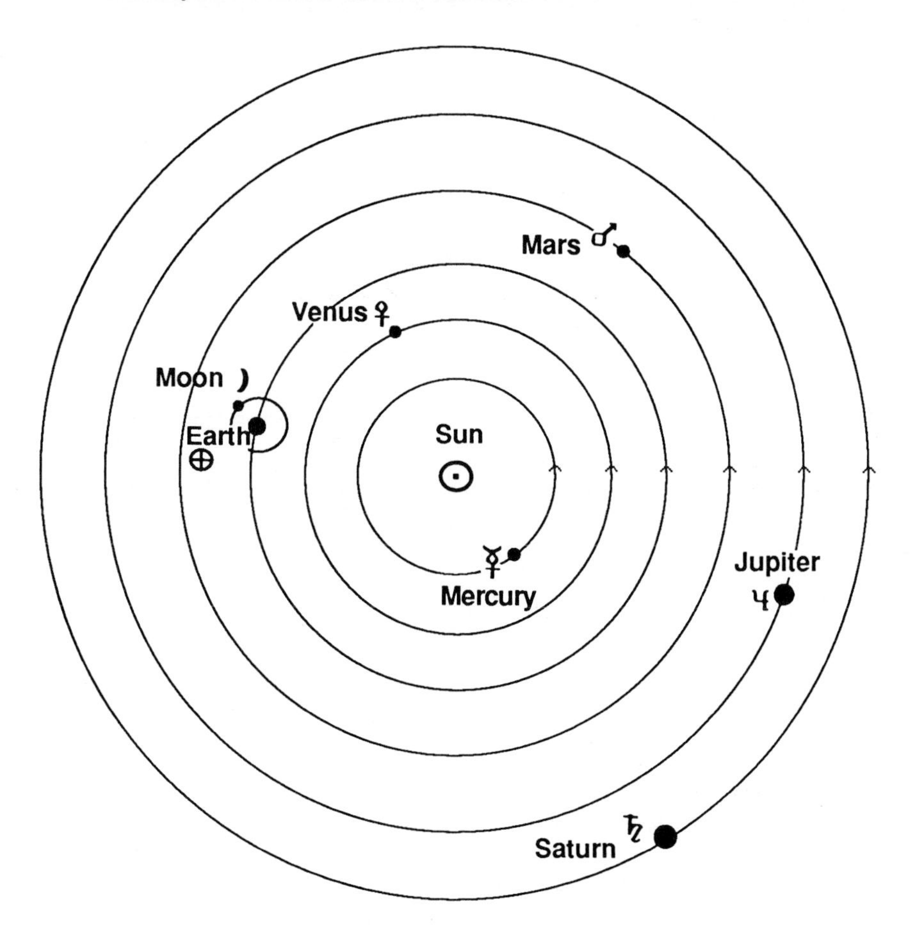

first astronomer to use a telescope. He was the first person to see sunspots, mountains on the moon, and the moons of Jupiter.

Recent scholarship indicates that one of Galileo's students first suggested how the telescope might be used in an attempt to determine whether the Ptolemaic or Copernican picture of the universe is correct. In the Ptolemaic picture, Venus revolves in a small orbit (an epicycle) centered on a line connecting the earth and the sun, as shown in Figure 3.3. It follows that, when viewed from the earth, Venus is mostly dark since it is always illuminated from behind by the sun. At most, only a small crescent-shaped part of Venus will be illuminated, depending on which side of its epicycle Venus happens to be. In the Copernican picture, by contrast, there will be times when Venus is almost fully illuminated. This would happen whenever Venus is on the opposite side of the sun from the earth, as shown in Figure 3.4. One cannot detect these

Figure 3.3

The Ptolemaic model predicts that one cannot observe Venus fully illuminated.

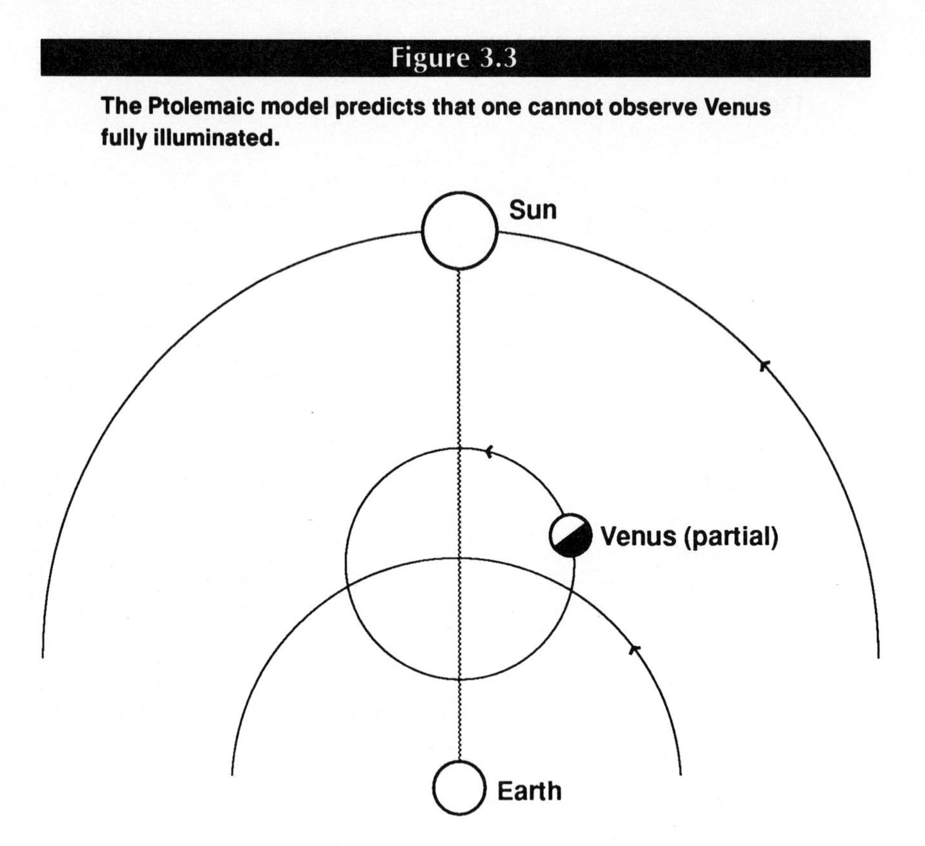

differences with the unaided eye. With a telescope, however, perhaps one could see whether Venus ever exhibits a full Venus the way the moon exhibits a full moon.

There is some historical evidence to the effect that when the theoretical significance of using his telescope to study Venus was first suggested to Galileo, Venus was, in fact, below the horizon for a number of weeks. Being anxious to claim the discovery for his own, Galileo, nevertheless, reported having made the requisite observations. He did not acknowledge the help of his student. Only later, in the latter part of 1610, did he, in fact, observe Venus go from a relatively dark sphere, to a crescent, to a fully lighted sphere, then back through a crescent phase (pointing the opposite direction from before), and finally returning to the original dark phase. James Watson was hardly the first major scientist to go out on a limb in pursuit of an important discovery.

Analysis

This episode sets up quite nicely as a prime example of a crucial experiment. There are two rival models with opposed predictions. The data clearly favor one model over the other. It is natural to identify the predictions before the data. Step 3 and Step 4 are, therefore, reversed.

Figure 3.4

The Copernican model predicts that one can observe Venus fully illuminated.

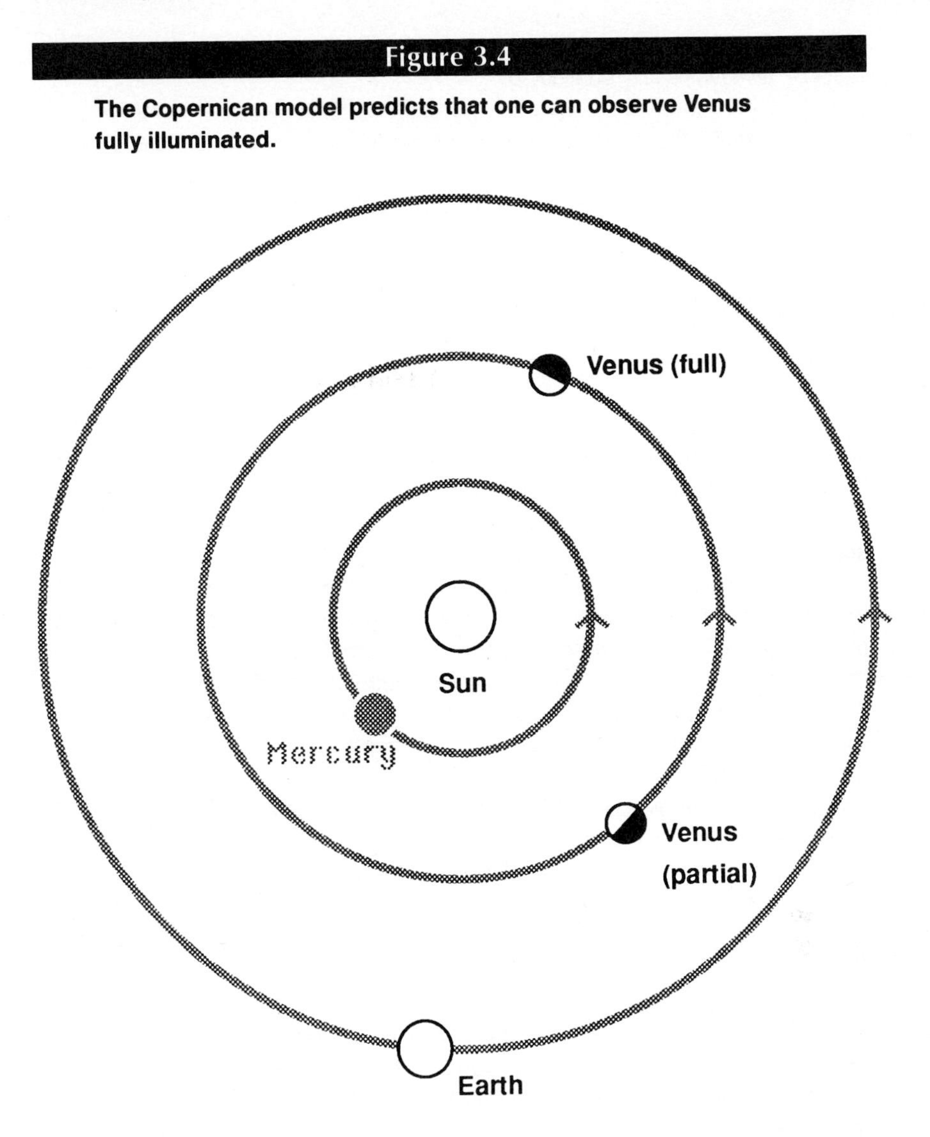

Step 1. The real world objects are the known heavenly bodies near the earth. The theoretical issue is how they are arranged and move relative to one another.

Step 2. There are two models: (1) the Ptolemaic model, as shown in Figure 3.1; and (2) the Copernican model, as shown in Figure 3.2.

Step 4. There are likewise two predictions: (1) the Ptolemaic prediction that Venus can never be seen fully illuminated, as shown in Figure 3.3; and (2) the Copernican prediction that Venus can be seen going through a complete set of phases, including being fully illuminated, as shown in Figure 3.4.

Step 3. The data are that Venus exhibits a complete set of phases and is sometimes fully illuminated.

Step 5. The data (1) disagree with the Ptolemaic prediction, and (2) agree with the Copernican prediction. The data, thus, provide evidence that the Ptolemaic model does not adequately represent the actual universe.

Step 6. The data would be impossible on the Ptolemaic model, and no other plausible models have been mentioned. So the data must be taken as providing positive evidence that the Copernican model provides a good fit to the actual universe.

Popular folklore often portrays Galileo as having slain the dragon of the Ptolemaic system with a single shot from his mighty telescope. Nothing could be further from the truth. Some people were initially convinced of the superiority of the Copernican system, but many more were not. Many questioned whether what Galileo reported seeing were not just reflections in his telescope. Few other people had access to any telescope at all, other than those built by Galileo himself. Also, many people wondered why Venus did not appear, to the unaided eye, to vary as much in its distance from the earth as required by the Copernican picture. Many were unimpressed with Galileo's explanation that, because Venus was more fully illuminated when it was on the other side of the sun, it appeared closer than it really was. And, of course, many people continued to find it unimaginable that the earth really could be both revolving around the sun and spinning on its axis at what had to be tremendous speeds. Why do we not feel any such motion? Indeed, why are we not simply thrown off into space by the rapid spinning motion? Because of these questions, and similar ones, the "scientific revolution" was not completed for fifty to 100 years after Galileo first turned his telescope to the heavens.

Finally, there is the issue of opposition by the Catholic Church to Galileo's views. Again, popular folklore portrays Galileo as the valiant defender of free scientific inquiry against religious dogmatism. The truth is far more complex. Galileo was a deeply religious man. With his blessings, his favorite daughter spent her life in a religious order. For a long time, Galileo thought his work supported the power of the Church. Initially, many religious scholars and authorities of the Church welcomed his findings. Only later, partly because of the continuing threat of the Reformation begun by Luther and others, did the Church attempt to prevent Galileo from teaching, or publishing his views. Nor did Galileo do all that he might have to escape the power of the Church. In 1610, he used his growing fame to secure a position in Florence, where the influence of the Church was much greater than in Padua, which was part of the Venetian Republic. He could later have sought safety and freedom to publish in Venice, or even in Germany, but chose to remain in Florence.

3.2 ISAAC NEWTON AND HALLEY'S COMET

Isaac Newton was born in 1642, the year Galileo died. He brought together the scientific work of the previous century by creating a set of theoretical models that could be applied both to terrestrial objects, such as cannon balls and swinging lamps, and to celestial objects, such as the moon and the planets. So successful were these models that Newtonian science became one of the main inspirations for the Age of Enlightenment, which lasted for most of the eighteenth century.

Newtonian Models

The objects in Newtonian models are called, simply, "bodies." Bodies have a number of important properties, some more familiar than others.

Position. At any moment of time, every body is said to be located at some point in space. That point is its position at that moment.

Velocity. Velocity is change in position per unit of time. One part of velocity is what we commonly call speed, for example, 50 miles per hour. The other part of velocity is direction. A body moving 50 miles per hour heading north and a body moving 50 miles per hour heading south would have the same speed, but opposite velocities, because they are moving in opposite directions.

Acceleration. Acceleration is change in velocity per unit of time. A body going from rest to 60 miles per hour in one minute would be accelerating at an average rate of one mile per hour per second. A body standing still has zero acceleration. So does one moving at a constant 50 miles per hour. A body slowing down has negative acceleration.

Mass. Newton was among the first to realize that there is a distinction between mass and weight. Since the advent of space flight, there have been many dramatic demonstrations of this fact. An astronaut has the same mass both on earth and on the moon, but weighs less on the moon.

Momentum. Momentum is defined as the product of mass times velocity. Whatever amount of mass an object has, its momentum is that number multiplied by its velocity. An object standing still has no velocity and thus no momentum, even though it may have a large mass.

Force. This is the most difficult concept in classical physics. Whole books have been written just on Newton's conception of force, and many others on the concept of force in general. We will restrict our attention to gravitational force. Gravitation is an attractive force produced by the mass of any body. It is because the moon has less mass than the earth, and thus exerts a smaller gravitational force, that one weighs less on the moon than on Earth.

Let us turn, now, to what Newton said about how bodies behave. This is summarized in Newton's three *Laws of Motion* and his *Law of Universal Gravitation.*

First Law of Motion. If there is no force acting on a body, the momentum of that body will remain constant.

This law tells us that if any body is moving with a constant velocity—that is, at a fixed speed in a straight line—then it will just keep moving at that velocity as long as no force interferes. A special case of this law occurs if a body is at rest, that is, not moving at all. According to the first law, it must just stay at rest until some force acts on it.

Second Law of Motion. If there is a force acting on a body, that body will accelerate by an amount directly proportional to the strength of the force and inversely proportional to its mass.

This law tells us that the more force you apply to a body, the faster it will accelerate. It also tells us that a given force will accelerate a lighter mass more than a heavier one.

Third Law of Motion. If one body exerts a force on a second, then the second exerts on the first a force that is equal in strength, but in the opposite direction.

Notice that these laws all tell us what a particle should do if it is subject to certain forces, but none of them tells us what the forces themselves would be. That is the job of the gravitation law.

Law of Universal Gravitation. Any two bodies exert attractive forces on each other that are directed along a line connecting them and are proportional to the product of their masses divided by the square of the distance between them.

This law tells us that the gravitational force between two bodies increases with increasing mass, and decreases rapidly with increasing distance. Dividing by the square of the distance means that doubling the distance between two masses will reduce their mutual force to only one-fourth of what it was. Figure 3.5 illustrates these relationships (insofar as they can be pictured) for the simple case of a system consisting of only two bodies.

Before proceeding, you should make special note of the fact that the above renditions of Newton's laws are given totally in words. No mathematical symbols are used. Yet everyone knows that physics is highly mathematical. What is going on here?

The answer is that, in fact, there is nothing that can be said using mathematical symbols that cannot be said in words. What matters are the underlying relationships, and these can be described in either words or special symbols. The difficulty is that the use of words is so cumbersome that by merely using words it would be practically impossible to do the calculations required to solve particular problems. For example, all the steps in a simple problem in long division could be written out in words, but you could hardly imagine actually figuring out the answer that way. It takes a special symbolism, really a special language, to do that easily. To achieve a fairly good intuitive understanding of the fundamental relationships, however, a special symbolism is rarely necessary. A reasonable, intuitive understanding of what is going on is generally good enough to be able to evaluate the relevance of new data.

Newton's work was so successful because so many things in the world, in fact, behave like systems of Newtonian bodies. The motions of falling

Figure 3.5

A Newtonian model for two bodies with gravitational attraction.

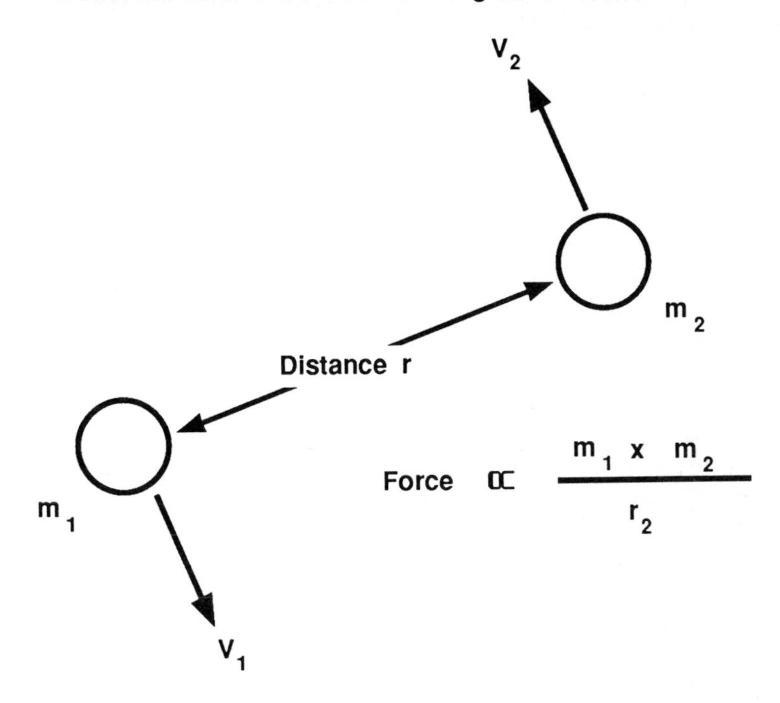

weights, the paths of cannon balls, the swings of a pendulum, the revolutions of the moon, and the motions of the planets. Because they were describable in clear mathematical form, Newtonian models made possible precise predictions of all these things, and many more.

Halley's Comet

Newton's work was first published in 1687. Around 1695, Edmund Halley, an English astronomer and friend of Newton, began applying Newtonian models to the motions of comets. He was probably acting on Newton's own suggestion that comets may be like small planets with very large elliptical orbits. In any case, comets were very interesting phenomena because they had always been viewed as mysterious, even ominous. Their appearances certainly exhibited no apparent regularity. If Newton's suggestion was correct, however, the behavior of comets would exhibit a great deal of underlying regularity.

Halley began investigating a comet that he himself had observed in 1682. The 1682 observations gave a quite precise location, relative to the sun and the background stars, of the comet's path when it was within sight. However, because only a small part of the total orbit could be observed, it was not possible to determine how big the orbit might be and, thus, how long it would take for the comet to return. Indeed, it was impossible to determine whether the orbit was an ellipse, as Newton suggested, or a parabola. Newton's theory allowed the possibility of a parabolic orbit, but such an orbit would mean that the comet would come by only once and then leave the solar system forever. If, however, the orbit was elliptical, as Newton suggested, the comet would have traveled that same path many times before.

Halley began digging into the records of observations for previous comets. He found twenty-four cases, going back roughly 150 years, for which the observations were precise enough to compare with the observations of 1682. For two of these, one in 1606–1607 and one in 1530–1531, the recorded orbits were very close to that of the 1682 comet. Halley argued that it was extremely unlikely that three different comets should have such similar orbits, and concluded that these were three appearances of the same comet in an elliptical orbit with a period of roughly seventy-six years. He speculated, but could not prove, that the slight discrepancies in the three orbits were caused by gravitational influences from the planets, particularly Jupiter.

But Halley did not stop here. Using the data from all three cases, together with the hypothesis that he was dealing with a system represented by a Newtonian model, Halley calculated the time of the next return. He boldly predicted that the comet would be seen again in the latter part of December 1758. Figure 3.6 will help you to keep in mind the relevant details of this example.

Halley published his work on comets in 1705. It was well received by Newton and the growing band of English Newtonians. It did little, however, to convince the French. Halley himself died, a respected scientist, in 1743, fifteen years before the predicted return of the comet. By this time, even the French were coming around to the Newtonian way of thinking, and Halley's prediction was remembered. In 1756, the French Academy of Science offered a prize for the best calculation of the return of the comet. The comet

Figure 3.6

Halley's comet.

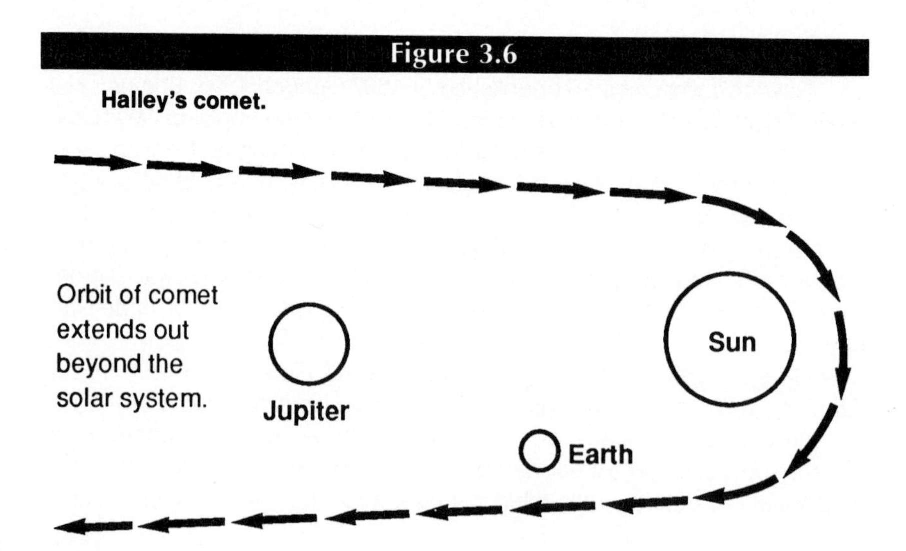

reappeared, as predicted, on Christmas Day in 1758 and was officially named "Halley's Comet."

Analysis

As presented above, the episode involving Halley's comet includes two different sets of data. It is instructive to do two analyses because, although both sets of data provide evidence for Halley's hypothesis, the evidence provided by the second set of data is superior. Again, Steps 3 and 4 are reversed.

> ***Step 1.*** The real world object of most direct interest was the comet observed in 1682. Of course, Halley was interested in other comets as well.
>
> ***Step 2.*** The model was a Newtonian model for two bodies in an elliptical orbit attracting one another by the force of gravity. The two bodies were identified with the sun and the comet, respectively.
>
> ***Step 4.*** The first prediction was that one would be able to find records of a comet similar to the comet of 1682, and recurring at roughly equal intervals.
>
> ***Step 3.*** Among twenty-four records of comets, those for comets in 1530–31 and 1606–07 had recorded orbits similar to the comet of 1682 and were spaced at about seventy-six-year intervals.
>
> ***Step 5.*** The data clearly agreed with the prediction.
>
> ***Step 6.*** The only alternative hypothesis was that there were three different comets with similar orbits at seventy-six-year intervals. Halley himself claimed that this was unlikely. Given that the documented incidence of similar comets was only twenty-four in 150 years, Halley's claim seems plausible. The data did provide evidence that the comet of 1682 fits the Newtonian model.

Before commenting on this analysis, let us analyze the data from the return of the comet in 1758.

> ***Step 1.*** The real world object of most direct interest is the comet observed in 1682.
>
> ***Step 2.*** The model was a Newtonian model for two bodies in an elliptical orbit attracting one another by the force of gravity. The two bodies were identified with the sun and the comet, respectively.
>
> ***Step 4.*** The second prediction was that the comet would return near the end of 1758.
>
> ***Step 3.*** A comet with the requisite orbit did appear as predicted.
>
> ***Step 5.*** The data clearly agreed with the prediction.

Step 6. The only alternative hypothesis was that another comet with the same orbit just happened to appear in the predicted interval seventy-six years later. That seemed to everyone extremely unlikely. So the data provided very good evidence that the Newtonian model fits.

The main difference in the two sets of data is that, in the second case, it was less likely that one should have found data agreeing with the prediction if the suggested model, in fact, did not fit the real world. In the first case, Halley predicted only that there might be some record of comets like the comet of 1682 with equally spaced intervals between them. To satisfy this prediction it was sufficient to find records of two equally spaced comets with roughly the orbit of the observed comet of 1682. The time interval was originally left open. It could have been five years, or fifty. In the second case, the prediction restricted the return of the comet to a single, several-week period seventy-six years later. That was much less likely to happen if what they were seeing were, in fact, different comets. So the evidence that the Newtonian model is correct was better in the second case than in the first.

Note, finally, that in the first case the "prediction" was of something in the past rather than the future. Note further that this prediction was not that comets had appeared at earlier intervals. That could no longer have been observed by Halley. What he predicted was that he would be able to find the appropriate records. Those he could observe.

3.3 THE DOWNFALL OF THE PHLOGISTON THEORY

In science, as in everyday life, one tends to remember the winners. The losers are usually forgotten. Some losers, however, are remembered if only because they had a period of considerable success before their downfall. A famous loser is the phlogiston theory.

Fire, like the motions of the heavens, has always fascinated people. In the Western world, recorded speculation about the nature of fire goes back to the Greeks. To them, we owe both the myth of Prometheus and the view that the world is made up of a few separate elements: earth, air, fire, and water. All four are present in the process of combustion. The common-sense view of combustion is that something is driven out of the burning object, leaving only ashes behind.

By the eighteenth century, this something had a well-established name, phlogiston—the fire stuff. Assuming that combustible material contains phlogiston explains most of the obvious facts about combustion. Heating drives off the phlogiston into the air; cooling makes it less volatile; smothering holds it in. The well-known fact that a burning candle placed in an enclosed container soon goes out was explained by saying that the enclosed air gets saturated with phlogiston so that the phlogiston remaining in the wax has nowhere to go.

Phlogiston accounts not only for combustion but also for the very important process of smelting. This is the process by which crude ores are turned into more refined metals. Generally this is done by carefully heating the ores, together with a measured amount of charcoal, to a controlled temperature. It was claimed that the charcoal contains an excess of phlogiston, which, at moderately high temperatures, leaves the charcoal and combines with the ore to form the metal. This hypothesis was substantiated by the fact that further heating at higher temperatures returns the metal to its original state. The phlogiston is driven out of the metal by the higher temperature. Even rusting was explained as the result of the phlogiston slowly escaping from the metal.

These claims may be taken as the "laws" that characterize phlogiston models. Such models lay behind many hypotheses about systems undergoing combustion, rusting, or the process of smelting. The phlogiston theory included the general hypothesis that this sort of model fits most cases of combustion, smelting, rusting, and so on. In what follows, we will concentrate on combustion.

Combustion is very difficult to study. Most things we commonly burn are made of many substances and give off many gasses when burned. Moreover, combustion generally is rapid and violent. Progress in such studies required finding some simple, well-controlled subjects for experimentation. In the 1770s, chemists developed a number of techniques for performing such experiments. The leaders were Joseph Priestly in England and Antoine Lavoisier in France. Priestly supported the phlogiston theory; Lavoisier led the revolution that overthrew it.

In the 1770s, utilizing techniques first developed by Priestly, Lavoisier performed a number of careful experiments with mercury. In one of these experiments, he floated a precisely measured amount of mercury on a liquid and covered it with a glass jar, thus enclosing a known amount of air (see Figure 3.7). The mercury was then heated, using the rays of the sun focused by a powerful magnifying glass (a burning glass). In such circumstances, as Lavoisier well knew, a red powder, or ash, forms on the surface of the mercury. Some of the mercury undergoes a controlled burning.

Applying the phlogiston model to this experiment, one would expect two things. First, the resulting mercury plus red ash should weigh less than the original sample of mercury alone. This is because some phlogiston must be driven off, leaving the ash behind. And the volume of air inside the jar should increase because it now contains the phlogiston that was driven out of the mercury. This means that the level of the liquid inside the jar would drop to make room for the additional "air." When Lavoisier completed the experiment, the water level had gone up, not down, and the mercury/ash combination weighed more than the original mercury alone.

Analysis

Again, Steps 3 and 4 are reversed.

Step 1. The general real world phenomenon under investigation is the process of combustion. The specific case is the controlled combustion of mercury.

Figure 3.7

Lavoisier's experiment.

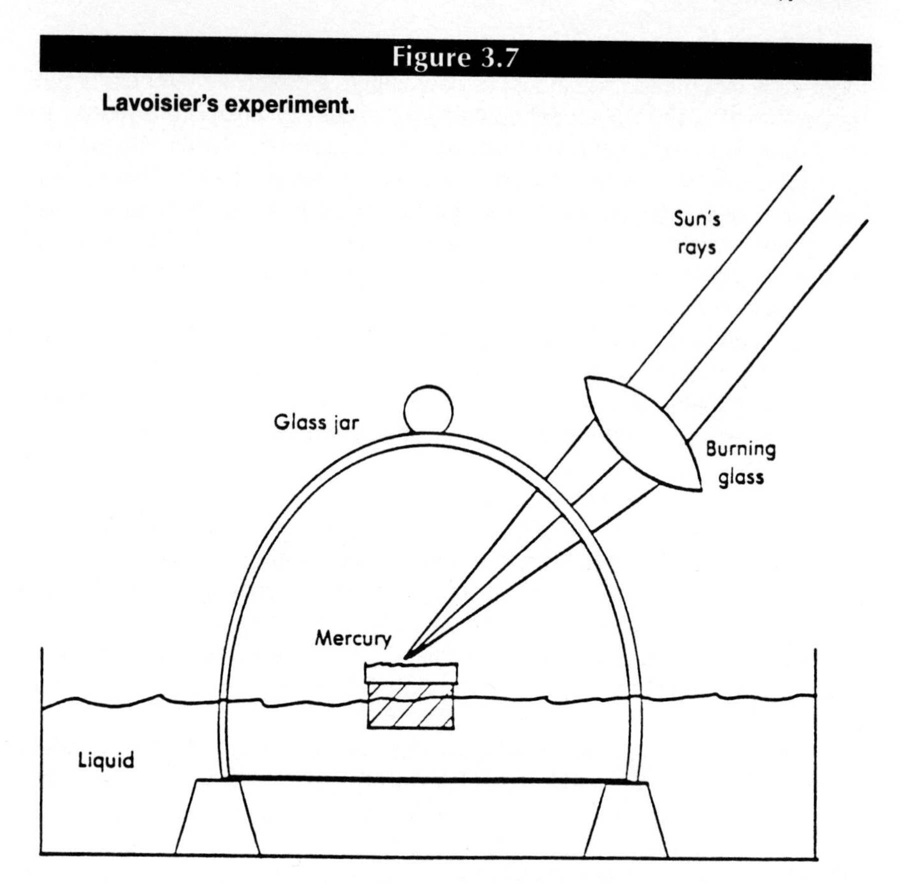

Step 2. The model is a phlogiston model as characterized by the stated laws of phlogiston.

Step 4. The predictions, for reasons explained in the discussion, are (1) that the water level under the bell jar should go down, and (2) that the mercury/ash combination should weigh less than the original mercury alone.

Step 3. The data are (1) that the water level under the bell jar went up, and (2) that the mercury/ash combination weighed more than the original mercury alone.

Step 5. The data and the prediction disagree. The data, therefore, provide evidence that the phlogiston model fails to represent the controlled combustion of mercury as carried out in this experiment.

3.4 MENDELIAN GENETICS

The scientific revolution of the seventeenth century changed biology as well as physics. But the real revolution in biology did not begin until the middle of the nineteenth century with Charles Darwin's discovery of evolution by natural selection and Gregor Mendel's investigations into the mechanisms of inheritance.

Mendel's Original Experiments

Mendel did most of his experimentation on a type of garden pea (*Pisum sativum*). Mendel noted that pea plants exhibit a number of characteristics that come in pairs, called "traits." The skins of the peas themselves (the seeds of the plant) exhibited the trait of being either smooth or wrinkled. Also, the color of the peas was either yellow or green. And the height of the plants was either tall or short. To keep things simple, we shall concentrate on only one characteristic, height.

Mendel discovered that when short plants are cross-pollinated with other short plants, the resulting seeds always yielded short plants. However, tall plants seemed to come in two types. Some pairs yielded seeds that led only to tall plants, but others yielded a mixture of tall and short offspring. The short plants, as well as the tall plants that produce only other tall plants, are called "true-breeding" plants, whereas the other tall plants are called "hybrids."

Experiment showed that if you cross-fertilize the true-breeding tall plants with the short plants (that were also true-breeding), the result was all tall plants. These tall plants, however, were all hybrids. If these hybrid plants were then self-fertilized, the result was a mixture of both tall and short offspring. So whatever was responsible for determining whether a plant is short or tall seemed to be transmitted through a generation of plants that were themselves all tall. The short plants showed up again only in the following generation.

Mendel's unique contribution was to isolate the different characteristics of his plants and, then, actually to count the plants exhibiting the different traits. He discovered that the ratio of tall to short in the second generation was roughly 3 to 1. This ratio occurred again and again, and with other characteristics besides height. Figure 3.8 shows Mendel's results.

Mendel's Model

Mendel sought to explain how a trait could be transmitted by plants that did not themselves exhibit that trait. Moreover, the explanation must account for the fact that the "hidden" trait appears in only one-fourth of the members of the succeeding generation. In an attempt to explain these things, Mendel constructed the following model.

Suppose that there are two things that determine any characteristic, such as height. Mendel called these things "factors." Factors correspond roughly to what we now call genes. Suppose further that one factor is associated with being tall and the other with being short. Also, suppose that each individual has some combination of two of these two types of factors. Let "H" stand for the factor associated with tall plants and "h" for the factor associated with short plants. Then there are really three different types of individuals: those with two "tall" factors, those with two "short" factors, and those with one of each. These three possibilities are shown schematically in Figure 3.9.

Figure 3.8

Mendel's original experiments.

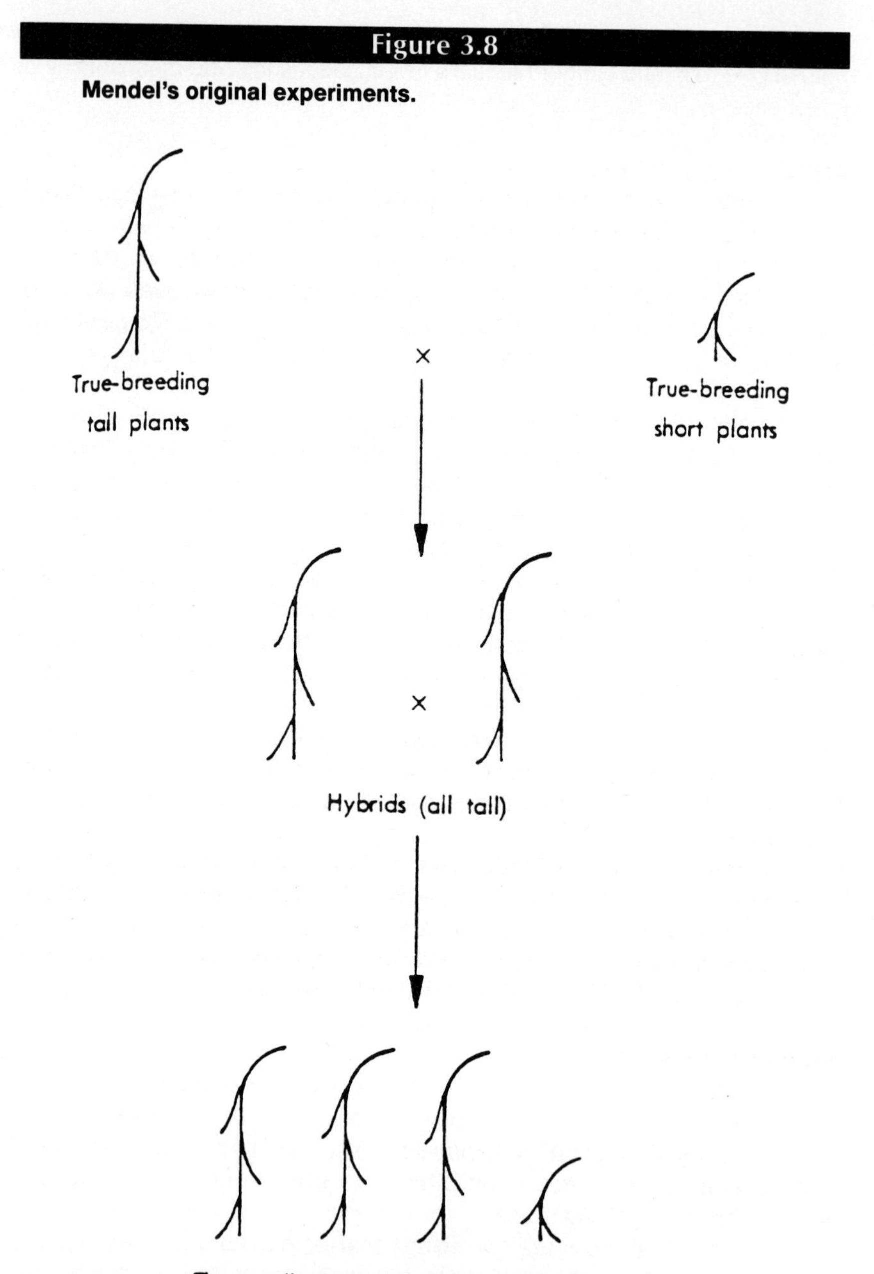

Next, assume that upon "mating" (pollination), the seeds for the next generation get one factor from each parent. Moreover, the selection of a factor from either parent is assumed to be a random process. This is the crucial part of what is usually called *Mendel's Law of Segregation.* In the formation of the seed that becomes a plant in the next generation, the four original factors (two from each parent) "segregate" themselves at random into groups of two, one from each parent.

Figure 3.9

How Mendel's model classifies individuals by combinations of factors.

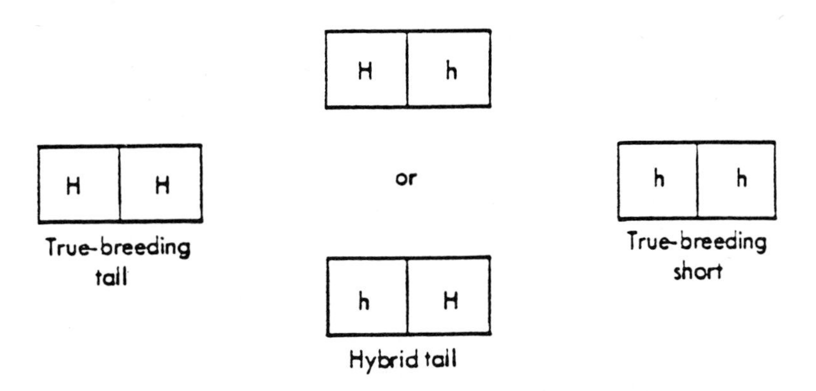

Finally, suppose that those individuals that have one factor of each type (the hybrids) exhibit the trait associated with only one of the two factors. This trait is called the dominant trait. The factor associated with this trait is called the dominant factor. That one of the factors should be dominant in this sense is known as *Mendel's Law of Dominance.* The other trait and its corresponding factor are called recessive.

Now let us take a new look at Mendel's experiment as pictured in Figure 3.8. This time, however, let us represent each individual by its factors alone. The result is shown in Figure 3.10. Let us go through this diagram carefully. The original parents consist of true-breeding tall and true-breeding short plants. Their union must produce hybrids because the offspring must get one factor from each parent, and each parent has only one type to give.

When both parents are hybrids, however, there are four different ways the offspring can get its two factors: a tall from each parent, a tall from the first and a short from the second, a short from the first and a tall from the second, or a short from each. Because, by the law of segregation, the combinations are selected randomly, there should be equal numbers of each of these four types of offspring. However, by the law of dominance, only one of these four types will actually be short. The other three—the one true-breeding tall and the two hybrids—will all be tall. We, thus, obtain the observed result that tall and short plants occur in the third generation in a ratio of 3 to 1.

The Backcross Test

Mendel's model suggests another experiment. Why not try crossing the second-generation hybrids with the true-breeding, short plants of the parental generation? A quick review of Figure 3.10 will reveal what kind of mating this backcross represents. It seems that no one had ever done this experiment before. Even if it had been done, no one had ever reported having actually counted the results to see what ratio of dominants to recessives would occur.

Let us figure out what ratio Mendel's hypothesis predicts. The experiment is represented symbolically in Figure 3.11. The numbers 1, 2, 3, and 4

Figure 3.10

Mendel's model applied to his original experiments.

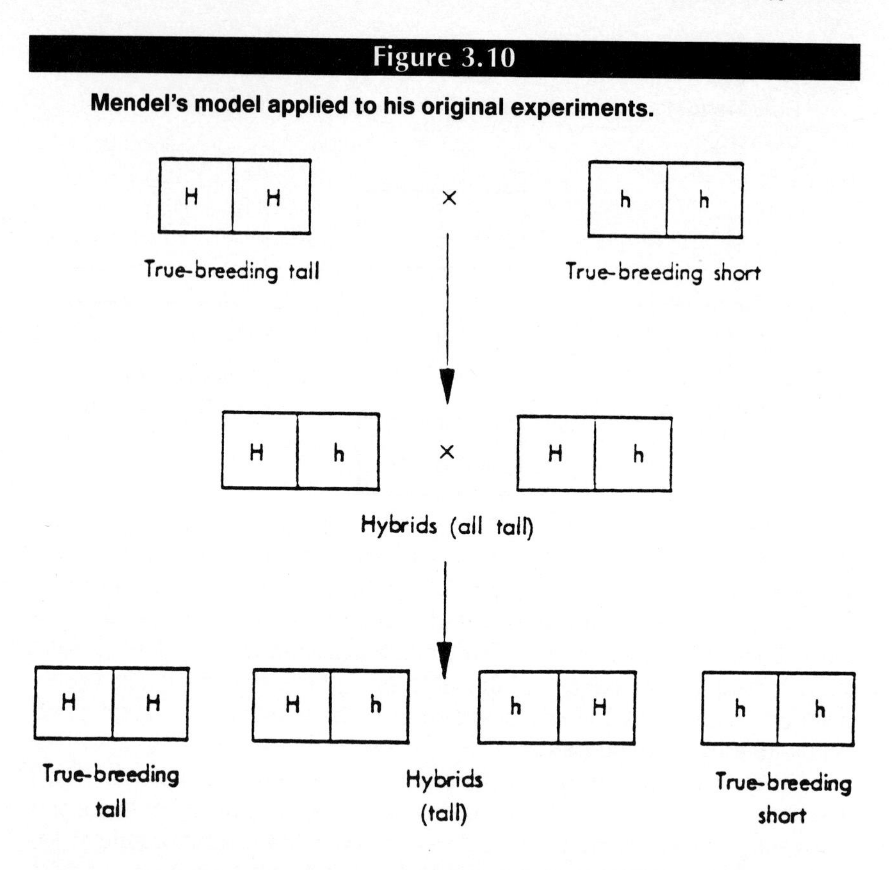

Figure 3.11

Mendel's model applied to the backcross experiment.

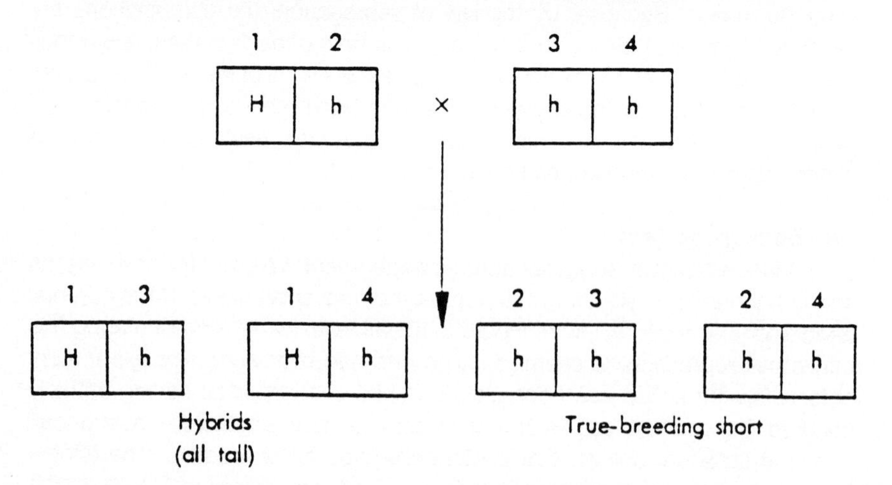

are attached to help keep straight the possible combinations of factors in the offspring.

According to Mendel's model, there are four possible ways an offspring could get its complement of two factors. According to the law of segregation, each of these possibilities is equally likely. So, on the average, there should be equal numbers of each type. Now, as you can see, two of the possibilities yield hybrids, which by the law of dominance will be tall. The other two are true-breeding short plants. So the ratio of tall to short plants in this sort of experiment should be 1 to 1. That is, on the average, half should be tall and half short. Mendel reported producing 208 plants by this type of cross-pollination. Of these, 106 came out tall and 102 short. That is, surely, close enough to call the prediction true.

Analysis

There are two sets of data to consider, that from the original experiments and that from the backcross experiments. We will do two separate analyses.

> ***Step 1.*** The real world process is the inheritance of traits by offspring from sexually reproducing parents.
>
> ***Step 2.*** The model is Mendel's factor model including the law of segregation and the law of dominance.
>
> ***Step 3.*** The data are the 3 to 1 ratios of tall to short plants in the original plant-breeding experiments.
>
> ***Step 4.*** The prediction is that the ratio of dominants to recessives should be 3 to 1.
>
> ***Step 5.*** If being tall is identified as a dominant trait and being short as a recessive trait, then the data and the prediction agree.
>
> ***Step 6.*** Was the prediction likely to agree with the data even if Mendel's model did not provide a good fit to the real world? Answering this question requires considerable thought.

Standard presentations of Mendel's theory, as found in college textbooks for example, do not explicitly refer to alternative models that might equally explain the data. None, for example, were mentioned in the above presentation. Non-specialists coming to this subject for the first time are, thus, left to wonder just how likely it would be to observe a 3 to 1 ratio if Mendel's model did not fit the case. There are two general points to consider.

One point is that 3 to 1 is a very simple ratio. So, even without knowing much at all about genetics, one could suspect that there must be many fairly simple mechanisms that could produce that ratio. Thus, taking the agreement between Mendel's prediction and the data as evidence for his hypothesis could fairly easily lead one to supporting a false hypothesis.

A second point concerns how Mendel constructed his model. He clearly discovered the 3 to 1 ratio before he constructed his model. That means he was searching for a model that would yield a prediction of three tall to one short. It is very difficult to say how likely it is that such a search process should lead to a well-fitting, rather than a poorly fitting, model. The only thing we know for sure is that the process is very likely to lead to a model that yields the correct prediction, regardless of how well the model fits. That is because Mendel set out to find a model that would yield the 3 to 1 ratio and would clearly have rejected any model that did not do so. We do not know how many, if any, such models there were. Nevertheless, whatever model he ended up presenting for public scrutiny, it would be sure to yield the correct prediction even if it did not fit the case in other respects.

The backcross experiment is significantly different. For completeness we will run through the whole analysis. Again, Steps 3 and 4 are reversed.

> ***Step 1.*** The real world process is the inheritance of traits by offspring from sexually reproducing parents.
>
> ***Step 2.*** The model is Mendel's factor model including the law of segregation and the law of dominance.
>
> ***Step 4.*** The prediction is a 1 to 1 ratio of tall to short offspring.
>
> ***Step 3.*** The data were 106 tall plants and 102 short plants out of a total of 208 offspring.
>
> ***Step 5.*** The data are in close agreement with the prediction.
>
> ***Step 6.*** Was the prediction likely to agree with the data even if Mendel's model did not provide a good fit to the real world? Here again the question requires extra thought.

The worry that the model was itself selected especially because it predicted the data is here eliminated. It is clear from all accounts that the model was developed prior to the backcross experiment. We still face the fact, however, that there would seem to be numerous different, but equally plausible, models that would predict a simple 1 to 1 ratio for the backcross experiment. The case for Mendel's hypothesis is somewhat strengthened by the fact that the model must correctly predict both the original 3 to 1 ratio in the standard breeding experiment and the 1 to 1 ratio in the backcross experiment. Nevertheless, it does not seem that all the data were very unlikely to occur if Mendel's model did not fit the real genetic mechanisms. So we cannot take the agreement between prediction and data as providing very strong evidence in support of Mendel's hypothesis.

Mendel worked on *Pisum* from roughly 1856 to 1866. At the time, he was a monk in a monastery in the city of Brno in what is now Czechoslovakia. His account of his discoveries was published in an obscure journal, which mailed out 115 copies. Mendel corresponded with one renowned scientist who understood his work. This scientist, however, was studying another species of plant that, for complex reasons, did not yield Mendel's simple ratios. Indeed, few traits with which we are familiar in humans and other mammals show such simple patterns of inheritance. In 1868, Mendel was appointed abbot of his monastery. His duties so absorbed his energies that by 1871 he was no longer able to continue his experiments. His work was largely forgotten until 1900 when several people rediscovered his results. Then began the chain of discoveries that led up to the work of Watson and Crick in the 1950s.

3.5 THE REVOLUTION IN GEOLOGY

At the beginning of the twentieth century, the generally-accepted hypothesis about the overall structure of the earth was that it was formed as a molten sphere that gradually cooled and formed a crust—on which we now live. As the sphere continued to cool, it contracted. As the whole sphere contracted inward, the crust cracked and deformed, yielding continents, mountain ranges, and the other major geological features now observed. This is the "contractionist hypothesis." It is a consequence of this hypothesis that the continents have always been in more or less their present positions. Once the crust solidified, there was no way for oceans or land masses to move around on the surface.

We can describe this situation by saying that geologists had developed a theoretical model of the earth based on an analogy with a molten sphere suspended in space. Physicists and geologists had a fairly good idea of what such a sphere would do. The question was just how similar our actual Earth is to such a model.

Until about 1950, most of our knowledge of the earth's geology could be explained using a contractionist model, but there were some difficulties. (1) There are remarkable similarities in the coastlines, and other geological features, of widely separated land masses. The fit between the coasts of eastern South America and western Africa is particularly striking. It is difficult to see why, on a contractionist model, this should be so. (2) There are relatively new mountains running all along the west coast of North and South America, from Chile to Alaska, but only relatively old, and less extensive, mountains on the east coast. There are also volcanoes and earthquakes along the west but not the east coasts. Why the asymmetry? (3) There are many similar species of plants and animals in places separated by oceans, as in Africa and South America. How could this have happened?

It was not that contractionists lacked possible answers to these and similar difficulties. The addition of no-longer-existing land bridges to contractionist models, for example, would allow animals and plants to have migrated from one continent to another. But why, then, are there no remaining traces of these bridges?

Drift Models

In 1915, a German scientist, Alfred Wegener, developed an entirely different model of the earth in a book entitled *The Origins of Continents and Oceans*. In contractionist models, the major movements are up and down—oceans sinking and mountains being pushed up. In Wegener's model, the major motions are horizontal. In particular, Wegener argued that the original cooling resulted in one large land mass, which he called Pangea. This mass, he claimed, subsequently broke up into continents, which then drifted to their present positions—and they are still drifting, though very slowly. Wegener was by no means the first person to hold such a hypothesis, but at the time his was the most elaborate and systematic treatment of this view.

With this model, one can easily account for many of those features of the earth that troubled contractionists. (1) The match in the coastlines and general geology of Africa and South America would be the natural result of South America having split off from what is now Africa and drifted to its present position. (2) The mountain ranges on the west coast of North and South America are the natural result of the two continents pushing their way west. (3) The similarities in plants and animals now on different continents resulted from their having originally all been together on Pangea before it broke up. Wegener regarded these consequences of his model as providing good evidence for the hypothesis that the development of the earth really did fit something like his model.

Very few scientists in 1915 regarded Wegener's hypothesis as even remotely plausible. The main reason was simply that it seemed impossible that the continents really could move. In the first place, there seemed no horizontal forces to push them. Secondly, the continents are composed of softer material than the ocean floors—the lighter materials having floated to the top when the earth was still molten. How could this softer material push through a harder material? Thus, even if there were something pushing the continents, they would just break apart, crushed against the edges of the more solid ocean floors.

If geologists were to be convinced that the continents move, they would have to be given both a better model and more powerful data than Wegener had provided. In fact, better models were devised by the time of Wegener's death in 1930, but it took until the 1960s before convincing data could be found.

Sea Floor Spreading

One key to understanding the structure of the earth was the discovery of radioactivity by Marie and Pierre Curie in the 1890s. Radioactivity results from the spontaneous decay of atoms of a few heavy elements, such as uranium. When this happens, energy is given off and transformed into heat. If there were a reasonable amount of radioactivity taking place within the earth, the inside of the earth would be much hotter than contraction models generally assume. This heat could produce more molten material much closer to the surface than was previously thought possible. And this, finally, allows the possibility of large-scale, horizontal movement driven by slow-moving, convection currents of molten rock. These implications of radioactivity for drift models were recognized by the late 1920s, but there was no way then to test such speculative ideas. The secret to testing them lay at the bottom of the oceans.

It was not until the 1950s that oceanographers obtained detailed knowledge of the large ridges running along the ocean floor. One such ridge runs roughly down the center of the Atlantic Ocean and another lies off the coast of the United States in the eastern Pacific. These ridges were discovered to have a number of special characteristics. For example, there is a depression running down the middle of the ridges. The area within the depression, and along the ridge generally, is warmer than the surrounding areas. Also, the magnetism of the floor material alternates as one moves at right angles away from the ridge. Figure 3.12 is a schematic cross section of a ridge showing these features. What could have produced such ridges?

In 1960, Harry Hess, a respected geologist with a reputation for having a good scientific imagination, proposed a model that he himself called "geopoetry." In Hess's model, there are convection currents of molten material (produced by radioactive heating) under the ocean floors. In this model, ridges are produced by a rising convection current that then spreads out forming the ocean floor. The spreading floor then descends back into the core at a trench that might be several thousand miles away (see Figure 3.13). This model immediately accounts for the existence of ridges, their central depression, and the temperature differences.

Hess was aware from the beginning that his model fell into the continental-drift tradition. An easy elaboration of the model makes possible the horizontal movement of the continents. One need only assume that the continents ride along on a piece of the earth's crust, which is pushed by the expanding ocean floor. This model eliminates the need to explain how the continents could plow through the harder ocean floors. Indeed, they do not. They ride along on top of the moving, harder material.

As an anonymous reviewer of a later paper by another drift supporter put it, this is the kind of thing one might talk about at a cocktail party, but it should not be published in a respectable, scientific journal. Most scientists regarded the application of Hess's model to the earth as being utterly implausible. How, indeed, could one ever test such a model? The secret lay in the alternating magnetic zones lying along the ridges.

Figure 3.12

Schematic cross section of an oceanic ridge.

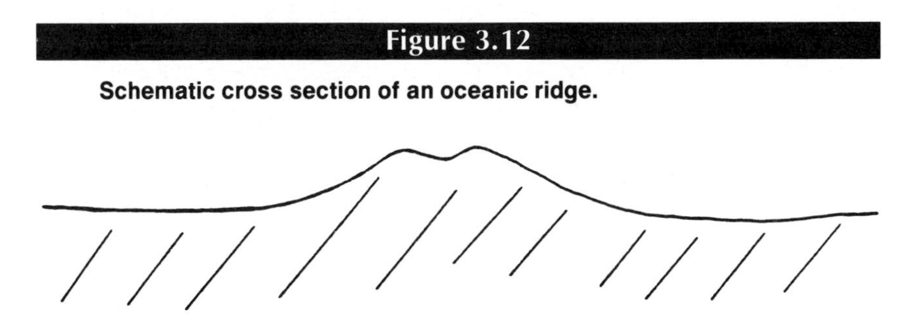

Figure 3.13

Hess's model of sea floor spreading.

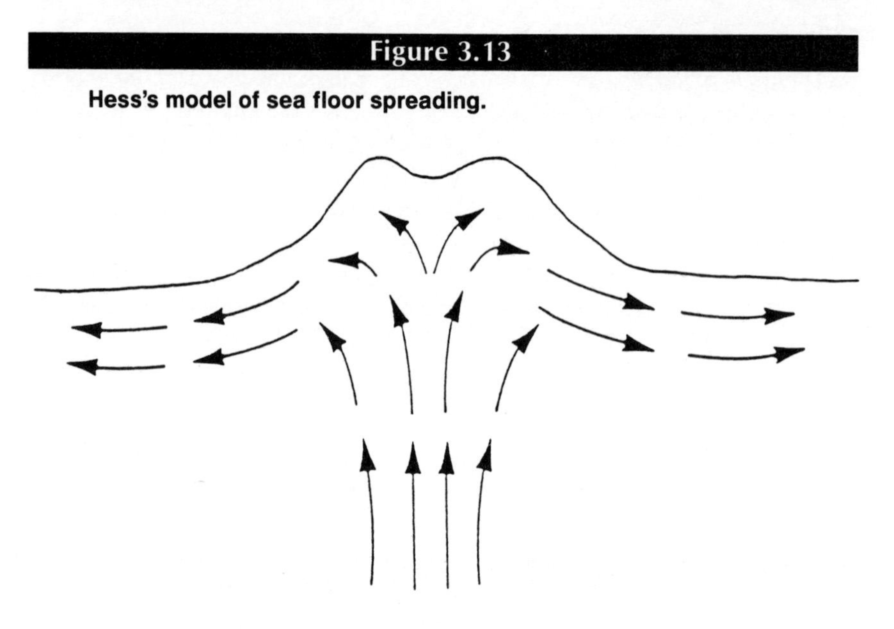

Magnetism, Geomagnetism, and Paleomagnetism

That the earth itself is a large magnet was known during the scientific revolution in the seventeenth century. In the 1950s, many scientists were studying the magnetism of the earth (geomagnetism), including the history of the earth's magnetism (paleomagnetism). Magnetic material, such as iron, that was originally free to move but then solidified into position, maintains the orientation it had when it solidified. By examining material formed over long periods of time without being disturbed, it is possible to determine how the magnetic field of the earth has changed over millions of years. Samples of such materials, around volcanoes for example, seemed to exhibit a complete (180 degree) reversal of orientation at various times. This indicated that perhaps the earth's magnetic field itself had actually reversed poles a number of times. That is, the north magnetic pole became the south magnetic pole and vice versa.

What has this to do with sea floor spreading and continental drift? It took several years for anyone to see the connection. Then, in 1963, working completely independently, a geologist in Canada (Lawrence Morley) and a geology graduate student in Cambridge, England (Fred Vine), came up with a model for explaining the alternating strips of differently magnetized materials along the ridges. The iron-like material in the rising molten material is free to orient itself with the earth's magnetic field at that time. When it reaches the surface, however, it cools and the magnetized material gets locked into place. If the earth's magnetic poles reverse, then material coming along later will be oriented the other way.

If this model is correct, the history of the earth's magnetic field, for millions of years, would be laid out on the ocean floor in strips parallel to the ocean ridges. Moreover, there should be a correspondence between these alternating, magnetic strips and the alternating, magnetic layers found in lava flows near volcanoes (see Figure 3.14). All would provide records of the same sequence of global magnetic-pole reversals.

Vine (together with his supervisor, Drummond Matthews) and Morley published their views in 1963 and 1964, respectively. (Morley actually had the idea first, but his paper was turned down by two journals before it was finally published.) The reception by most geologists and geophysicists ranged from complete disinterest to outright hostility. That there actually should be such a pattern was generally regarded as fantastically improbable.

In the fall of 1965, an oceanographic research ship operated by the Lamont Geological Observatory at Columbia University was recording magnetic effects across the Pacific-Antarctic Ridge in the southeastern Pacific. By dragging a magnetometer near the ocean floor, they measured the changes in magnetic intensity in the floor as one moves perpendicularly across and away from a major ridge. When the data was analyzed early in 1966, it revealed a

Figure 3.14

Predicted correspondence between magnetic records on land and along the ocean floor.

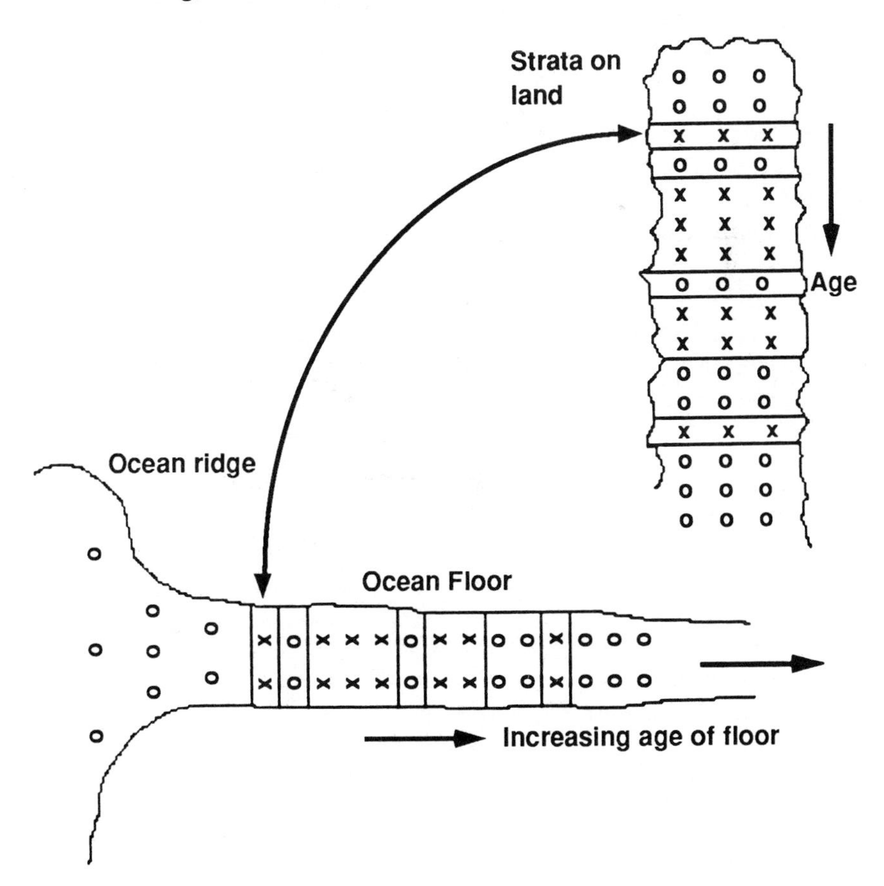

pattern of increasing- and decreasing-magnetic intensity matching the pattern of magnetic reversals found in minerals on land.

At the same time, other researchers at Lamont began to study the magnetism in core samples obtained from the sediment layer at the bottom of the ocean near the southern tip of South America. Sediment is formed by soft material piling up on the bottom of the ocean and compressing the material underneath, which eventually solidifies. Because such sediment contains traces of magnetic materials, it, too, should exhibit the reversals in the earth's magnetic field. Indeed, they found the same pattern of alternating, magnetic orientation in the sediment.

The reaction of the geological community was immediate and dramatic. Everyone embraced drift models. During the next few years, geologists developed the idea of sea-floor spreading into what is now called plate tectonics. According to the theory of plate tectonics, the earth's surface consists of a number of plates that slowly move about the surface, constantly changing the configuration of continents and oceans. Much of geological research now consists of working out the details of this general theory.

Analysis

There are two analyses to consider. One focuses on the data Wegener used to support his model of continental drift in the 1920s. The second focuses on the magnetic data first gathered in the 1960s.

Step 1. The real world object is the structure and history of oceans and land masses on the surface of the earth.

Step 2. The model is Wegener's model of a planet with drifting continents.

Step 3. The data include (1) the matching coastlines of Africa and South America, (2) the mountain ranges on the western edges of North and South America, and (3) the similarities among plants and animals in places like Africa and South America.

Step 4. Wegener's model predicts the three features noted above.

Step 5. The data and the predictions agree.

Step 6. Were these predictions likely to agree with the data even if Wegener's model did not provide a good fit to the real world? As supporters of contractionist models argued, the data Wegener cited could plausibly be accounted for either by chance or by versions of a contractionist hypothesis. The matching of the coastlines and the location of mountain ranges were generally ascribed to chance. The similarities in plants and animals were said to have resulted from the existence of large, land bridges that have since sunk or washed away. So Wegener's data were not regarded as very improbable if his model did not fit the actual history and structure of the earth's surface. The data were thus inconclusive regarding the applicability of Wegener's drift model to the actual history of the earth.

The situation in the 1960s was significantly different. Again, Steps 3 and 4 are reversed.

Step 1. The real world object is the structure and history of oceans and land masses on the surface of the earth.

Step 2. The model is the drift model developed by Hess, Vine, and others.

Step 4. The prediction is that one will find alternating strips of normally aligned and reversed magnetic material parallel to oceanic ridges. And the pattern of these strips will match the pattern of magnetic reversals found in successive layers of rocks such as those near volcanoes. It would also match the pattern of magnetic materials in deep-sea sediments.

Step 3. The data consist of measurements of magnetic material on land, along oceanic ridges, and in sediment cores.

Step 5. The data and the predictions agree in great detail.

Step 6. Were these predictions likely to agree with the data even if Hess's and Vine's model did not provide a good fit to the real world? No. Proponents of contractionist models all along regarded it as extremely unlikely that any such data could ever be found. There were no plausible contractionist models that predicted such data. So the data provided very good evidence that the new drift models do fit the actual structure and history of the earth's crust.

EXERCISES

Analyze these reports following the six-point program for evaluating theoretical hypotheses developed in the text. Number and label your steps. Be as clear and concise as you can, keeping in mind that you must say enough to demonstrate that you do know what you are talking about. A simple "yes" or "no" is never a sufficient answer. Some of these reports are taken directly from recent magazine or newspaper articles and are presented here with only minimal editing.

Exercise 3.1

The Discovery of Neptune

During the first half of the nineteenth century, astronomers were still working out tables and charts giving the positions of the various planets. In this, they were aided by Newtonian, theoretical models. But then the outermost planet, Uranus, caused some difficulties. Its observed orbit differed from what it should have been according to the then best-fitting Newtonian models. The difference was much too great to be attributed solely to inaccuracies in measurement. They were forced to conclude that their current models were not correct. They did not, however, give up Newton's theory of celestial mechanics. By that time, there had been so many successful predictions using Newtonian models that they were reluctant to conclude that the general theory could be wrong. Around 1843, the English astronomer J. C. Adams and, somewhat later, the French astronomer Leverrier, independently calculated that the observed orbit of Uranus could be explained if there were an additional planet beyond Uranus whose gravitational force produced the deviations from the earlier Newtonian predictions—which, of course, assumed no such planet. Using this more elaborate Newtonian model, Adams and Leverrier were able to calculate just where the new planet should be at any particular time. The planet, named Neptune by Leverrier, was observed in 1846, just where it was predicted to be.

Exercise 3.2

The Missing Planet: The Story of Vulcan

Like the orbit of Uranus, the observed orbit of the innermost planet, Mercury, failed to fit Newtonian models by amounts that could be reliably measured. Fresh from their discovery of Neptune, many astronomers immediately assumed that there must be yet another planet, closer to the sun than Mercury. Leverrier even named the new planet Vulcan and calculated just where it should be. Although several people claimed to have seen Vulcan, these reports were never substantiated.

Exercise 3.3

The Wave Theory of Light: 1818

After a century in the shadow of Newtonian ideas, according to which light is composed of small particles, wave models of light were revived around 1800, first by an Englishman, Thomas Young, and then by a Frenchman, Augustin Fresnel. Fresnel's model was submitted for a prize offered by the French Academy of Science. One of the judges for the Academy, S. D. Poisson, deduced that, according to Fresnel's model, the shadow of a small, circular disk produced by a narrow beam of light should exhibit a bright spot right in the center of the shadow. Poisson and the other judges are reputed to have thought that this refuted Fresnel's hypotheses because they had never heard of there being such a phenomenon and regarded it as highly unlikely to exist. No known particle models predicted such a spot. When the experiment was carried out in carefully controlled circumstances, however, there was the spot, just as required by Fresnel's model. Fresnel received the prize in 1818.

Exercise 3.4

The Oxygen Theory

Antoine Lavoisier was not only interested in refuting the phlogiston theory, he sought also to establish his own oxygen theory. In many respects, oxygen models have just the opposite structure of phlogiston models. According to Lavoisier's oxygen model, when combustion occurs there is a component of normal air (about one-fifth by his measurements) that combines with the combustible material. This component of air came to be called "oxygen." Analyze Lavoisier's experiment with mercury as a crucial experiment applied to both phlogiston and oxygen hypotheses. It may help you to know that the red powder that formed on top of the mercury is mercuric oxide.

Exercise 3.5

Childbed Fever

During the nineteenth century, many women died from what was then called "childbed fever." The first person who had any success in discovering the cause of this disease was a young Viennese doctor, Ignaz Semmelweis. In the years 1844 to 1846, the death rate from childbed fever in the First Maternity Division of the Vienna General Hospital averaged roughly 10 percent. Curiously, the rate in the Second Division, where women were attended by midwives rather than doctors, was only around 2 percent.

Semmelweis tried in vain for two years to discover why the rate should be higher in the "better" division. Then, one day a fellow doctor received a small cut on the finger from a student's scalpel during an autopsy. The colleague died exhibiting symptoms very much like those of childbed fever. Semmelweis wondered whether the disease might be caused by something in "cadaveric matter" that was being transmitted to the women during childbirth on the hands of the doctors and medical students, who spent every morning in the autopsy room before making their rounds.

Semmelweis reasoned that if this idea were right, the death rate could be cut dramatically simply by requiring the doctors and students to wash their hands in a strong cleansing agent before examining their patients. He, then, insisted that no doctors enter the maternity ward without first washing their hands in a solution of chlorinated lime, which he assumed must be strong enough to remove whatever it was that caused the disease. It worked. The death rate in the First Division for 1848 was less than 2 percent.

Exercise 3.6

The Wave Theory of Light: 1849

During the first half of the nineteenth century, there was a long controversy over the nature of light. According to Newtonians, light consists of small particles moving at high velocities, so that what we call light rays are really Newtonian particles. The competing theory, advocated mainly by French physicists, was based on the idea that light is really an example of a system of waves, like waves on the surface of a calm lake, in a bowl of jelly, or on a vibrating, stretched string.

Using standard Newtonian models, it was calculated that light should travel faster in water than in air, and by a precisely determined amount. Similarly, using wave models, it was calculated that light should travel slower in water than in air, also by a precisely determined amount.

It took until 1849 for anyone to design instruments that could measure the velocity of light accurately enough to detect the predicted differences. When the experiment was finally done, it was found that the velocity of light is lower in water, and by the amount claimed by the wave theorists.

Exercise 3.7

The Digestive System

At the beginning of the twentieth century, it was known that the pancreas, a small organ near the stomach, secreted digestive juices into the duodenum, which connects the stomach with the small intestine. This happened whenever partially digested food entered the duodenum from the stomach. The question was whether the signal that started the pancreas working was transmitted by the nervous system or by a chemical substance carried by the blood. At the time, there were known examples of both nerve-stimulated organs and chemical-stimulated organs. No other stimulating mechanism was known.

To decide the issue, two physiologists, W. M. Bayliss and E. H. Sterling, cut all the nerves going to and from the duodenum of an experimental animal. All blood vessels to and from the duodenum were left in place. They also inserted tubes to detect flow of digestive juices from the pancreas to the duodenum.

The result of this experiment was that, when food entered the duodenum of the experimental animal, the pancreas secreted digestive juices in the normal way.

Exercise 3.8

Pasteur and Anthrax

Louis Pasteur, who died in 1895, is best remembered as inventor of the process we now call "pasteurization." For example, carefully heating milk to about 63 degrees Celsius for 30 minutes and then cooling it rapidly allows the milk to be stored below 10 degrees Celsius for several weeks rather than several days. Pasteur's explanation for why the process works is that the heating debilitates the small organisms, "microbes," that produce the substance that causes milk to go sour.

By 1800, the English physician Edward Jenner had shown that inoculation with relatively harmless cowpox provides humans with immunity to deadly smallpox. Pasteur theorized that smallpox and many other diseases are produced by microbes. He suspected that debilitating, but not killing, the microbes would allow one to generalize Jenner's process of "vaccination" (from vacca, *meaning cow) to other diseases.*

In 1880, nearly 10 percent of the sheep, cows, and goats in France were dying of anthrax. Pasteur extracted from dead animals material containing the anthrax bacillus and experimented with ways of attenuating it. Eventually, he discovered that heating for eight days at between 42 degrees Celsius and 44 degrees Celsius would do the trick.

To convince the general public of his result, Pasteur arranged a public demonstration experiment. On May 5, 1881, at a farm near the town of Pouilly-le-Fort, twenty-four sheep, six cows, and one goat were inoculated with an attenuated anthrax bacillus. On May 31, all thirty-one inoculated animals, plus a similar assortment of twenty-nine other healthy animals that had not been inoculated, were injected with fully active anthrax bacilli. By the end of the day on June 2nd, all the unvaccinated sheep were dead and all the unvaccinated cows were sick. All the vaccinated animals were still perfectly healthy. Pictures of healthy animals surrounded by the corpses of those that had not been inoculated made very good press.

By June of the next year, roughly 300,000 animals, including 25,000 cows, had been vaccinated with Pasteur's vaccine, and the mortality rate from anthrax had dropped to nearly 0.5 percent.

Exercise 3.9

Island Biogeography

In the years 1831–36, the young Charles Darwin sailed around the world aboard the British Navy ship H.M.S. Beagle. The Beagle was engaged in a survey of South America as well as various Atlantic and Pacific islands. Darwin went along as an unpaid naturalist. Among the many islands they visited were the Cape Verde Islands and the Galapagos Islands. The Cape Verde Islands lie in the Atlantic Ocean roughly 300 miles west of the coast of what is now Senegal. The Galapagos Islands lie in the Pacific Ocean roughly 650 miles west of Ecuador. The two sets of islands are very similar, being geologically relatively young archipelagos of volcanic origin. Yet, the bird species on the two sets of islands are very different. The bird species on the Cape Verde Islands, however, are similar to species on the nearby African continent. Likewise, bird species on the Galapagos Islands are similar to species on the nearby South American continent. The species on the islands, however, are definitely different species than those on the corresponding mainlands.

It was some years later that Darwin constructed his theory of evolution by natural selection, which he claimed explained these earlier observations. He suggested that a few birds managed to migrate to the islands from the mainland. Their offspring exhibited natural variations in traits, such as the shape of their beaks. Birds exhibiting variations better suited to the new environment would tend to have more offspring. Those variations would thus be transmitted to relatively larger numbers in succeeding generations. In time, the population would evolve into a new species while, nevertheless, retaining many similarities with the original founders of the population. That, Darwin claimed, is how things came to be as he had found them on these islands in the early 1830s.

Exercise 3.10

A Confirmation of Continental Drift

Some of the best tests of hypotheses involving continental drift have become possible only since the development of accurate methods of determining the age of rocks, methods based on the examination of the products of radioactive decay. The following two paragraphs are taken from an article, "The Confirmation of Continental Drift," that appeared in the Scientific American *for April 1968. The particular hypothesis at issue is that Africa and South America are part of the system of land masses that have broken up and drifted apart. The investigators write:*

Of special interest to us at the start was the sharp boundary between the 2000-million-year-old geological province in Ghana, the Ivory Coast, and westward from these countries, and the 600-million-year-old province in Dahomey, Nigeria, and east. This boundary heads in a southwesterly direction into the ocean near Accra in Ghana. If Brazil had been joined to Africa 500 million years ago, the boundary between the two provinces should enter South America close to the town of Sao Luis on the northeast coast of Brazil. Our first order of business was therefore to date the rocks from the vicinity of Sao Luis.

To our surprise and delight, the ages fell into two groups: 2000 million years on the west and 600 million years on the east of a boundary line that lay exactly where it had been predicted. Apparently, a 2000-million-year-old piece of West Africa had been left on the continent of South America.

Exercise 3.11

Clues to the Drift of Continents and the Divergence of Species

In 1910, Alfred Wegener, a German meteorologist and explorer, began a long quest for evidence supporting the seemingly preposterous notion that the continents drift hither and yon. He eventually found it in a strange guise—the global distribution of marsupials, animals such as kangaroos and opossums that carry their young in a pouch. Marsupials, Wegener pointed out, are largely confined to Australia and South America, which are separated by thousands of miles of ocean. Yet, "even the parasites of the Australian and South American marsupials are the same," he wrote. The phenomenon, he concluded, "dates back to the time when Australia was still joined to South America via Antarctica."

That Antarctica provided the link between the Americas and Australia 65 million years ago, long after the great southern continent of the era, Gondwanaland, began breaking up to form Africa, South America, India, Australia, and Antarctica, has now been dramatically confirmed by the discovery of fossil marsupial bones in the onetime land bridge. . . .

When Antarctica linked Australia and South America 80 million years ago, it lay in a more hospitable latitude than now, facilitating passage of temperate-zone animals. This new search for fossil evidence of the link, financed by the National Science Foundation, focused on Seymour Island off the Antarctic Peninsula, famous for the fossils of giant penguins. . . . A party led by Dr. William J. Zinsmeister of the Institute for Polar Studies at Ohio State University spent four weeks combing the area. They found nothing until the end of their stay, when they returned to a site rich in penguin remains. Michael O. Woodburne of the University of California at Riverside spotted a marsupial jawbone. The scientists soon found four fragments from two animals, which resembled a species living during the same era—40 million years ago—at the southern end of South America. The teeth characterize berry-eating marsupials, says Zinsmeister. He believes the animals, about seven inches long, lived in vegetation near what was then shoreline.

Exercise 3.12

Latest Fossil Finds Link Ancient Africa to the Carolinas

Sara L. Samson, a graduate student in geology at the University of South Carolina, felt her shoelace come loose while on a field trip near Batesburg, South Carolina. What she saw when she bent to tie it was the first in a series of finds at that site that won over many geologists to the controversial view that a broad chunk of North America, stretching from Alabama through Georgia and both Carolinas, is a foreign body—a visitor that came from across an ancient sea about 400 million years ago, was thrust up onto this continent, and stayed behind when the sea opened again to form the modern Atlantic Ocean.

The object that attracted Ms. Samson's attention was a stone with what she suspected might be a fossil imprint. She showed it to a fellow student who pronounced it of no interest and threw it away. Ms. Samson, however, retrieved it and showed it to Dr. Donald T. Secor, Jr., professor of geology at the university, who recognized it as the partly preserved head of a trilobite, a many-segmented sea creature that lived more than 500 million years ago.

Since that March day, Ms. Samson and her colleagues, scouring the area, have found the source of that first fossil in bedrock in which many further specimens are embedded. At least five separate species have been identified, including prehistoric aliens—natives of the waters that fringed the far side of the ocean that separated America from Africa and Europe before the great collision. Such finds are evidence not only for that collision, but also strengthen the general case for plate tectonics, the theory that the continents assumed their present shape as a result of the wanderings of the great plates of which the earth's crust is composed. . . .

Ms. Sampson presented her findings at an annual meeting of the Geological Society of America in New Orleans. Her specimens have been examined by Dr. A. R. Palmer of the society, one of the world's leading authorities on trilobites and a former skeptic as to "foreign" segments in the American landscape.

"It is really exciting stuff," he said in a telephone interview. "They are very definitely un-American." Farther north, other non-American fossils had previously been discovered in formations 450 million years old along the Atlantic rim of New England and Canada. It was Dr. Palmer's examination of these fossils that won him over. Not only did they include trilobites once native to the English coast, but the specimens were in the same sequence of rocks. "It is the whole picture," he said, "that is convincing."

It was just such fossils that first led Dr. J. Tuzo Wilson of the University of Toronto to propose, in 1966, that those coastal areas once lay on the far side of an ancient ocean.

The closing up of this ocean, Dr. Wilson suggested, produced the Appalachian Mountains. As the continental masses collided, their surfaces were folded and piled miles high in spectacular contortions of landscape of which today's comparatively mild mountains are only a faint suggestion. Later, Dr. Wilson said, the continents separated, slowly producing the present Atlantic Ocean, but leaving part of Europe and Africa glued to New England. Other fossils in Scotland and coastal Norway, he said, indicate that those areas were once part of America, but were carried off by Europe as the new Atlantic widened. Now the trilobite finds in the South seem to have confirmed the long-held suspicions that part of that region also originated abroad. . . .

Perhaps the chief iconoclast, where proposals that sections of the American landscape came from far away are concerned, has been Dr. Anita Harris of the Geological Survey, a specialist in the microscopic fossils called conodonts. Dr. Harris does not deny that the great plates into which the earth's surface is divided are moving, but she believes those motions explain less of the earth's history than has been claimed for them. Other forces, such as currents, have been ignored by the plate-tectonic theory, she says.

Reached by telephone Dr. Harris conceded, on the authority of Dr. Palmer, that the trilobites in South Carolina were probably of European-African affinity. But, she added, this did not prove their home territory lay across a wide ocean or had been attached to Africa.

Exercise 3.13

Project

Look up an historical account of a scientific episode involving some theoretical hypothesis or theory. You might begin with an account in a standard textbook for a science course. Then, you could look up the same episode in an encyclopedia or in a book on the history of science, perhaps a whole book on the episode in question. Analyze the case following the six-step program developed in the text. Depending on how extensive an example you have uncovered, you may have to do several analyses involving different sets of data or different experiments.

CHAPTER 4

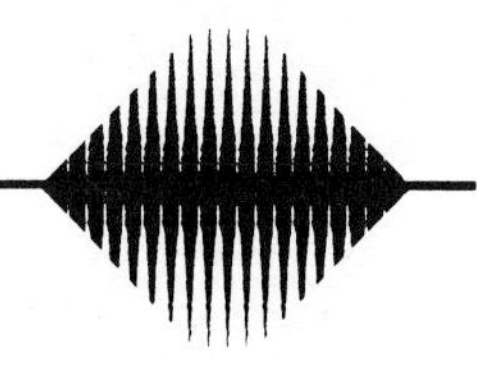

Marginal Science

Up to now, the examples we have studied clearly belong either to the history of science or to contemporary science. They are all part of straight, mainstream science. In this chapter, we will examine a number of examples whose very status as science would often be questioned. It is important to gain experience in evaluating such examples because the popular media contain many instances of purportedly scientific reports whose scientific status is, in fact, highly questionable. Often, no indication is given that these instances are any different from cases clearly within the mainstream practice of modern science.

Many textbooks on philosophy, logic, or science attempt to provide simple criteria that one could use to distinguish genuine science from pseudoscience. I will not follow this strategy because I think it is impossible to devise the required criteria. For example, it is often said that science involves experimentation and pseudoscience lacks experiments. However, there are no experiments in astrophysics, which everyone admits is a genuine science. Moreover, parapsychologists do perform experiments, and many people would regard parapsychology as a pseudoscience. I suspect there are similar counterexamples for any rigid criterion that might be proposed.

Rather than seeking special principles to draw a sharp line between science and pseudoscience, we will simply apply to these new cases the program we have already developed. So the question for us will not be whether a particular inquiry is genuinely scientific. The question is simply whether the data cited do or do not provide evidence for a proposed model. We will find that there are inquiries for which the data cited turn out typically to be inconclusive. Such inquiries, we will say, are instances of marginal science. This leaves open, at least, the possibility that they might one day move into the scientific mainstream.

This approach presumes only that there is some sort of model being proposed, and that some data are being cited in support of the hypothesis that the proposed model represents the real world. So the only inquiries excluded from our approach would be those that either (1) deny using models to represent the world, or (2) deny the relevance of an appeal to data in support of models proposed. In the first case, we are clearly not dealing with science. In the second case, we would have to ask what could possibly be the basis for claims that the proposed models do represent the world if one makes no appeal to data about what has actually been experienced in the world. In the past, people have claimed to have ways to authenticate models as representing the world without any appeal to data. Some people still do. This dispute is an interesting one, but we will not pursue it in this text. Science has become so prominent in Western culture that even advocates of the most outlandish hypotheses typically cite data in support of their claims. In all of these cases, we can apply our standard program for evaluating the proposed hypotheses.

4.1 THE CASE OF LITTLE HANS

Our first case, Freudian psychology, is clearly a marginal science. A century after Freud's early work, the categories of his theory have become part of everyday language. Yet, the scientific status of his theories continues to be a subject of much debate within the scientific community.

Sigmund Freud was born in 1856 and lived most of his life in Vienna. He studied medicine and then spent more than a decade conducting research in neurophysiology. Unable to continue making a living doing research, he began attempting to treat people who, because he knew something about the brain and nervous system, came to him seeking help for various psychological disorders. He spent the rest of his life developing a theory of the mind and a technique (psychoanalysis) for treating people with psychological problems. Freud's theory and the technique developed together over many decades until his death in 1939. His evidence for the correctness of the theory came primarily from his clinical practice. What follows is a partial sketch of Freud's mature theory.

Freud's Theory

For Freud, the mind has several structures, each with distinct characteristics and functions. One structure consists of three parts, the unconscious mind, the preconscious mind, and the conscious mind. The difference is a matter of awareness. We are not aware of things in our unconscious mind, but are aware of things in our conscious mind. Objects of the preconscious mind can, with special efforts, sometimes be brought to awareness.

Another structure also consists of three parts, the id, ego, and superego. The id consists of primitive, primarily biological, desires whose fulfillment produces pleasure. The superego contains the rules of proper behavior that are learned from one's parents and others. It corresponds to what we normally call our "conscience." The ego is the part of the mind that engages the world and attempts to balance the demands of the id with the requirements of the superego. The id is mostly unconscious, but the superego is mostly conscious. The structures are pictured in Figure 4.1.

Freud also developed a model of psychological development to go along with his model of the mind's structure. A young child, he said, is ruled primarily by its id. Early in childhood, boys and girls develop Oedipal, or Electra, complexes in which they form an emotional attachment to the parent of the opposite sex and reject the parent of the same sex. Thus, for example, little girls love their fathers and hate their mothers. Successful ego development requires adapting to the demand of the superego that one identify with the parent of the same sex.

Caught between the conflicting demands of the id and the superego, the ego develops a number of defense mechanisms for coping with the stress. One

Figure 4.1

The elements of Freud's model of the human mind.

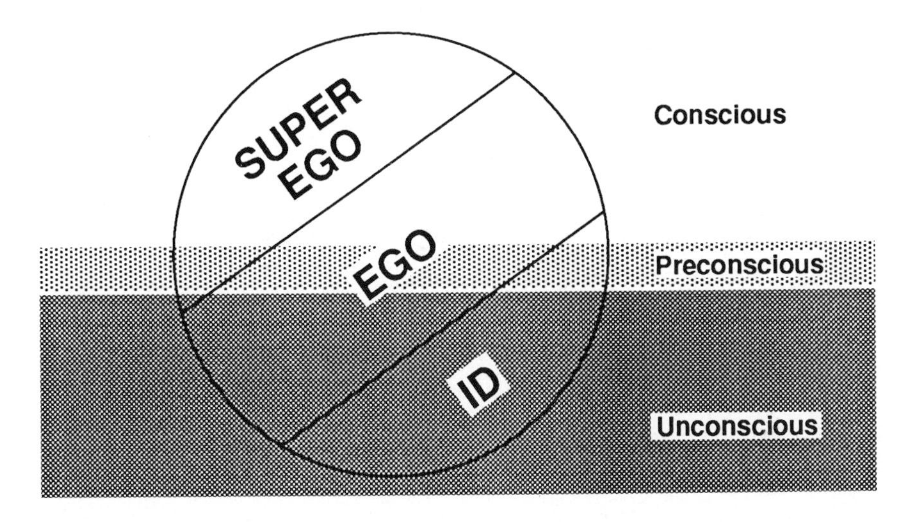

is to repress the demands of the id by pushing them down into the unconscious. Another is to project thoughts, or feelings, from their real object onto another object. Operating defense mechanisms requires much psychic energy. In some cases, the ego may simply not be strong enough successfully to cope with the demands placed upon it and lose touch with reality, directing the person to perform highly inappropriate, "psychotic," actions.

Finally, Freud thought that psychological disorders could be cured if the patient could bring the source of the conflict into consciousness and come to understand its true causes. It is the job of the therapist, he thought, to help the patient perform this task.

Little Hans

Little Hans was the son of a Viennese physician who was also a friend of Freud and a follower of his teachings. In the fall of 1907, four-year-old Hans developed an extreme fear of horses, so much so that he refused to go outside, where there were many horses in the streets drawing carriages. He even expressed the fear that a horse might come into his bedroom and bite him. His parents were much perplexed and his father brought him to see Dr. Freud.

Freud learned that Hans had spent the previous summer in a resort some distance from Vienna. There he had many playmates and spent much time alone with his mother during periods when the father was back in Vienna tending to his patients. Sometimes, when Hans was anxious about being alone in the dark, his mother let him spend the night in her bed. Hans was also alarmed when he heard someone else being warned not to get too close to a horse because it might bite him.

Upon returning to Vienna in the fall, the family moved to a larger apartment where Hans had few friends and his own bedroom. Earlier, he had shared his parents' bedroom. Hans's father now objected to his coming into the parents' bedroom in the mornings. And Hans spent much less time with his mother, there now being a nursemaid who would take him for walks in the park. On one of these walks, Hans was very frightened by a horse falling down in the street. Shortly after this episode Hans developed his extreme fear of horses. He expressed particular fear of the things the horses had around their eyes and mouth.

Freud's Hypothesis

Freud hypothesized that Little Hans was undergoing an Oedipal crisis. His id missed his mother's companionship and attention. But his ego was unconsciously afraid his father would be angry with him for wanting to be with his mother more. However, since Hans's superego had been taught that one should love one's parents, his ego repressed his fear of his father and displaced it onto horses. Why horses? Well, he had learned to fear being bitten by horses. And Freud also thought that Hans must be associating the blinders and muzzles of the horses with his father's glasses and beard.

Applying his own theories to the case, Freud asked the child if his father's glasses and beard did not remind him of the blinders and muzzles of the horses. Then, he suggested to the child that maybe he really was afraid his father would disapprove of his desire for more time with the mother. Freud, also, had the father reassure Little Hans that he was not really angry at him for wanting more of the mother's attentions. Little Hans engaged in some symbolic defiance of the father. Over the next few days it seemed to the father that the symptoms did lessen somewhat. It was several months later, however, before the symptoms disappeared.

Analysis

Steps 3 and 4 here are again reversed.

Step 1. The real world object of study is Little Hans and his fear of horses.

Step 2. The model is that developed by Freud as described above.

Step 4. The prediction is that the symptoms will disappear once the child is made aware of the real basis for his behavior and reassured that nothing bad will happen if he ceases to suppress his real feelings of fear of his father.

Step 3. The data are that the symptoms appeared to diminish and, finally, disappeared several months later.

Step 5. The data and the prediction do pretty much agree.

Step 6. Would the data have been likely to agree with the prediction even if Freud's model did not really fit the case? Are there other plausible models that would have yielded the same predictions?

The standard descriptions of this case do not mention other models, so one is left on one's own to evaluate the question in Step 6. But it requires only a rudimentary knowledge of young children to come up with very plausible rival models.

Everyone knows that children desire attention and affection. By expressing fear of the dark, Hans gained the privilege of sleeping in his mother's bed. That tactic did not work with the father. Developing an extreme fear of horses, however, did gain Little Hans considerable attention, even from the father, who took him off to see his friend Freud. Having gained the attention he desired, the symptoms gradually declined.

That the symptoms should go away in several months was, in any case, quite predictable simply because children around the age of four are typically quite changeable. They will outgrow most "phases" in a few months no matter what one does. Hans's professed fear of horses can be seen as a somewhat extreme phase, but hardly that unusual for a four-year-old.

The conclusion is that the results of this clinical case do not really provide much evidence for the correctness of Freud's theory (or for the efficacy of his method of treatment). Neither do they refute it. The data cited are inconclusive regarding Freud's hypotheses.

4.2 ASTROLOGY

Most scholars and scientists regard astrology as a marginal science at best. Yet, polls indicate that as many as a third of all Americans believe that there is something to the practice of astrology. Millions read their horoscopes in the daily newspaper. Special monthly magazines with more detailed, daily horoscopes are available at newsstands. Also, there are in the United States roughly ten thousand "professional" astrologers who make at least part of their living casting horoscopes and giving advice based on astrological lore.

The practice of astrology goes back to the Babylonians (roughly 2000 B.C.) to whom we also owe the beginnings of astronomy. The versions of astrology practiced today derive mainly from classical Greece (400 B.C. to A.D. 200). Thus, the fact that the familiar signs of the zodiac are divided into earth signs, fire signs, and so on, is based on a correspondence with the Greek theory that everything is made of four basic elements: earth, air, fire, and water. The major classical works on astrology were written by the second-century Greek astronomer Ptolemy, the same person who developed the theory of the universe that Copernicus and Galileo challenged fourteen centuries later. Even some of the major figures in the scientific revolution of the seventeenth century, such as Johannes Kepler, combined the study of astrology with work in astronomy.

At a time when the sun, moon, and planets were associated with gods (Jupiter, Mars, and so on) who were thought to influence human life, it was not unreasonable to suppose that the positions of the sun, moon, and planets at one's birth might have some influence on one's subsequent life. It also helps to think that the earth is the center of the universe, and not merely a middle-sized planet in orbit around a middle-sized star.

These beliefs have long since been given up. It is now well known that the gravitational force of the planets on the earth is minuscule. This book exerts a much greater gravitational force on you right now than the planet Mars ever could. Where you now sit, the radio waves picked up by your portable radio are many times stronger than radio waves from Jupiter, and the magnetic field of the magnet that drives the radio's speaker is many times stronger than the magnetic field originating in any astronomical body other than the earth itself. There is just no good reason to believe that there are any forces of the type assumed by astrology. Yet, people continue to believe in astrology. Why?

In interviews with people who put at least some stock in astrological descriptions and predictions, the most common answer to the question why they believe is simply that it works. In saying that it works, people seem, at least, to be claiming that the descriptions and predictions based on astrological theories are borne out in practice. What descriptions and predictions are these? Are they generally true? Do they provide evidence that the models of astrological theory do fit the real world?

Looking up my own sun sign (Sagittarius), in an astrological magazine, I find the following description:

> It represents the intended capture of unknown knowledge, wisdom, truth, and perfection in the philosophies. It represents all things dealing with higher thought, which bring people greater awareness of themselves and their place in the universe. Education, courts, religion, and literature are represented here.

The daily "prediction" for a typical day for my sign reads something like this:

> Your personality will glow, you'll attract much attention with charm and pleasing ways. Make work enthusiastic, especially if it is linked with younger people. Use ingenuity and skills, which are at their height of performance. Be generous financially, contribute to a worthy cause. Health is excellent and travel aspects are good. Carry out a plan you know is valuable in practical ways. Love breakups are possible, some dishonesty in another's attitude.

As a matter of fact, I find it fairly easy to interpret much of both the description and the prediction as being true of me in general and of my actions on a particular day. So let us analyze the relevance of this data for the hypothesis that astrological theories do fit the real world.

Analysis

Step 1. The subject matter is the attitudes and everyday actions of ordinary people, myself included.

Step 2. The models are those provided by traditional astrological theory in which the positions of the sun, moon, planets, and stars at the time of people's births determine (or at least influence) their later attitudes and everyday actions.

Step 3. The data, in my own case, are my attitudes and the actions I performed on the day in question.

Step 4. The predictions are those given in the excerpts quoted above.

Step 5. The data and the predictions do agree pretty well.

Step 6. Were the predictions likely to agree with the data even if the models of astrological theory do not provide a good fit to the real world? Yes. For reasons to be discussed below, these predictions were fairly likely to agree with the data, in any case. So the agreement between the data and the predictions fails to provide any evidence in favor of the hypotheses that make up astrological theory.

There are two general reasons why the predictions based on astrological models were likely to agree with the data concerning my attitudes and activities, even if these models have little resemblance to the real world. These reasons deserve special attention because they highlight typical features of the predictions associated with many marginal sciences.

The Predictions Are Vague

The descriptions and predictions derived from astrological theory are stated in quite general terms. Anyone with a little imagination can easily see themselves in the descriptions. The description for Sagittarians, for example, seems to fit well a university professor. But read the description for Gemini:

> It is the symbol of the negative and affirmative mental aspects; the combination of different ideas that are joined in an earthly and idealistic decision; an adaptable memory for facts; communication and the search for knowledge.

That fits me as well as the description for my own sign.

Similarly, the daily predictions and advice are formulated in a vague way that makes it easy to interpret them as either true, or relevant. "You will attract much attention with charm and pleasing ways." How much attention? From whom? What kind of charm? What sorts of pleasing ways? Because answers to these questions are missing, this prediction is very likely to be classified as fulfilled for anyone on almost any given day.

Here we have an alternative explanation for the agreement between the predictions and the data. They agree because the predictions are framed in a vague form that can be readily interpreted as being true of almost anyone. So a fairly good agreement between data and prediction is very likely quite apart from any fit between astrological models and the real world. There is, therefore, no basis for taking the agreement between data and predictions as providing evidence that the models of astrological theory fit the real world.

There Are Multiple Predictions

As exhibited in the above excerpts, astrological descriptions and predictions generally have several parts. The typical daily horoscope contains at least a half-dozen predictions or recommendations. This increases the chances that one or two will stand out as particularly correct, or relevant. In general, it is fairly likely that one or two out of a half-dozen vague predictions should be easily interpreted as having come true—whether or not the models of astrological theory bear any relation to the real world. There is only one chance in fifty-two of getting the ace of spades in a single draw from a standard deck of cards, but the chances of getting the ace increase as the number of draws increases.

The fact that there are multiple predictions, in addition to each prediction being vague, strengthens the alternative explanation of the agreement between the predictions and the data. If there are many predictions, that increases the chances of one or two being clearly correct. If they are all vague, that makes it difficult to argue that any particular prediction is clearly incorrect. The result is a strong impression that the predictions are, by and large, correct. But this impression may be solely a response to the vagueness and multiplicity of the predictions and have no bearing whatsoever on the fit between astrological models and the world.

Astrology as an Interpretative Framework

When people say that astrology works, they may mean much more than simply that the descriptions and predictions of astrological theory seem to come true. They may regularly use astrological categories to interpret both their own actions and the actions of other people, and may find these interpretations highly satisfying. "Why is Linda always so self-assured? Oh, she's an Aries, and Aries are like that." The fact that astrological lore can be used in this way may have nothing to do with how well astrological models, in fact, fit the real world. It may simply be due to the extreme flexibility of the framework and to the fact that people who use it this way seldom have any interest in noting, or remembering, cases in which specific predictions seem to fail. It is an interesting, and somewhat distressing, fact that an interpretative framework such as astrology can be highly satisfying even though the models that underlie it bear little resemblance to the real world.

4.3 PSYCHICS

If you regularly shop at supermarkets or walk past newsstands, you know that there are numerous publications that feature predictions by contemporary psychics. Among the best known is Jeane Dixon. As several biographies attest, many people believe that Ms. Dixon has special abilities to foresee the future. Moreover, it is explicitly argued that her powers are proved by her many, past successes.

Ms. Dixon, and others in her profession, regularly issue whole sets of predictions. Thus, you may in January see the headline: "Jeane Dixon's

Predictions for the Coming Year." The predictions concern all types of subjects, but most deal with the lives of celebrities, that is, well-known people in entertainment, politics, and the arts. For example, it is widely claimed that Ms. Dixon predicted the assassination of John Kennedy in 1963. What she really predicted, it appears, was that he would die in office, which is far less dramatic. In any case, does the fact that this and other predictions have been correct provide any basis for believing in her reputed special powers to see the future? Let us simply apply our program for evaluating theoretical hypotheses.

Analysis

Step 1. The real world subject is Jeane Dixon's activity as one who writes about the lives of celebrities.

Step 2. The model is, understandably, not spelled out in much detail. But it involves some kind of special psychic ability to "see" the future, an ability not possessed by most, normal human beings.

Step 3. The data are that she correctly predicted that John Kennedy would die in office.

Step 4. If the model fits, that is, if Ms. Dixon really does have psychic powers, then one would presume that she could successfully predict Kennedy's death while still in office. (Here I use the word "presume" rather than "predict" because the data involve another, quite different, prediction.)

Step 5. There is agreement between the data and the presumption that Ms. Dixon could correctly predict Kennedy's death in office.

Step 6. Was the presumption likely to agree with the data, even if Ms. Dixon possesses no special psychic powers? Are there any other plausible hypotheses that could account for her success in predicting Kennedy's death?

Was her prediction vague enough to make it likely that it would be successful in any case? Not really. When she made her prediction, Kennedy was a vigorous man in his mid-forties. He had from two to six years remaining in office—depending on whether he sought and won a second term. That such a man would die in the next six years would not be fantastically improbable, but it is not very probable either. So we could not fully explain her success by the fact that this particular prediction was one that would quite likely have been successful regardless of her having any special psychic powers.

There is, however, another possible explanation for the success of her prediction—multiple predictions. Jeane Dixon makes a great number of predictions, as many as a hundred or two hundred a year. The Kennedy prediction was just one that turned out to be successful out of a large number of predictions. What, then, is the chance that a few of her many predictions would come true if the hypothesis that she has psychic powers is false? Pretty high. Even if no single prediction has a very high chance of coming true, the chances that a few of them will come true are high.

So here we have a plausible explanation for Jeane Dixon's success in the Kennedy case. Of the many predictions she has made, this one happened to come true. If it had not, then some other comparably dramatic, successful prediction would be used as data by those who support her claims to special psychic powers. The answer to the question in Step 6, then, is "yes." The success of the Kennedy prediction provides no support for the hypothesis that Jeane Dixon has psychic powers.

What About Failed Predictions?

If Jeane Dixon makes many predictions, and they are not all so vague as to guarantee success, then there should be some that have failed. What about these failed predictions? Ms. Dixon and her supporters obviously do not talk much about these. Here are two that have appeared over the years. The *Star* headline for July 12, 1977, was "Jeane Dixon: Predictions." Among the predictions in that issue, there are a number about Elvis Presley, for example, that the following year he would give a tremendous benefit performance. Elvis Presley died several months later, leaving most of Ms. Dixon's predictions unfulfilled. Again, in the issue for December 29, 1981, the headline reads, "Jeane Dixon: 101 Predictions for 1982." On the front page, we see "Di: She'll Have a Girl." Inside it says only that she probably will have a girl. In 1982, the Princess of Wales, Lady Diana, gave birth to Prince William.

These cases should count as evidence that Jeane Dixon has no special psychic powers. If she had such powers, she certainly would have been exercising them to their utmost when she made these predictions. Elvis's death and Prince William's birth were major events in the celebrity world. Ms. Dixon specializes in such events. If she can see anything, she should have seen these. But she did not. There is no need to go through the entire analysis over again. At Step 5, we note the clear disagreement between the data and the presumption, based on the hypothesis of her psychic powers, that these predictions by Ms. Dixon should have been successful. That is evidence against the hypothesis that she has psychic powers.

In response to this evidence, Ms. Dixon's supporters would still insist that she has psychic powers, but would probably be willing to admit that her powers are not perfect. Indeed, they are so imperfect that she sometimes completely misses major events. Nevertheless, these supporters would still have to maintain that her rate of success is better on the average than that of an ordinary gossip columnist who claims no special psychic powers. This presumption could be verified by comparing a large number of her predictions with those made by ordinary gossip columnists. The analysis of such tests requires the techniques to be developed in Part Two of this book. To my knowledge, however, no such experiment has ever been undertaken. My personal opinion is that any such data would show that, on the average, Jeane Dixon is no better than her competitors.

4.4 CHARIOTS OF THE GODS?

It is a recurring theme in American popular culture that the earth has been, or is now being, visited by aliens from another world. There are many variations on this theme. Within the community of people familiar with these matters, this is known as the hypothesis of extraterrestrial visitation, or ETV. Strictly speaking, ETV is a special version of the broader view that there are genuine unidentified flying objects, or UFOs. Those who believe in ETV think they have an explanation for the existence of UFOs. One might think there are genuine UFOs, however, without believing they have an extraterrestrial origin.

As an example of this genre of theory, let us look at Erich von Daniken's *Chariots of the Gods?* This book is reputed to have sold over 30 million copies in thirty-five different languages. Whatever its scientific success, it has certainly been a commercial success.

Von Daniken claims that the ETV hypothesis is true and, moreover, that the visitations took place early in human history. He thus proposes to explain various happenings in the early history of the human race as being the direct result of actions by these early, extraterrestrial visitors. In support of his claims, he produces numerous examples of archaeological, or anthropological, findings that otherwise seem difficult to explain.

Many different cultures have myths about gods who came from the heavens and performed great feats. How is it that so many cultures, widely separated in place and time, have similar myths?

In Egypt and in Peru, there are huge temples, pyramids, and other structures constructed out of cut stones weighing many tons. How could ordinary men have constructed such things with only the simple tools then available?

The plain of Nazca in Peru exhibits what resemble roads cut in patterns. It is difficult to imagine why, or how, these were constructed.

On Easter Island in the Pacific Ocean, there are hundreds of stone statues weighing up to 80 tons erected a fair distance from the source of the rock from which they were carved. Who could have carved them? How could they have been moved and set up on a remote island that could not support more than a few thousand people?

Von Daniken has an easy answer for all of these questions. These things are all explainable as the direct result of the activities of ancient astronauts. Von Daniken assumes that any beings capable of reaching Earth from somewhere else in the universe would have no difficulty performing feats, such as building pyramids. Now let us subject von Daniken's claims to a standard analysis.

Analysis

Step 1. The subject matter encompasses parts of human history on Earth.

Step 2. Von Daniken's model of human history includes visitations by ancient astronauts who performed various feats while on Earth.

Step 3. The data include the discovery of pyramids in Egypt and Peru, the statues on Easter Island, and other things like those noted above.

Step 4. The prediction is that we can now find all the things noted above.

Step 5. The data and the prediction do agree. We have indeed found all the things von Daniken's model says we should find.

Step 6. Would our finding all the things von Daniken discusses be likely if there had been no visitations by ancient astronauts? Are there other plausible explanations for the existence of the data as cited? As usual, everything depends on how one answers this question.

An Alternative Model

There is one obvious alternative explanation for the agreement between von Daniken's predictions and the data that is quite independent of any similarity between von Daniken's model and the real world. This is that von Daniken deliberately constructed his model so as to yield predictions agreeing with the known data. Such a process of model construction requires only imagination. Almost anyone could do it. Even you or I. Let me show you how.

Imagine a model in which there once was an advanced race of humanoids similar to us who evolved on the earth several thousands of years before the earliest peoples now known to anthropologists and archaeologists. According to this model, these early humanoids developed a fairly advanced technology that made them superior to other evolving prehuman primates. Using their advanced technology, these humanoids moved out from their original home in the Middle East and established outposts in Asia, South America, Easter Island, and so on. We can imagine that they even developed large ships and airplanes. It is this race of early humanoids that is responsible for the statues, temples, and other unusual things that we now find in isolated parts of the world. Moreover, it is the memory of this race that is recorded in the myths of various peoples around the world. Unfortunately, when it came to social organization, these advanced humanoids were no wiser than present-day humans and began warring among themselves. Eventually they destroyed themselves and most of their artifacts. All that remains are the few traces that current anthropologists and archaeologists cannot explain. My hypothesis makes all these things easily understandable.

My terrestrial model, which does not require ETV, predicts exactly the same data as von Daniken's model incorporating ETV. And my model is initially no less plausible than his. So the agreement between von Daniken's predictions and the data is not necessarily unlikely if his hypothesis is false. Thus, the agreement between the data and his predictions does not provide evidence that his model fits, no more than it provides evidence that my model fits. Here it is useful to remember Halley's successful prediction of the return of a comet seventy-six years later. In this case, ensuring the agreement between prediction and data was beyond the power of anyone's imagination. The only alternative to Newton's model was the improbable supposition that another, very similar, comet just happened to come by at just the right time seventy-six years later.

What About the Data?

Writers like von Daniken often have a good intuitive sense of how scientific data is evaluated. He, therefore, devotes many pages to convincing the reader of the fantastic nature of the artifacts he describes. This is to make it seem that the existence of such things would be highly unlikely unless something like his hypothesis is accepted. The trouble is that average readers have little detailed knowledge of the things he describes. They are dependent on the author himself to supply the data. In this situation, the author may be tempted to distort or withhold known facts in order to make his case seem better than it really is. It seems clear that von Daniken has succumbed to this temptation.

In his discussion of the statues on Easter Island, von Daniken cites the work of anthropologist-adventurer Thor Heyerdahl. In the book cited, Heyerdahl reports actually observing some modern-day inhabitants of Easter Island carving, moving, and setting up statues similar to those originally found on the island. Using only a crude wooden sled, it took fewer than 200 men to move a 12-ton statue. Using large poles as levers and stones to prop it up, it took twelve men only eighteen days to set up a statue weighing around 25 tons. Heyerdahl's book even contains a picture showing the statue in the process of being set up. Thus, most of what von Daniken says about this case is just plain false. There is a relatively simple, perfectly natural, explanation of how these statues could have come to be where they are.

The moral for our purposes is that it takes more than good reasoning not to be misled into thinking that there is real evidence for a hypothesis when in fact there is none. You have to learn when to be suspicious of the data and when to check further on your own. Then, you have to take the time to do it. There are no good rules that you can use here except to maintain a healthy skepticism regarding things that seem to go against most of what you already know. Sometimes you even have to consider the authors' motives. Are they interested in learning the truth, or are they primarily concerned to attract attention and to sell books?

4.5 WHEN PROPHECY FAILS

As part of a study concerning the dynamics of belief in groups, some social psychologists once infiltrated a number of small cults. Among the groups observed, there was one led by a woman who was given the fictitious name of Mrs. Marion Keech. The events described, however, really took place.

Mrs. Keech claimed to have made contact with extraterrestrial beings and she regularly revealed to her followers what she claimed were messages from her contacts. One day, she said that her contacts had told her that they were going to destroy the world. She even gave a specific day on which this was supposed to happen. These extraterrestrial beings also told her, she claimed, that they would rescue her and anyone else who wished to be saved if they would wait at a designated location.

On the appointed day, Mrs. Keech and her followers were at the designated place. The day wore on, but the world did not end. Finally, Mrs. Keech informed her followers that she had received a telepathic message from her contacts informing her that they had decided to spare the world from destruction because she and her followers had been so strong in their faith that the rescue would, indeed, take place. The psychologist who was the undercover observer reported that the assembled followers returned to their homes, apparently strengthened in their belief that Mrs. Keech was indeed in contact with powerful extra terrestrial beings.

Analysis

This case is best understood by realizing that there are, in fact, two different models involved. In the first model, Mrs. Keech's extraterrestrial contacts are going to destroy the world no matter what. In the second model, they change their minds and do not destroy the world. Although it is a bit tedious, we will do two separate analyses, one at a time. Again, Steps 3 and 4 are reversed in the first analysis.

Step 1. The real world object of study is Mrs. Keech and her cult.

Step 2. The first model has Mrs. Keech in contact with extraterrestrials who say they will destroy the world on the designated day but will rescue her and her followers if they will congregate at a designated place.

Step 4. The prediction is that the world will be destroyed and the followers will be saved.

Step 3. The data are that the world is not destroyed and Mrs. Keech's supposed contacts do not appear.

Step 5. The prediction and the data do not agree. The model should be rejected.

In fact, it seems that Mrs. Keech's followers were quite willing, and maybe even relieved, to reject this model. But this was partly because they were provided with another, similar model, which they immediately adopted. Here the analysis proceeds as follows:

Step 1. The real world object of study is Mrs. Keech and her cult.

Step 2. The new model has Mrs. Keech in contact with extraterrestrials who say they have decided not to destroy the world because she and her followers were so steadfast in their belief that they would be rescued.

Step 3. The data are that the world is not destroyed and Mrs. Keech's supposed contacts do not appear.

Step 4. Now the prediction is that the world will not be destroyed and there will be no rescue.

Step 5. The prediction and the data now agree.

Step 6. That the data would agree with the new prediction is just what everyone should have expected in the first place. Just about every model of the world known to mankind yields the prediction that the world was not going to be destroyed that day. So the data provide no evidence that the new model fits the real world.

From the standpoint of the psychologists studying this cult, the remarkable fact that requires some psychological explanation is that a group of people will believe something quite unusual with no real evidence whatsoever. All they really had to go on was Mrs. Keech's pronouncements. How this might come about is another story.

In analyzing such cases it is very important to keep in mind that both the prediction and the data refer only to things, or processes, that can be observed in the real world. The world being destroyed is obviously such a process. So is the nonappearance of extraterrestrials. What cannot be part of the prediction, or the data, are the nonobservable activities of the supposed extraterrestrials, such as changing their minds about destroying the world. That happens only in Mrs. Keech's model, and there is no evidence that this model is exemplified in the real world.

4.6 AN EXPERIMENT WITH TIME

The 1960s, the Age of Aquarius, gave birth to the New Age of the 1980s. New Age culture is sympathetic to many marginal beliefs, such as those in clairvoyance, telepathy, out-of-body experiences, and reincarnation. Here is one theory offered to explain these supposed phenomena.

First, some data. Get yourself in a relaxed position and place a clock, or watch, with a sweep-second hand in your line of sight so that you gaze naturally on it without any effort. Gaze at the clock for a while, absorbing the rhythm of the second hand. Then close your eyes and imagine yourself in a relaxed position somewhere else, such as lying on a familiar, quiet beach. Imagine as many details as you can to make the imagined scene as realistic as possible. Pretend you are really there listening to the waves. Now, very slowly, open your eyes and just let them gaze straight ahead. Do not attempt to focus on the clock. Just let it be unfocused in your line of sight. If you do it right, you may have the experience that the second hand seems to skip a beat, or even seems to stand still for a second or two. Do not think about it or focus your eyes on the clock. That will destroy the desired effect.

Although this effect is confidently reported, I myself have not yet experienced it. That some people should sometimes have this experience seems plausible enough, so let us suppose for the sake of argument that it does sometimes happen. What are we to make of this?

This phenomenon, assuming it exists, is offered as evidence that a person's psyche, sometimes called the "observer," can, at least briefly, leave the body and travel at the speed of light, or maybe even faster, to distant places. So if you imagine the scene realistically enough, your psyche actually is thought to be there observing the scene at the beach. While it is gone, your body remains home functioning normally except that there is no consciousness receiving the signals from your senses. The reason the second hand seems to stand still, on this theory, is that your consciousness has momentarily left the body, and when it returns it picks up where it left off. Your body, meanwhile, follows the physical rhythm of the second hand. Thus, the anomalous experience of seeming to see the second hand stand still while at the same time feeling that it should have moved forward.

There is, of course, much more to the theory than has been presented above. The broader theory portrays, for example, both the body and psyche as having natural rhythms. In this broader theory, therefore, the observer's motions away from and back to the body are said to oscillate like a pendulum. There is even some dispute as to whether the motion of the psyche is limited by the speed of light. Perhaps it can travel infinitely fast, thus being able to cover infinite distances and, in effect, be everywhere at once. We need not, however, go into these details in order to analyze what we already have before us.

Analysis

Step 1. The real world objects of study are ordinary humans who are conscious and have physical bodies.

Step 2. The model is that of a person's psyche as being a distinct, though immaterial, thing that is able to leave the body and travel rapidly to distant places.

Step 3. The data is the supposedly common occurrence of someone experiencing a second hand as seeming to skip a beat or two.

Step 4. It is part of the model that experiences such as that of the skipped beats of a second hand are possible.

Step 5. The data and the prediction agree.

Step 6. The agreement between prediction and data is, however, plausibly to be expected even if the model fails completely to represent the real world. One need not know the mechanism of the experience. Everyone has experienced a variety of optical illusions, such as lines of equal length drawn in such a way as to appear to differ in length. It is a very plausible, alternative hypothesis that the purported phenomenon is something similar. In any case, there are many completely natural phenomena similar to this one. So the experiment fails to provide any evidence that the model is correct.

4.7 THE BURDEN OF PROOF

By and large, the results of our evaluations in this chapter have been inconclusive. In most cases, we have concluded that the data cited do not provide evidence for thinking that the hypotheses put forward are true. It may seem to some that this is unsatisfying. They would like to be able to conclude that the hypotheses in question are mistaken—the proposed models do not fit. Although understandable, this feeling is based on a misunderstanding both of our objectives and of the real power of the program for analyzing reports that we have developed.

In debating, as in the law, there is a concept of the burden of proof. When a claim is disputed, the question is who has the responsibility for proving it true or false. That party carries the burden of proof. As a consumer of scientific information, one should always take the position that the burden of proof is on the producers, or purveyors, of that information. They have to convince you that they are right. That means they have to provide you with data that constitute good evidence for their claim. You, like a jury, need only evaluate what they provide. Thus, if you conclude that the data supplied do not provide good evidence for their claims, you have done your job. You have reached the useful conclusion that you need not, at least for the moment, pay any more attention to their claims.

Being able to conclude that the data presented are irrelevant to whether the proposed model is correct is, in fact, a very powerful conclusion. It means that one can safely ignore the hypothesis in question. One need not take it seriously. Of course, this leaves open the possibility that the hypothesis might be true and that other data might provide strong evidence to that effect. Until presented with such data, however, one may remain uncommitted.

In written presentations and conversations, advocates of a particular point of view will often attempt to shift the burden of proof onto you, the audience. They will challenge you to prove that they are wrong. You should not give in to the temptation to take up the challenge. They want you to think that unless you can prove that they are wrong, you must agree that they are correct. But this is simply not so. Your being unable to show that they are wrong does not mean that they are right. That is, your inability to produce evidence against their claims does not constitute evidence for their claims. It may only be evidence of your general ignorance of something about which you have every right to be ignorant. So remember, it is not your job to prove that they are wrong. It is their job to show that they are right.

In debate and the law, as in warfare, the advantage always lies with the defense. If you allow the burden of proof to shift to you, that will put you on the offensive side of the debate. This requires, typically, that you be able to produce data disagreeing with the predictions of their models. Not being especially up on the subject, you are unlikely to have such data on the tip of your tongue. Even if you can come up with some relevant data, advocates of the position are likely to reveal still more details about the models that render your data irrelevant to, or maybe even supportive of, their position. Far better to stick with the defensive position. Keep the burden of proof where it belongs.

EXERCISES

Analyze these reports following the six-point program for evaluating theoretical hypotheses developed in the text. Number and label your steps. Be as clear and concise as you can, keeping in mind that you must say enough to demonstrate that you do know what you are talking about. A simple "yes" or "no" is never a sufficient answer. Some of these reports are taken directly from recent magazine or newspaper articles and are presented here with only minimal editing.

Exercise 4.1

UFO Update

Five TBM Avenger torpedo bombers left the Naval Air Station in Fort Lauderdale, Florida, on December 5, 1945, for a routine training mission. But less than three hours later, all five planes and their crew of fourteen were missing. A Martin flying boat (a plane that can land on water) was immediately sent to the rescue; but that craft, along with thirteen crew members, also disappeared.

Cryptic messages attributed to the flight instructor, including the cry "Don't come after me; they look like they're from outer space," have long fueled speculation that aliens were involved in the planes' mysterious disappearance. Now a Poway, California, man named Wesley Bateman claims he has evidence that extraterrestrials hurled the planes into space. In fact, Bateman asserts, at least one of the Avengers is entombed in ice about 6000 miles above the earth, just over the Bermuda Triangle.

Bateman, who says he conducts privately funded, UFO research "for a group of people who want to remain anonymous," made his discovery while watching a videotape about UFOs. After watching the tape several times, Bateman says, he noticed an object allegedly photographed by Apollo 11 astronauts. What's more, he declares, he realized that the object resembled a TBM Avenger. "The heaviest part of the plane—the nose—is pointed toward the earth," he states. "And you can easily recognize the bubble turret and the tail."

But how did a 1945, propeller-driven airplane get into orbit? Bateman theorizes that depth charges dropped by the Avengers may have damaged an alien craft under the sea. "When the UFO rose out of the water to avoid further damage," Bateman says, "its rapid departure created a propulsion vortex, sucking up a lot of seawater and the planes along with it."

As far as aviation journalist and UFO skeptic Philip J. Klass is concerned, however, Bateman's theory is full of holes. "I've personally spent days in the Navy archives going through the records of Flight 19, and there is simply no mystery to it," he insists. Klass says the craft originally headed east, toward the Bahamas. The flight instructor became disoriented and decided that the aircraft were over the Florida Keys. He thus ordered the planes north, toward what he thought was the mainland. But since the planes were over the Bahamas, heading north pointed them toward Greenland. Greenland was just too far away, and the aircraft simply ran out of gas.

But how does Klass explain the photo of an Avenger some 6000 miles above the earth? "I could say that the object looks like the face of God," he says. "Besides, our radar monitors every object in Earth orbit. It's powerful enough to detect a six-inch-long metal strap that has fallen off a satellite. An object the size of an Avenger would long ago have attracted attention." NASA spokesman Ken Atchison adds, "I was around during the Apollo days, and I've never heard of the astronauts seeing or photographing anything like what Bateman is talking about." Still, Bateman insists that seeing is believing. "I can't imagine anyone looking at this photo," he asserts, "and saying it's not an Avenger. This is conclusive proof that UFOs exist."

Exercise 4.2

Transactional Analysis

Transactional Analysis (TA) is a form of psychological therapy based on the theory that each person has a composite personality consisting of three parts called "child," "parent," and "adult," respectively. A person's parent personality is made up from that person's experiences with his or her own parents (or substitutes). It includes many "don't's" and some "do's." The child personality is made up of a person's earliest (mostly nonverbal) feelings, such as feelings of helplessness, or curiosity. The adult personality develops as one learns to control one's own life. Associated with each component personality are particular ways of interacting with others. According to the theory, mismatches between people's various personalities lead to conflict and unhappiness.

Experience with patients is often cited as evidence for this theory. For example, a couple comes to a TA therapist. The therapist has them discuss some problem and observes how they interact. The therapist then formulates a hypothesis about their interaction. The husband, for example, might be seen as responding with his child personality, while the wife is seen as responding with her parent personality. Neither adult personality, the therapist reasons, is happy with the responses from the other person. The therapist infers that they will both be happier if he can eliminate the mismatch by getting them both to respond directly to each other's adult personality.

In a number of succeeding sessions, the therapist intervenes in discussions attempting to bring out the adult personality in both husband and wife. Typically, the couple benefits from the therapy in that, afterwards, they are somewhat better able to communicate and are generally happier.

Exercise 4.3

Levitation

A recent article in the student newspaper of a Big Ten university discussed a student who claimed that he was able to reach a deep state of meditation in which he could levitate—that is, float above the ground (or floor) with literally no support. The reporter inquired whether he meant that he really was levitating or that it merely felt like it. No, the student insisted, he was really doing it. At that point, the reporter asked for a demonstration. Later, the reporter watched as the student meditated, but observed no levitation. Informed that he had not levitated, the student explained that having somebody watching made him too nervous to achieve a sufficiently deep state of meditation. The reporter apparently accepted the explanation because he wrote up the story in a straightforward—not tongue-in-cheek—manner.

Exercise 4.4

Predicting the Stock Market

When the stock market crashed in 1929, an economist with unorthodox views claimed that the crash proved his theory was right. He had predicted the crash a year before. It turned out that every year since 1921 he had been predicting a crash for the following year. This did not, however, prevent him from continuing to cite the 1929 crash as evidence for his views.

Exercise 4.5

Aquarians

According to astrological theory, Aquarians, those born between January 21 and February 18, are said to be generally scientific but eccentric, that is, brilliant but unconventional. In support of this hypothesis, it is often noted that a number of famous scientists, such as Copernicus, Galileo, and Thomas Edison, were Aquarians.

Exercise 4.6

The Shocking True Story: Tabloid Predictions Fail

Professional psychics featured in the supermarket tabloids struck out in 1989, failing to foresee the collapse of the Berlin Wall, the San Francisco earthquake, the political unrest in China, or the U.S. invasion of Panama.

Few if any of the events they did predict, like the crash of a jetliner into the Vatican, the Statue of Liberty toppling over, militant antismokers rioting in Washington, D.C., and President Bush being impeached, came to pass.

Forecaster Jeane Dixon, whose claim that she predicted the Kennedy assassination has never been proven, did say the Berlin Wall would be sold brick by brick for souvenirs. But she said the sale would happen early in the twenty-first century.

One psychic predicted in the National Examiner *that Panamanian dictator Manuel Noriega would step down and move to New York, where he would seek to become a photographer's model for a pineapple juice distributor. In fact, he was ousted in a U.S. invasion.*

Forecasters said terrorists would blow up the wrong stadium when they tried to disrupt the Super Bowl, and that Soviet leader Mikhail Gorbachev would flee to the United States.

The National Enquirer's *ten leading psychics forecast that in 1989 Madonna would have a baby, Jane Fonda would gain ninety pounds, and the Vatican plane crash would kill hundreds, destroying priceless masterpieces.*

Psychics said in the Globe *that 1989 would be the year an alien spaceship was revealed to the public, Priscilla Presley announced that Elvis was still alive, and scientists proved that food preservatives were addictive.*

In Weekly World News, *psychic Sophia Sabak predicted that a doctor would discover a pill that eliminated the need for sleep, the lost continent of Atlantis would reappear, and Soviet cosmonauts would be rescued by American astronauts.*

Exercise 4.7

Land of the Bird Men

The following passages are taken from the opening paragraphs of Chapter 8 of Erich von Daniken's Chariots of the Gods? *The chapter is titled "Easter Island—Land of the Bird Men." These passages were originally intended to provide evidence for the hypothesis that the earth was long ago visited by ancient astronauts who left behind various artifacts scattered around the world.*

The first European seafarers who landed on Easter Island at the beginning of the eighteenth century . . . saw hundreds of gigantic statues, some of which are between 33 and 66 feet high and weigh as much as 50 tons. Originally these colossuses also wore hats . . . which weighed more than 10 tons apiece.

Easter Island lies far away from any continent or civilization. The islanders are more familiar with the moon and the stars than any other country. No trees grow on the island, which is a tiny speck of volcanic stone. The usual explanation, that the stone giants were moved to their present sites on wooden rollers, is not feasible in this case, either. In addition, the island can scarcely have provided food for more than 2000 inhabitants. (A few hundred natives live on Easter Island today.) A shipping trade which brought food and clothing to the island for the stonemasons is hardly credible in antiquity. Then, who cut the statues out of the rock, who carved them and transported them to their sites? How were they moved across country for miles without rollers? How were they . . . polished and erected? How were the hats, the stone for which came from a different quarry from that of the statues, put in place?

Exercise 4.8

Debunking the Shroud of Turin

Since the Middle Ages, multitudes have believed that a piece of linen enshrined in Turin, Italy, is the burial shroud that Jesus Christ left in the tomb when he rose from the grave. But Turin's Anastasio Cardinal Ballestrero has announced that scientific testing proves the yellowing 14-foot-long fabric is only six or seven centuries old and could not have dated from the time of Jesus. Thus ended the most intense scientific study ever conducted on a Christian relic.

The new findings may please skeptics, but the shroud saga is not a major embarrassment for the Roman Catholic Church. Shortly after the earliest known exhibit of the shroud, in 1354, a French bishop declared it to be a fraud. Through subsequent centuries the church refused to confirm its authenticity. The examination that finally discredited the shroud was conducted with the full blessing of the church, in an unusual alliance between honest faith and objective science. When Pope John Paul was informed of the negative report, he ordered, "Publish it."

Ballestrero had initially agreed to an extraordinary series of scientific tests on the shroud in 1978, but refused to permit carbon 14 testing, which was crucial to determining the fabric's age. Handkerchief-size samples needed to be cut out, which, to Ballestrero, was unthinkable for such a revered historical item.

After technical improvements made it possible to use samples the size of postage stamps, however, the Cardinal allowed cuttings to be taken.

Testing was done simultaneously at the University of Arizona, Britain's Oxford University, and Switzerland's Federal Institute of Technology in Zurich. Each laboratory received four unmarked samples: a shroud cutting and three control pieces, one of which dated from the first century. The samples were chemically cleaned, burned to produce carbon dioxide, catalytically converted into graphite, and, then, tested for carbon 14 isotopes to fix the date by calculating the amount of radioactive decay. Only London's British Museum, which coordinated the testing, knew which samples were which.

Arizona's physicist Douglas Donahue says that the three laboratories reached a "remarkable agreement," all estimating dates within 100 years of one another. Averaging of the data produced a 95 percent probability that the shroud originated between 1260 and 1380 and near-absolute certainty that it dates from no earlier than 1200. However, some Catholics held out the slim hope that there was a scientific oversight and the shroud might be redated someday.

Exercise 4.9

Thousands Plan Life Below, After Doomsday

At least 2000 followers of a religious group have streamed into majestic Paradise Valley from around the world, preparing to spend years in underground shelters safe from the nuclear Armageddon they believe is imminent.

Heeding a call from their leader, members of the Church Universal and Triumphant have sold their homes, closed their bank accounts, bade goodbye to relatives, and headed to the valley and the elaborate system of concrete and steel shelters that the church has built here. Many have paid up to $10,000 to reserve a spot guaranteeing them a role in planning the future after most of the world is dead.

"This is the time we have been told by Mother to be prepared," said Audrey Duplessy as she loaded a U-Haul truck full of rolled oats and dried beans for sustenance in the underground shelter.

Elizabeth Clare Prophet, known to her followers as Guru Ma, which in church language means "the teacher who is mother," says the world is in a dangerous period.

Followers of the 50-year-old native of Red Bank, New Jersey, say she is referring to nuclear war, an interpretation that church leaders do not dispute.

Murray Steinman, a church spokesman, said that Mrs. Prophet is not convinced that recent international events mean tension is abating in the world; rather, she believes they are a result of deception by the Soviet Union to lure the United States into complacency. . . .

This is not the first time that Mrs. Prophet has predicted the end of the world. Most recently, she had spoken of a possible nuclear conflict for last October.

Her husband, Ed Francis, who is a church vice president, said in a statement this week, "The shelters will be used only if there is a nuclear war." To move underground before war, he said, would be "simply preposterous."

He also said, "We are not saying goodbye to the world."

However, many of the church members, who have come here from Europe, South America, and parts of North America, talk as if they are saying goodbye.

Like many church members, Mrs. Duplessy, who came with her college-age daughter, said she withdrew all her money from the bank and is now moving her most important personal possessions into her assigned space in a shelter 12 feet underground.

She paid $6000 for the space in a bunker she will share with 125 people. The price, she said, includes seven months of community food.

"I'm a faithful cheles of Mrs. Prophet," she said, first explaining that the word meant "slave," then correcting herself to say "devotee."

If nuclear war does not come, Mrs. Duplessy and the other church members will live in mobile homes, cabins, trailers, and other residences that Mrs. Prophet's followers have purchased from the church. Church members who have journeyed here from far away have been housed by members who live in the area.

"To me the whole thing is a big scam," said Carlo Cieri, a county commissioner who has been a frequent critic of the church. "If we had stricter laws, this would have never happened."

Today, at one of the church housing complexes, members who were stocking the shelters said they are praying intensively to lessen the chance of Armageddon.

If nuclear war does not come, Mrs. Prophet will not lose any credibility, said several members; they said they will attribute the lessening of tensions to their prayers.

"But we must be prepared now because Mother says this is a dangerous time," said Emma Issa, who came here from Lawrence, Massachusetts. . . .

The church is built around the belief that Mrs. Prophet

receives, and passes on to members of the church, messages from a group of souls known as the Ascended Masters, who convinced her to lead her flock to Paradise Valley, an area best known for its trout fishing and large elk herds.

Exercise 4.10

Pyramid Power

The following case has been cited in support of the hypothesis that pyramids have special powers. A young woman who was having difficulty with her complexion was told to keep a pitcher of water under a pyramid and then wash her face in that water, with only the mildest soap, once in the morning and once in the evening. She was also told to put nothing else on her face, no creams or medications of any kind, and no makeup. Although she has been in the habit of using quantities of makeup, she agreed to the experiment. Within two weeks there was a clearly noticeable improvement in her complexion.

Exercise 4.11

Reincarnation

The headline in a recent issue of a national weekly "newspaper" states that there have been hundreds of cases of reincarnation in the United States. These cases, the headline proclaims, provide convincing evidence that there is life after death. Turning to the inside pages one finds accounts of patients who have been cured of various complaints by being hypnotized and then "regressed" to a previous life. The following cases are typical of those presented.

A fifty-year-old woman claimed to have suffered from severe headaches, several a week for over thirty-five years. She claimed to have seen ten different doctors who had prescribed various pain killers and other drugs—none of which worked. Then, in a single two-hour session under hypnosis, she "discovered" the true cause of her headaches. In an earlier life, she had been a young man in nineteenth-century New England. One day, while on the way to visit his fiancee, the young man fell into a gully, hit his head on a rock, and was killed. A year and a half after her session with the therapist, the woman claimed she had since suffered only one or two headaches.

A twenty-five-year-old real estate dealer complained of several serious allergies, including a very strong reaction to corn. Under hypnosis, he was "regressed" back to an earlier life as a commander in a Mongolian army. In one campaign, the commander refused to order his men to kill innocent women and children. Because of this disobedience, his superiors had him tortured by being force-fed corn and water, which caused him to bloat so much that he died. After learning of his earlier life, the realtor claimed to be rid of most of his allergies and to be able to eat corn with no ill effects.

Several of the psychologists engaged in this sort of therapy are quoted as being convinced that their work provides scientific evidence that reincarnation does occur and that there is indeed life after death.

Exercise 4.12

Project

Find an article or book on a popular scientific topic that interests you—not a topic that comes out of a formal science source. Evaluate the material following the standard program for evaluating theoretical hypotheses. You may turn this into a more extensive project by looking up other articles or books on the same subject. In particular, you may be able to find articles or books that explicitly set out to refute the views in the materials you had read first. In this case, you can evaluate the opposing hypotheses as well.

PART TWO

Statistical and Causal Hypotheses

CHAPTER 5

Statistical Models and Probability

This chapter has two objectives. One is to acquaint you with a few basic types of statistical models. The second is to provide you with some knowledge of probability models so that when you reach Chapter 6 you will be able to understand how statistical data may be used as evidence for statistical hypotheses. We will take up causal models and causal hypotheses in Chapters 7 and 8.

5.1 WHY STATISTICAL AND PROBABILISTIC MODELS ARE IMPORTANT

There are many reasons why it is important to be able to understand and evaluate statistical and probabilistic models. The main general reason is that such models are widely used in science, particularly in the social, behavioral, and biomedical sciences. These are the sciences whose results, typically, are most relevant to the everyday concerns of many people.

The reason these sciences use statistical and probabilistic models is that such models are appropriate for the kinds of questions scientists in these fields try to answer. One kind of question concerns the prevalence of various characteristics in large populations, such as the population of women in the United States. For example, what percentage of American women between twenty and thirty years of age hold full-time jobs? The answer to that question, as we shall see, is a statistical hypothesis. It would be exceedingly difficult, and very costly, to determine the age and employment status of every woman in the United States. Indeed, that would be like performing a full-scale census, something that even the federal government attempts only once every ten years. The way such questions are typically approached is by examining only a small sample of the whole population of interest. Information about the sample is then used as evidence for a particular hypothesis about the population. But evaluating the data is not a simple matter. It requires the use of probabilistic models. So one needs to know about those as well.

A second kind of question typical of the behavioral and biomedical sciences concerns the causes of characteristics exhibited by individuals. Does the dietary intake of cholesterol cause heart attacks in adult males? Not nearly enough is known about the biological and chemical mechanisms involved to answer that question for any particular person. By studying several large groups of men, however, one might be able to answer the question without knowing so very much about the precise, biological mechanisms involved. Evaluating the evidence, again, requires the use of probabilistic models.

For the moment, we will concentrate on statistical and probabilistic models. What we learn now will have direct application later to the understanding and evaluation of causal hypotheses.

5.2 THE ELEMENTS OF A STATISTICAL STUDY

We will begin by outlining the basic elements of any statistical study. The rest of this chapter will be devoted to just two of these elements.

The Real World Population

In any statistical study, there must be some population in the real world that is the object of the investigation. For example, American women between twenty and thirty are such a population. But populations in the scientific sense do not have to be made up of individual people. They do not even have to be made up of biological individuals, such as penguins. The Census Bureau and the Internal Revenue Service study households, which technically are not people, but small groups of people. The FBI publishes statistics on homicides in the United States. For them, homicides form a population. But homicides are not people; they are incidents involving people—a killer and a victim, for example.

Whatever the nature of the individuals making up a population under investigation, those individuals will have various characteristics, or properties, whose incidence in the population is of interest. The properties can be almost anything: holding a full-time job, consisting of three, unrelated people, or being committed with a gun. Just keep in mind that the characteristics of members of the population are also features of the real world.

The Sample

As already noted, most populations of interest are just too big to examine every member. The best that can be done is to examine some selected members of the population, called a sample. How the sample is selected turns out to be crucial. For the moment, it is only necessary to understand that a sample, like the population, is made up of real-world individuals with real properties. So selecting a sample is a physical process that has to be carried out by someone, somewhere, at some time.

A Model of the Population

As was emphasized in Part One of this text, thinking scientifically about anything requires constructing a model. In the present case, the model will be a statistical model of the real world population under investigation. Whether the model is abstract, like a set of numbers, or real, like a jar full of marbles, the model is always something distinct from the population being studied.

A Model of the Sample

Finally, corresponding to the sample from the real population is a model of that sample. The model of the sample is to be regarded as a sample from the model population.

Figure 5.1

The four elements of a statistical study.

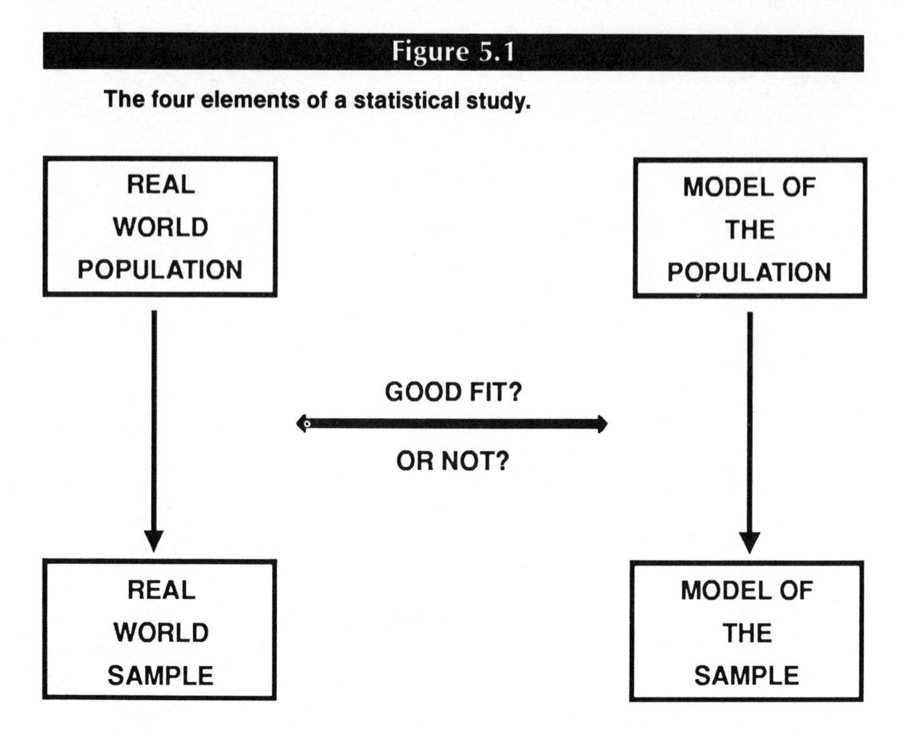

Putting the Elements Together

Figure 5.1 pictures the four elements of a statistical study in their proper relationship to one another. Looking across the top, there is the real world population pictured, on the left, with its corresponding model, at the top right. Looking across the bottom, there is the real world sample pictured, on the left, with its corresponding model, at the bottom right. The arrow on the left going down from the real world population to the sample represents the physical process by which the sample is selected from the population. The corresponding arrow on the right represents a model of that process, which may be a physical process but may also be purely imaginary. Overall, the whole right-hand side of the diagram represents models of the real world population and sample pictured on the left.

For the rest of this chapter, we will be exploring relationships on the right-hand side of the diagram. That is, we will be exploring possible statistical models of populations and idealized relationships between models of populations and models of samples. In the following chapter, we will begin using these models to understand the relationships in the real world pictured on the left-hand side of the diagram.

5.3 PROPORTIONS AND DISTRIBUTIONS

Current thinking about statistical models and probability originated in studies of various games of chance 300 years ago in the late seventeenth century. One of the models used in those earliest studies was that of a jar filled with marbles of various colors and sizes. We will take marbles in a jar as our standard model for thinking about populations and samples. In principle, this is a physical model, although we may do no more than talk about the model and never actually pick any marbles out of any jars. The closest we will come to a real jar of marbles is the picture in Figure 5.2.

Proportions

One of the simplest, statistical models consists of a jar full of marbles, some of which are red and some of which are any other colors but red. The number of red marbles divided by the total number of marbles in the jar is the proportion of red marbles in the jar. For example, if there were 100 marbles in the jar, and 60 of them were red, the proportion of red ones would be $^{60}/_{100}$, or $^{6}/_{10}$, or .60, or 60 percent.

In what follows, I shall assume you know how to go back and forth between fractions, decimals, and percentages. If for any reason this makes you uneasy, now is the time for a quick review. We are going to be seeing many simple proportions in various guises.

Figure 5.3 is an example of how we shall represent a proportion. The letters "MJ" at the top describe the population, which is Marbles in the Jar. The red marbles are pictured as being in the bottom part of the box and the nonred marbles are in the top part. The proportion, in this case 60 percent, is written off to the right. Note that all of the marbles are accounted for because every marble in the jar is either red or nonred.

Variables

Before we can proceed to more complex statistical models, we need to introduce some new technical ideas. These ideas are standard in the

Figure 5.2

Our standard model of a population.

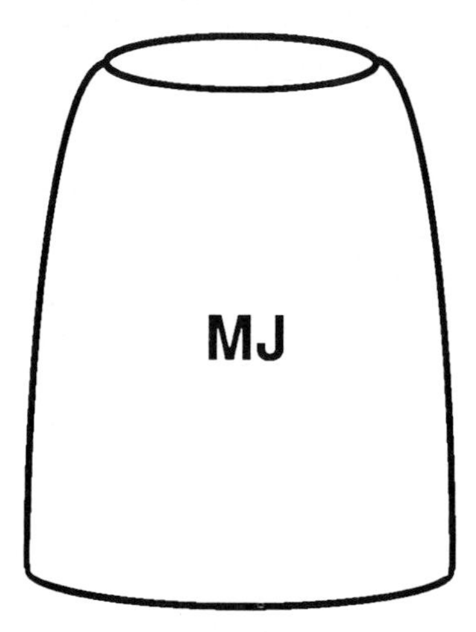

Figure 5.3

A representation of a proportion.

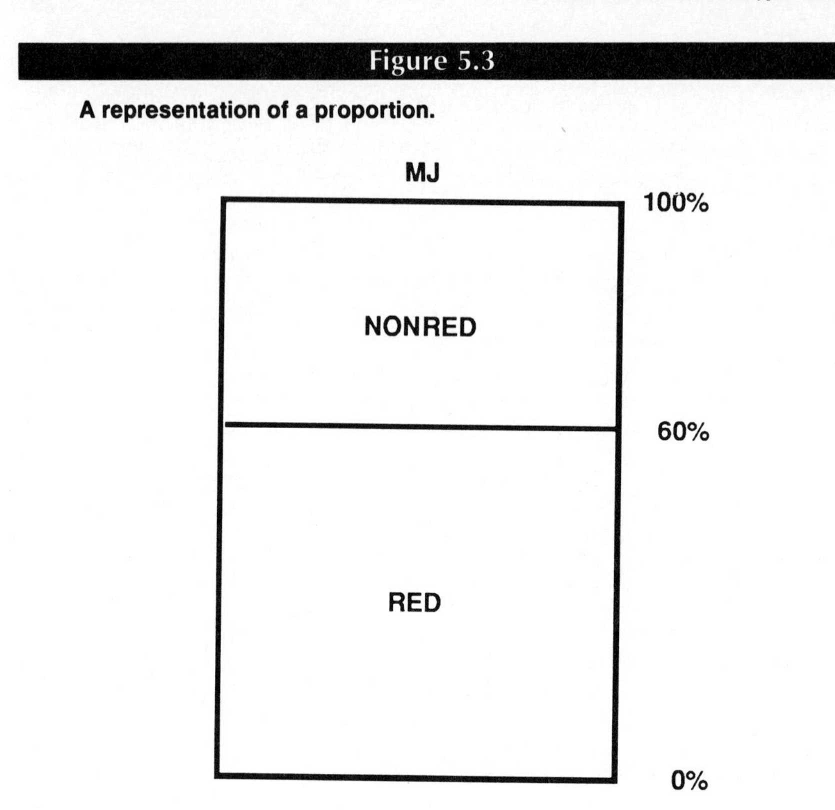

mathematical study of statistics and probability. They are more easily grasped, however, by means of examples than formal definitions.

The first idea is that of a variable. A variable is a general property that may exhibit different specific forms. For example, color may be thought of as a variable. Among its specific forms are red, green, and blue. Similarly, size could be a variable. Its specific forms could be, for example, large and small. The specific forms of a variable are called the values of the variable. Thus, in the example just given, the variable, color, has three possible values: red, green, and blue.

There are two important restrictions on the set of possible values of a variable: they must be exclusive and exhaustive. That the set of possible values be exclusive means that a member of the population can exhibit only one value. Each value of the variable excludes all others. In our example, no marble could be both red and green. That the set of possible values be exhaustive means that every member of the population must exhibit some value. Thus, the set of possible values must exhaust the possibilities for values of the variable. In the example, if red, green, and blue are the possible values of the color variable, then there cannot be any yellow marbles in the jar.

The language of variables applies primarily to statistical models, which are idealized representations of real populations. Among ordinary real objects, one might have great difficulty determining whether something should be classified as red or not. Maybe it should be called orange, or purple. When talking about statistical models, all such ambiguities are explicitly eliminated. Among the marbles in our imaginary jar, there are by stipulation none for which the determination of red, green, or blue is the least bit problematic.

Distributions

A statistical distribution is defined by a single variable with any number of possible values. For each possible value of the variable, there will be a definite percentage of the population assigned that value. For example, suppose our jar of 100 marbles has 50 red marbles, 30 green ones, and 20 blue. Here the variable is color and the three possible values of the variable are red, green, and blue. The corresponding percentages are thus 50 percent, 30 percent, and 20 percent, respectively. Note that because the values of the variables are exclusive and exhaustive, the percentages given must add up to 100 percent. Each marble is counted once and only once. Figure 5.4 exhibits the kind of diagram we will typically use to picture distributions. An equivalent representation would be a pie diagram as shown in Figure 5.5. These should be familiar to users of computer spreadsheets.

Figure 5.4

A representation of a distribution.

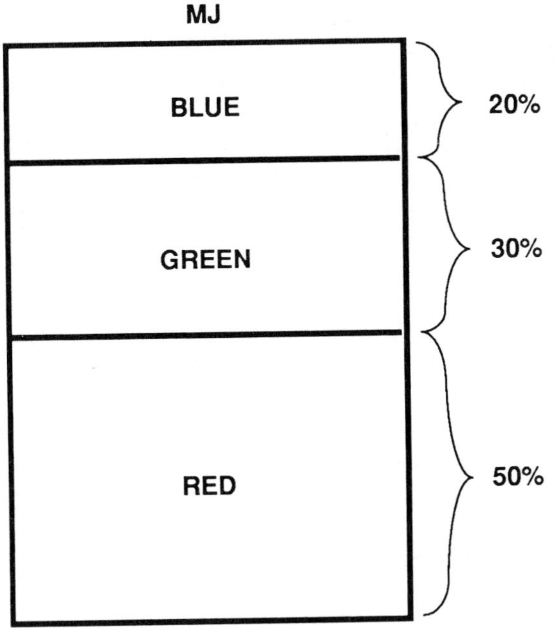

MJ

BLUE 20%

GREEN 30%

RED 50%

Figure 5.5

Representing a distribution with a pie diagram.

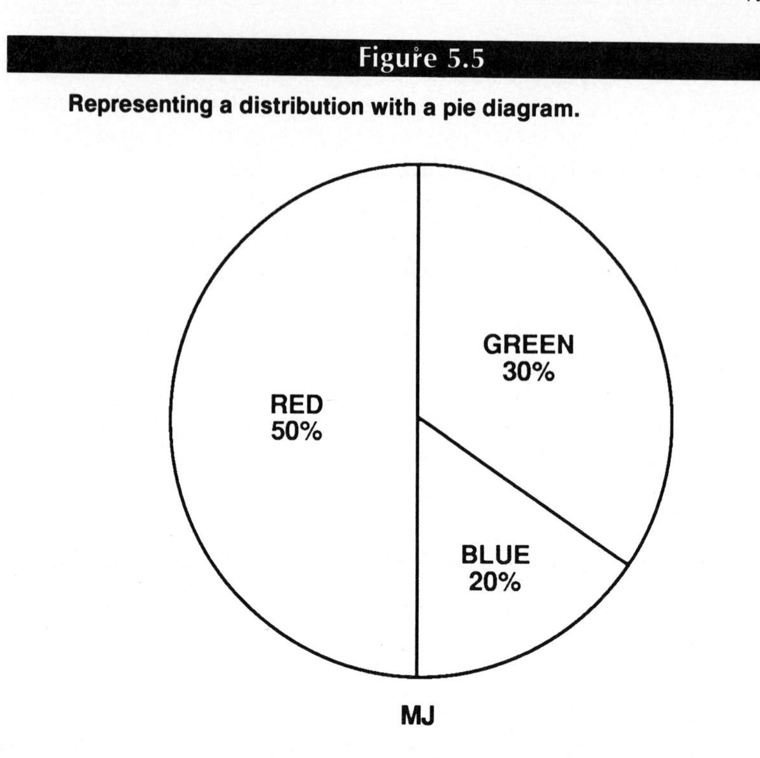

In our earlier example of a simple proportion, the variable was color and it had only two values, red and nonred. Thus, proportions are really special cases of distributions in which the variable has only two possible values. In general, the variable in a distribution can have any number of possible values.

5.4 SIMPLE CORRELATIONS

Correlation is a relationship between two variables, each with various numbers of possible values. We will concentrate on the simple case in which each of the two variables has only two possible values. As before, we will take color as one variable and suppose it has only two possible values, red and green. To deal with correlations, we will have to expand our model to include a second variable. Let us suppose, therefore, that marbles come in two sizes, large and small. So our second variable is size, and its two possible values are large and small. Figure 5.6 pictures our model for this case.

Variables Not Correlated

For our first case, imagine that our model population has 60 red and 40 green marbles. Among the red, we will suppose that there are 45 large and 15 small. Among the green, we will imagine 30 large and 10 small. Thus, among both the red and the green marbles, three quarters, or 75 percent, are large. Figure 5.7 exhibits the form of the diagrams we will use to represent correlations.

Figure 5.6

A model of a population with a possible correlation.

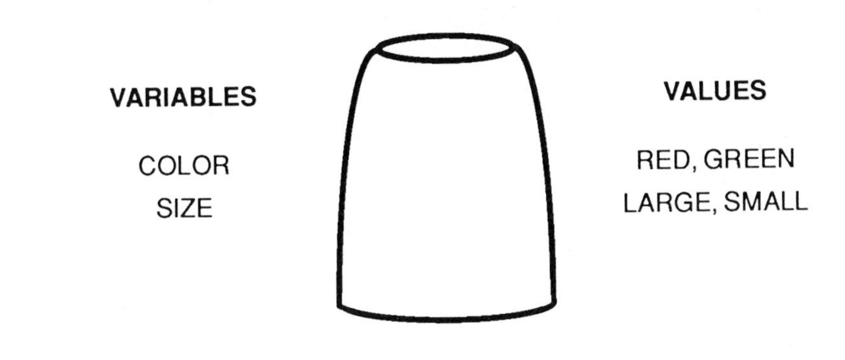

Figure 5.7

A model of a population in which size and color are not correlated.

MJ

	RED	GREEN
% LARGE	45/60 or 75%	30/40 or 75%

Having the percentage of large marbles being the *same* among both the red and the green marbles represents the special case in which we say that, in the model population, the two variables, color and size, are *not correlated.* What this means intuitively is that if one is interested in the proportion of large marbles, it does not matter whether one looks among the red or the green. The proportion of large is the same in both.

Variables Correlated

If, in the model population, the percentage of large marbles among the red marbles is *different* from the percentage of large among the green, then we say that the two variables, color and size, *are correlated.* Here there are two possibilities. The percentage of large among the red can be either *greater* than or *less* than the percentage of large among the green.

If the percentage of large among the red is *greater* than the percentage of large among the green, we say that, for this population, large is *positively correlated* with red. For example, suppose that 45 out of the 60 red marbles are large but that only 10 out of the 40 green marbles are large. In that case, the percentage of large among the red is 75 percent, but the percentage of large among the green is only 25 percent. In this particular population, red and large tend to go together. The idea of positive correlation is one way of making precise the idea of two properties "going together" in a particular population. Figure 5.8 pictures this example in which large is positively correlated with red.

The other possibility, if color and size are correlated, is that there is a *smaller* percentage of large marbles among the red than among the green. For example, suppose that only 20 out of the 60 red marbles are large while 30 out of the 40 green marbles are large. In that case, the percentage of large marbles among the red is 33 percent while the percentage of large marbles among the green is 75 percent. In such a population, we would say that being

Figure 5.8

A model of a population in which large size is positively correlated with red color.

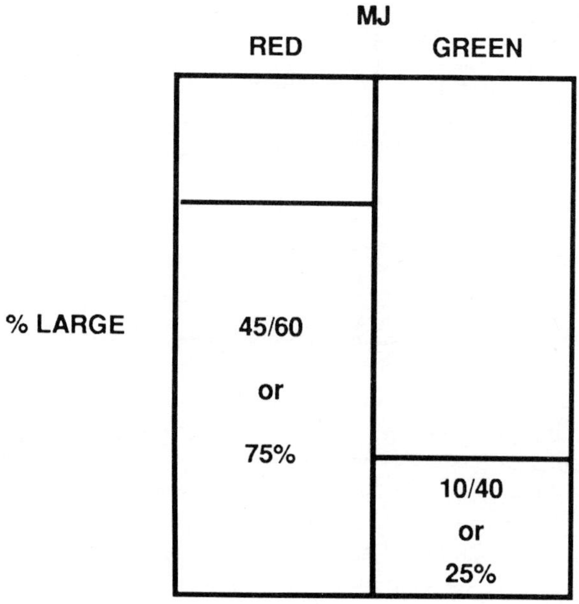

MJ

RED GREEN

% LARGE

45/60 or 75%

10/40 or 25%

large is *negatively correlated* with being red. The properties large and red tend not to go together. This case is pictured in Figure 5.9.

Once you understand what it is for there to be a positive, or negative, correlation between values of correlated variables, you should realize that there are many different ways to describe the same correlation among the variables. For example, Figure 5.8 pictures a model population in which there is a positive correlation between large and red marbles. That same correlation could be described as a negative correlation between large and green marbles. Or, it could be described as a positive correlation between small and green marbles. Or, it could be described as a negative correlation between small and red marbles. You should make sure that you understand why these are all just different ways of describing the same correlation among the variables color and size for the particular model pictured in Figure 5.8. If you do not understand this, you do not really understand what it is for two variables to be correlated.

One other general point. Whether two variables are correlated depends entirely on the particular population in question. Figures 5.7, 5.8, and 5.9 all concern the variables color and size, but the populations are different because there are different relative numbers of marbles assigned to the values of these two variables. Thus, whether the variables are correlated and, if so, how their values are related, differs from population to population. The idea

Figure 5.9

A model of a population in which large size is negatively correlated with red color.

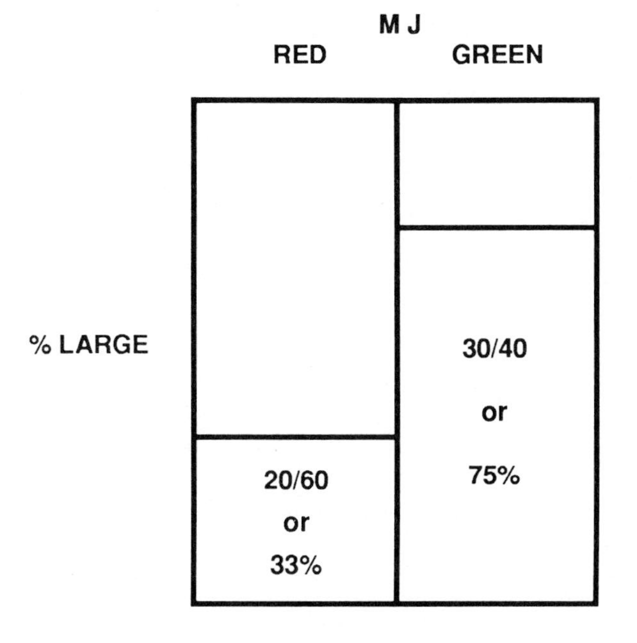

that somehow color and size are, in general, correlated, with no reference to a particular population, makes no sense.

Summary

The various correlation relationships for the model populations discussed above may be summarized as follows:

> If, for the population in question, the percentage of large marbles among the red marbles is the *same* as the percentage of large marbles among the green marbles, then the variables, size and color, are *not correlated* in this population.
>
> If, for the population in question, the percentage of large marbles among the red marbles is *greater* than the percentage of large marbles among the green marbles, then being large is *positively correlated* with being red in this population.
>
> If, for the population in question, the percentage of large marbles among the red marbles is *less* than the percentage of large marbles among the green marbles, then being large is *negatively correlated* with being red in this population.

5.5 SYMMETRY AND STRENGTH OF CORRELATIONS

Symmetry of Correlations

In all of the models of correlations discussed so far, we have focused on the color variable and, then, considered the percentages of large or small marbles among those of different colors. This focus is reflected in the fact that all of the diagrams of correlations exhibited thus far list the values of the color variable on top and represent the values of the size variable by a scale of zero to 100 percent on the side. But the association of two variables defined as correlation does not seem to favor one variable over another, and that is correct. So we should be able to represent the same relationships if we focus first on the size variable. That means listing the values of the size variable on the top of our diagrams and representing the values of the color variable along the side. Figure 5.10 provides two pictures of the model shown in Figure 5.7 in which there is no correlation between color and size. The picture on the left repeats Figure 5.7. The picture on the right shows the same population, but with the values of the size variable listed on top.

It is easy to derive the right-hand picture from the left-hand picture. Because there are 45 large red marbles and 30 large green ones, the total number of large marbles must be 45 plus 30, or 75. Likewise, there must be 15 plus 10, or 25, small marbles. Now among the 75 large marbles, 45 are red, for a proportion of $^{45}/_{75}$, or 60 percent, red. Similarly, among the small marbles, 15 are red, for a proportion of $^{15}/_{25}$, or 60 percent red. Thus, the percentage of red marbles is the same among both the large and small marbles. So the right-hand diagram tells us that there is no correlation between color and size in this population, which is just what the left-hand diagram told us. The corresponding percentages are different in the two diagrams, 75 percent and 60 percent,

Figure 5.10

Equivalent representations of the lack of correlation between size and color in the population pictured in Figure 5.7.

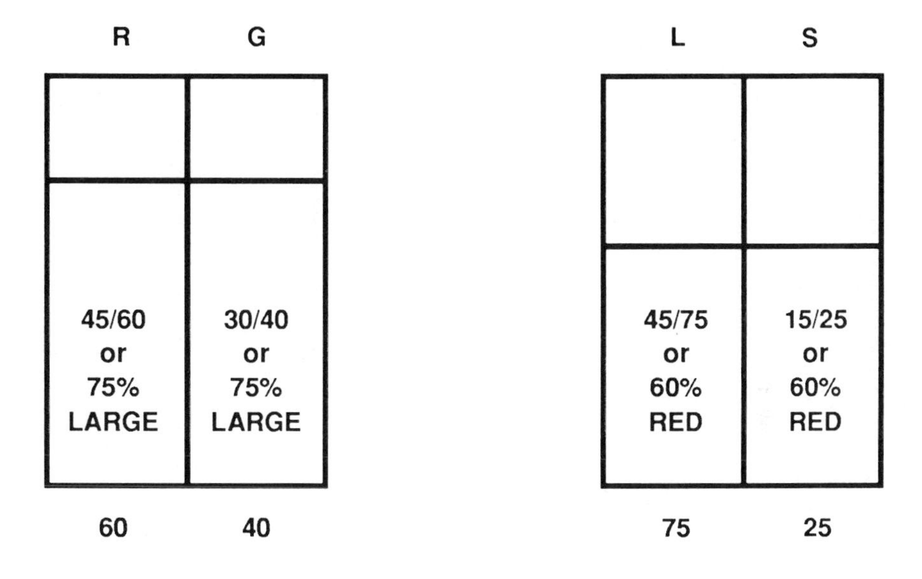

respectively, but what matters for the lack of correlation is the equality of the two percentages in each diagram.

We have worked out this equivalence only for one specific model, but the result is completely general. If representing a population using a diagram with the values of one variable on the top yields a result of no correlation between the variables, using a diagram with the values of the other variable on top will, likewise, yield a result of no correlation.

A similar result holds for positive and negative correlations among the various values of the two variables. For example, if the positive correlation between large and red marbles in Figure 5.8 is represented with the size variable at the top of the diagram, the result is a positive correlation between red and large marbles. This equivalence is shown in Figure 5.11. In general, if being large is positively correlated with being red in a population, then being red is positively correlated with being large in that population. That is what is meant by saying that positive correlation is symmetrical. The same holds, of course, for negative correlation.

The general symmetry of positive correlation can be demonstrated by examining Figure 5.10. Suppose that we were to exchange one small red marble for a large red marble in the jar. That would raise the proportion of large marbles among the red marbles to $46/60$ in the left-hand diagram. That indicates a (small) positive correlation between large and red. At the same time, however, the proportion of red marbles among the large marbles in the right-hand diagram in Figure 5.10 goes up to $46/76$, or 61 percent, while the proportion of red among the small marbles goes down to $14/24$, or 58 percent. That yields a (small)

Figure 5.11

Equivalent representations of the positive correlation between large size and red color in the population pictured in Figure 5.8.

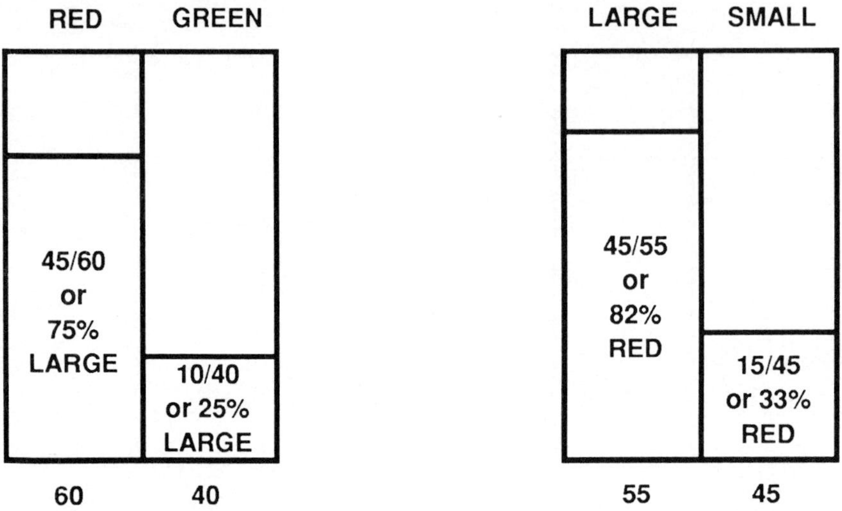

RED GREEN

45/60 or 75% LARGE

10/40 or 25% LARGE

60 40

LARGE SMALL

45/55 or 82% RED

15/45 or 33% RED

55 45

positive correlation between being red and being large in this population. Thus, a change in the population that produces a positive correlation between large and red automatically produces a positive correlation between red and large. As before, correlation is a symmetrical relationship.

Strength of Correlations

Thus far, we have largely treated correlations as just qualitative relationships. Either there is correlation between two variables, or there is not. If there is a correlation between the variables, then some values of the variables are positively or negatively correlated with others. But, clearly, correlation is also a quantitative relationship. Some correlations can be stronger than others. For example, we just considered a model in which the difference in percentages of red marbles among the large and small was just the difference between 61 percent and 58 percent. Intuitively, that is a small difference, corresponding to a weak correlation. In other models we have considered, the difference was that between 75 percent and 25 percent, which intuitively seems a much bigger difference, corresponding to a much stronger correlation.

For our purposes, we will take the difference between the two proportions expressed, however, as decimals rather than percentages, as a rough measure of the strength of the correlation between the values in question. More precisely stated, for our standard model:

> The strength of the correlation between large and red is the difference, expressed as a decimal, between the proportion of large marbles that are red and the proportion that are green.

For example, in the population shown in Figure 5.8, the strength of the correlation between large and red is (.75 – .25) = .50. That, by this measure, is a fairly strong correlation. The strongest possible correlation would have the value 1.00, which would happen if all the red marbles were large and all the green ones were small. There would be a perfect association between large and red. Similarly, the smallest possible correlation would have the value –1.00, which would happen if none of the red marbles were large and all of the green ones were large. If the variables are not correlated, the difference would, of course, be zero, yielding a strength of correlation of 0.00, which is just what one should expect.

A Correlation Coefficient

Measuring the strength of a correlation by the simple difference in proportions has the great advantage that one can perform the calculation in one's head. Thus, without resorting to a calculator, one can get a good idea of how strong a correlation really is. However, this is not the measure you would find in any textbook on statistics. For professional purposes, a difference in percentages is not a sufficiently good measure. One defect is that it is not symmetrical. That is, the difference in percentages may vary depending on which variable one represents on the top of the diagram. You can see this in Figure 5.11, where the strength of the correlation in the left-hand diagram is .50 and in the right-hand diagram, .49. Nevertheless, it is the same correlation.

For comparison, Figure 5.12 provides a more symmetrical representation of our standard model. This is the kind of representation one finds in standard statistics texts. Here:

a = the number of marbles that are red and small

b = the number of marbles that are green and small

c = the number of marbles that are red and large

d = the number of marbles that are green and large

The correlation coefficient is given by the following formula:

$$r = \frac{(bc - ad)}{[(a + b)(c + d)(a + c)(b + d)]^{1/2}}$$

As with the difference measure, this measure varies from 1.00 to – 1.00 and has value zero when the variables are not correlated. But unlike the difference measure, this one is also symmetrical. However, this measure is obviously too complex to compute in one's head. For our purposes, it is better to have a measure that one can use easily and quickly, even if it is not quite up to the standards of professional statisticians.

Figure 5.12

A representation of a correlation as found in typical statistics textbooks.

	RED	GREEN	
SMALL	a	b	a + b
LARGE	c	d	c + d
	a + c	b + d	

5.6 PROBABILITY MODELS

Statistical models are models of populations. As we saw at the beginning of this chapter, however, we need also to understand the process of selecting a sample from a population. For that, a fuller understanding of probability models is required. As it turns out, statistical models are a special case of probability models. So what we have just learned about statistical models will be useful in developing a fuller understanding of probability models themselves.

A Probability Model

We will begin with a more complex version of our model of marbles in a jar. This time we will assume that our jar contains 200 marbles of four colors (red, green, blue, and yellow) and two sizes (large and small) distributed as shown in Figure 5.13. In the context of probability models, statisticians talk about stochastic variables or random variables rather than just variables. So here we have two stochastic variables, color and size. As before, the possible values associated with any one variable must be both exclusive and exhaustive.

Probability is a measure associated with the values of stochastic variables defined for a particular population. The measure associated with any particular value has a numerical value equal to the proportion of members of the population assigned to that value. For example, in our model population, the proportion of marbles that are red is $^{100}/_{200}$, or $^{1}/_{2}$. So the probability of red in this population is .50.

Here it is helpful to introduce the standard notation used in the study of probability and statistics. Let **C** and **Z** stand for the random variables color and size, respectively. Let R, B, G, Y, L, and S stand for the values of these

Figure 5.13

The distribution of marbles in the population of probability model number one (Model 1).

COLOR		SIZE	
Red	100	50 L,	50 S
Green	50	25 L,	25 S
Blue	30	15 L,	15 S
Yellow	20	10 L,	10 S

variables. Then the expression **P**(**C** = R) = .50 is read as: The probability that the variable color has the value red is one half. Often the explicit reference to the variable is left out, yielding the abbreviated expression **P**(R) = .50.

Probability models have a definite structure, which is just the structure of proportions. The details of this structure are given by some possible ways of forming combinations of proportions. In fact, there are only two basic ways of forming combinations that matter for probability models.

Simple Addition Rule

One way of combining probabilities is to consider the probability of a member of the population exhibiting either one of two or more values of stochastic variables. For example, one could ask about the probability that a marble is either red or green. Examining the population of marbles we find that there are 100 red and 50 green for a total of 150/200 that are either red or green. Thus, **P**(R or G) = .75. Because **P**(R) = .50 and **P**(G) = .25, one might think that the probability of the combinations red or green is just the sum of their respective individual probabilities. But that suggestion is a bit hasty.

Consider the combination of marbles that are either red or large. What is the probability of that combination? That is, what is the probability of **P**(**C** = R or **Z** = L)? Looking at the composition of the population we see immediately that there are 100/200 red marbles and 100/200 large marbles. So **P**(R) = .50 and **P**(L) = .50. The sum of these two probabilities is one. But that cannot be right because that would mean that all of the marbles are red or large and we know that there are some that are not; for example, the 25 small green marbles. What has gone wrong?

The problem is that in counting up the red marbles and then counting up the large marbles we counted the large, red marbles twice. There are, after all, 50 large, red marbles in the population. Put in terms we have just learned, red and large are not exclusive values of stochastic variables. The simple addition rule suggested above works only for exclusive values.

Letting A and B stand for any values of random variables, we can now state this simple rule as follows:

Simple Addition Rule for Probabilities:

If A and B are exclusive values of random variables, **P**(A or B) = **P**(A) + **P**(B).

A little later we will learn what to do with values of variables that are not exclusive.

Simple Multiplication Rule

Another way of combining probabilities is to consider the probability of a member of the population exhibiting both of two values of stochastic variables. For example, one could ask about the probability that a marble is both red and large. From the descriptions of our model population, we see that there are 50/200 marbles that are both large and red. So **P**(R and L) = .25. We have already determined that **P**(R) = .50 and **P**(L) = .50, so it appears plausible that the probability of both red and large is just the *product* of the two individual probabilities. Again, this would be a somewhat hasty conclusion.

To see why the above simple rule is not quite right as it stands, consider the population with a composition as shown in Figure 5.14. The only difference between Model 2 and Model 1 is that the relative numbers of large and small among the various colored marbles have been changed. The total numbers of colored marbles and of large and small marbles remain the same. Thus, for the population in Model 2, **P**(R) = .50 and **P**(L) = .50. But the proportion of large red marbles is 80/200, which is .40, and not the product of .50 times .50, which is .25. What has gone wrong?

The problem is that in Model 1 the variables, color and size, were not correlated. In Model 2, these variables are correlated. This difference is pictured in Figure 5.15. The simple product rule works only if the variables corresponding to the respective values are not correlated.

Figure 5.14

The distribution of marbles in the population of probability model number two (Model 2).

COLOR		SIZE	
Red	100	80 L,	20 S
Green	50	10 L,	40 S
Blue	30	5 L,	25 S
Yellow	20	5 L,	15 S

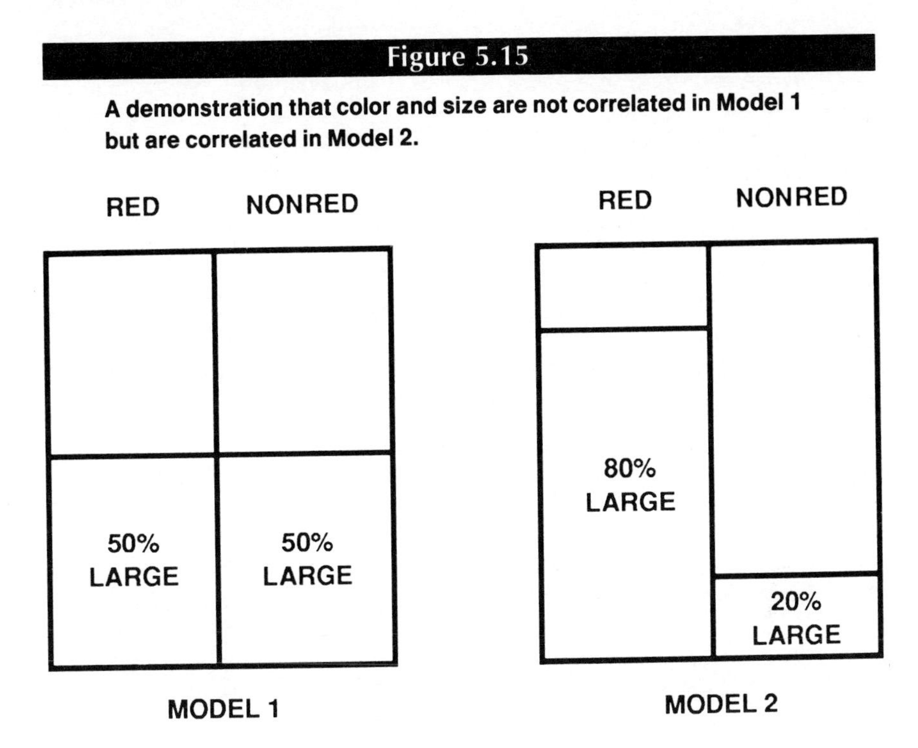

Figure 5.15

A demonstration that color and size are not correlated in Model 1 but are correlated in Model 2.

Simple Multiplication Rule for Probabilities:

If A and B are values of random variables that are not correlated, then
P(A and B) = **P**(A) × **P**(B).

We will now develop a rule that works even for the joint assignment of values of correlated variables.

Conditional Probabilities

Let us look at the probability of marbles that are both red and large marbles in a slightly different way. Think of picking out a large, red marble as a two-stage process in which you first select the color red and then look for large marbles only among the red ones. The probability of a marble being large, given that it is red, is known as a conditional probability and is symbolized as **P**(L/R). It can also be described as the probability of large conditional on red. Using this notation, we can generalize the multiplication rule as follows:

General Multiplication Rule for Probabilities:

If A and B are values of two random variables, then
P(A and B) = **P**(A) × **P**(B/A).

This rule applies whether or not the variables in question are correlated. If the variables are not correlated, then **P**(B/A) = **P**(B), and the general rule reduces to the simple rule.

Applying this general multiplication rule to our earlier example for large, red marbles in Model 2:

$$\mathbf{P}(\text{R and L}) = \mathbf{P}(\text{R}) \times \mathbf{P}(\text{L/R}) = .50 \times .80 = .40.$$

That, of course, is the correct answer because in Model 2 there are 80/200 marbles that are both red and large.

Now that we have a generalized version of the multiplication rule, we can produce a generalized version of the addition rule as well. This rule applies whether or not the values of the variables are exclusive.

General Addition Rule for Probabilities:

If A and B are two values of random variables, then **P**(A or B) = **P**(A) + **P**(B) − **P**(A and B).

Applying this rule to determining the probability of large red marbles in Model 1:

$$\mathbf{P}(\text{L or R}) = \mathbf{P}(\text{L}) + \mathbf{P}(\text{R}) - \mathbf{P}(\text{L and R}) = .50 + .50 - .25 = .75$$

This is correct because there are 150/200 marbles that are either large or red (or both) in Model 1. Thus, the generalized addition rule takes care of double counting by subtracting (once) the members of the population for which the two values overlap. If the values of the variables are exclusive, then **P**(A and B) = 0, and the general addition rule reduces to the simple addition rule given earlier.

5.7 THE FLIPPANT JUDGE

As an introduction to the study of sampling, we will now work through a simple problem whose solution requires the calculations necessary to understand the sampling process. As you will see, all that is required are the simple addition and multiplication rules.

The Problem

Imagine a jury consisting of three judges. The first two judges, through long experience, have established that their probability of reaching a correct decision is 3/4. That is, on the average, three-quarters of the times when they conclude that the defendant is guilty (or not guilty), the defendant is, in fact, guilty (or not guilty). The other one-quarter of the time they are mistaken one way or the other. The third judge is, unknown to anyone, irresponsible. For each case, he merely flips a coin. If the coin comes up heads, he says "guilty." If it comes up tails, he says "not guilty." The chances of this judge being correct are, of course, 1/2. The jury is pictured in Figure 5.16.

Figure 5.16

The jury.

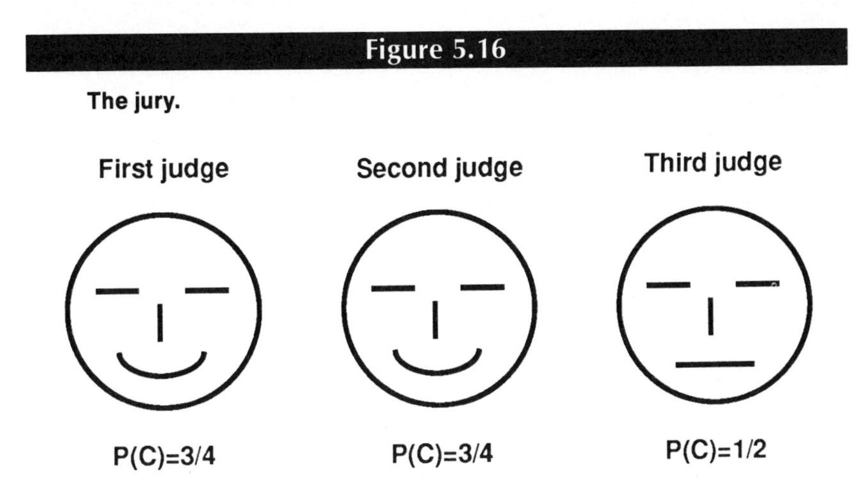

The final verdict is decided by a two-thirds majority. If any two of the three judges declare the defendant not guilty, that is the jury's verdict—and vice versa. Suppose that you have been brought before this jury. What is the probability that the jury reaches a correct verdict in your case? Before reading the solution, take a few minutes to see if you can solve the problem on your own. Nothing more than applications of the simple addition and multiplication rules are required.

The Solution

The key to the solution is to figure out precisely what probabilities to calculate and in what order. The only complication is that there are a number of combinations of correct and incorrect decisions by individual judges that leave the correctness of the jury's decision unchanged. So we need first to determine which combinations of individual, correct decisions constitute a correct verdict.

For each judge, there is a random variable indicating the status of his or her judgment. Let the letter C stand for the value of the variable indicating that the corresponding judge is correct. Likewise, let the letter M indicate that the corresponding judge is mistaken. Remembering that the jury reaches a correct verdict whenever two or more of the judges are correct, the combinations for which the jury reaches a correct verdict are all pictured in Figure 5.17.

The combinations that make the jury correct are, clearly, mutually exclusive. So to determine the overall probability of a correct verdict, we need only apply the addition rule to the probabilities of these combinations. We want to know the probability that one or another of these combinations obtains. But first, we need to calculate the probabilities of each of these combinations themselves.

The first possibility for a correct verdict, for example, is the combination of the first judge being correct, while the second and third judges are also correct. The second possibility is the combination in which the first and second judges are correct, while the third is incorrect, and so on. Moreover, because the judges operate independently, the variables corresponding to

Figure 5.17

Combinations of correct decisions by individual judges that yield a correct verdict by the whole jury.

First judge	Second judge	Third judge
C	C	C
C	C	M
C	M	C
M	C	C

decisions by individual judges are not correlated. For example, the percentage of correct decisions by the second judge is the same if we consider only cases in which the first judge is correct or only cases in which the first judge is incorrect. So we can apply the simple multiplication rule to each combination.

Thus, the probabilities corresponding to correct verdicts are:

$$\mathbf{P}(1C, 2C, 3C) = \mathbf{P}(1C) \times \mathbf{P}(2C) \times \mathbf{P}(3C) = 3/4 \times 3/4 \times 1/2 = 9/32$$
$$\mathbf{P}(1C, 2C, 3M) = \mathbf{P}(1C) \times \mathbf{P}(2C) \times \mathbf{P}(3M) = 3/4 \times 3/4 \times 1/2 = 9/32$$
$$\mathbf{P}(1C, 2M, 3C) = \mathbf{P}(1C) \times \mathbf{P}(2M) \times \mathbf{P}(3C) = 3/4 \times 1/4 \times 1/2 = 3/32$$
$$\mathbf{P}(1M, 2C, 3C) = \mathbf{P}(1M) \times \mathbf{P}(2C) \times \mathbf{P}(3C) = 1/4 \times 3/4 \times 1/2 = 3/32$$

The probability that the jury's verdict is correct is simply the sum of these probabilities, which is $24/32$, or $3/4$. The effect of the irresponsible judge, therefore, is to cancel out the judgment of one of the good judges. The jury could do equally well with only one of the good judges.

You should now calculate the probabilities for the combinations leading to incorrect verdicts. Their sum, of course, should be $1/4$.

5.8 SAMPLING

Whether we are talking about sampling from a real population, such as all Americans over eighteen years old, or a model population, such as our idealized marbles in a jar, the process of sampling can be pictured as in Figure 5.18. The idea is to learn something about the whole population by examining a small selection of individuals from that population. What can be learned, however, depends crucially on how the selecting is done.

Replacement

There is one preliminary distinction to be made before we can proceed. This is the distinction between sampling with replacement and sampling without replacement. The difference is whether an individual already examined is eligible to be selected again later in the selection process. Are the individuals who have already been examined replaced—that is, put back into the population—or are they not replaced?

Figure 5.18

A representation of a sampling process.

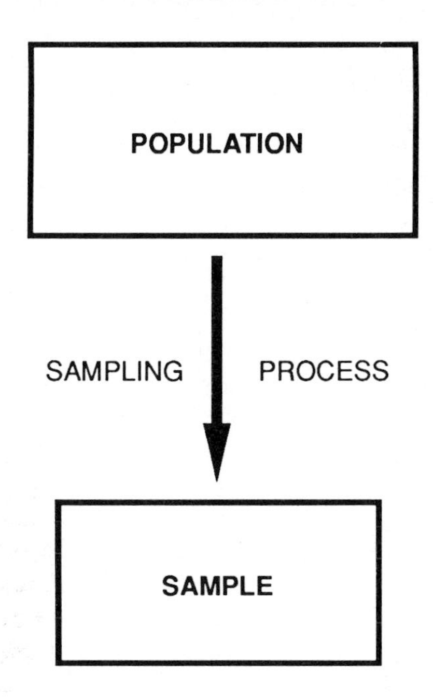

This distinction is important only if the sample constitutes a sizable fraction of the whole population. Suppose we were to select, without replacement, 50 marbles from among the 200 marbles of the population in Model 1. And suppose that, by coincidence, all 50 turned out to be red. That would mean that the proportion of red marbles in the population would be quite different when selecting the fiftieth marble than when selecting the first. In effect, the population sampled later would be a different population than that sampled earlier. That would make it very difficult to infer anything from the sample about the proportion of red marbles in the original population. This problem is eliminated if we assume that individual marbles are replaced immediately after each selection so that for the next selection the population is just the same as it was earlier. For this reason, we will always assume that sampling from our small, model populations is done with replacement.

When we turn to applying our models to real world populations, replacement is of little practical importance. Most real world populations are very much larger than any samples we might consider. For example, when sampling from a population of 100 million adult Americans, removing even a few thousand individuals from the population would make almost no difference in the proportions of individuals exhibiting properties of interest, such as being a smoker or a registered Democrat. Thus, for all practical purposes, sampling without replacement from a large population is like sampling with replacement from a small population.

What all this really means is that, from a theoretical point of view, the size of the population is of no importance in understanding the process of sampling. In practice, the size of the population makes a big difference. The actual process of sampling from a large population is always much more difficult than sampling from a small one. We will explore some of these difficulties when we come to applying our simple models to real world populations.

Random Sampling: Two Trials

In statistics texts, the selection of an individual from a population is often called a trial. As the smallest sample involving more than one trial is a sample with two, we will consider this case first. A sample of two is too small to be of much use in evaluating claims about a population, but it is easy to understand the probabilities involved. To keep things as simple as possible, let us consider sampling (with replacement) from our original model population of 200 marbles (Model 1). We will focus on just the color variable, and then only on two values of that variable, red (R) and not red (N).

The easiest way to think about sampling is to imagine a sequence of trials, each consisting of the selection of one member of the population. We can then represent each trial by a different random variable, conveniently labeled $\mathbf{X}_1, \mathbf{X}_2, \mathbf{X}_3 \ldots \mathbf{X}_n$. Each of these random variables has two possible values, R and N. The case of just two trials would, thus, be represented by the sequence $(\mathbf{X}_1, \mathbf{X}_2)$.

There are only four possible results of selecting two marbles from the Model 1 population of 200. We might get a red marble on both trials, a red marble on the first and a nonred marble on the second, a nonred marble on the first and red marble on the second, and, finally, a nonred marble on both trials. These four possible results may be symbolized as:

(R,R) (R,N) (N,R) (N,N)

A common way of thinking about sampling is to imagine not only the sequence of two trials actually carried out, but also a large hypothetical set of sequences of two trials each. If carried out, each member of this set would have one of the four possible results described above. This set is pictured in Figure 5.19. It is helpful to think of this set as itself a population, not of marbles, but of sequences of two selections of marbles from the population of Model 1.

We can now characterize the crucial process known as random sampling. There are two things necessary for the sampling process to be random. (1) On each trial the probability of red must equal the probability of red in the population, which in Model 1 is .50. This may be represented by the proportion of results of each trial in the large, imaginary set of two trials being .50. That is, half of the first trials in the set of two trials would yield an R, and half of the second trials in the set of two trials would also yield an R. (2) There is no correlation between the outcomes of the two trials. For example, if one looked at members of the set of two imaginary trials in which the first trial yielded an R, and then for that set counted up the proportion of results of the second trial that also showed an R, that proportion would again be .50.

Figure 5.19

A hypothetical population of pairs of selections of a marble from the population of Model 1.

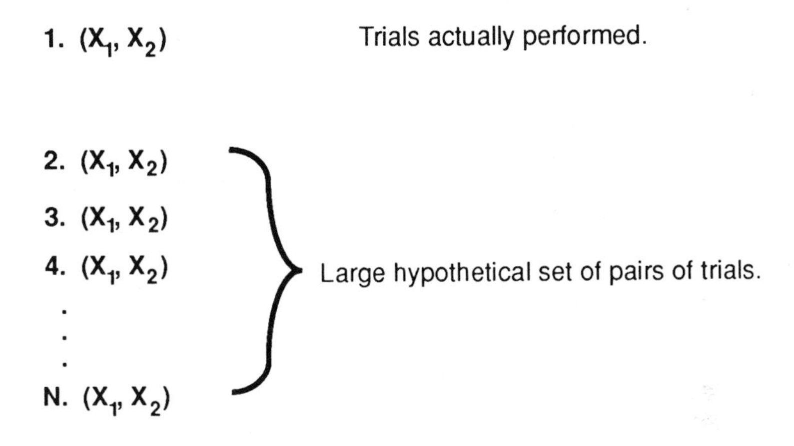

It is important to realize that randomness is not a feature of a sample by itself. You cannot tell a random sample just by looking at the sample. Randomness is a feature of the process by which the sample is selected. We have followed tradition in characterizing that process in terms of the results that would be expected if the whole sampling process itself were repeated many times. Later, we will learn something about the kinds of actual physical-selection processes operating on real populations that could be expected to achieve the desired results in repeated applications.

We can now proceed to the calculation of probabilities for the various results of selecting two marbles, at random, from our model population. The assumption of random sampling, together with the fact that the population of Model 1 contains 50 percent red marbles, means that in a large set of two trials each, the first trial would yield a red half of the time. That is $\mathbf{P}(\mathbf{X_1} = R) = .50$. Similarly, $\mathbf{P}(\mathbf{X_1} = N) = .50$. By the same reasoning, $\mathbf{P}(\mathbf{X_2} = R) = .50$ and $\mathbf{P}(\mathbf{X_2} = N) = .50$.

The assumption of random sampling also tells us that the two random variables, $\mathbf{X_1}$ and $\mathbf{X_2}$, are not correlated. Thus, if we wish to compute the probability of getting a red marble on both the first and second trials, we can use the simple multiplication rule. Thus:

$$\mathbf{P}(\mathbf{X_1} = R \text{ and } \mathbf{X_2} = R) = \mathbf{P}(R,R) = \mathbf{P}(\mathbf{X_1} = R) \times \mathbf{P}(\mathbf{X_2} = R) = .50 \times .50 = .25$$

By a similar calculation, $\mathbf{P}(R,N) = \mathbf{P}(N,R) = \mathbf{P}(N,N) = .25$.

You should realize that the possible results of multiple sampling do not have to have the same probability. They do in our example because half of the marbles in our model population are red. If we had considered the values green (G) and not green (N) instead, then the probabilities would not all be the same. For example, $\mathbf{P}(G,G) = 1/4 \times 1/4 = 1/16$, but $\mathbf{P}(G,N) = 1/4 \times 3/4 = 3/16$. It is a little easier to deal with the case in which the original population probabilities are 50-50, so we will continue to use that example for a while yet.

In most applications of sampling, one is not interested in the order in which the results occur in a set of trials. It is the relative number, or relative frequency, of types of individuals that is important for the evaluation of statistical, or causal, hypotheses. Thus, for example, it usually makes no difference whether a trial had the result (R,N) or (N,R). The important thing is that these two results both have one red and one nonred marble. That is, the relative frequency of red marbles is the same in both of these results, namely one out of two, or $\frac{1}{2}$.

The result of selecting two marbles might, therefore, be no R's, one R, or two R's. That is, the relative frequency of R's can have three different possible values: $\frac{0}{2}$ (0), $\frac{1}{2}$ (.50), and $\frac{2}{2}$ (1.00). Let us use the letter **Y** to represent the random variable with these three values. Now, what are the probabilities corresponding to **P**(**Y**)?

For no R's or two R's, the calculations are easy. **P**(0 R's) = **P**(N,N) = $\frac{1}{4}$. Similarly, **P**(2 R's) = **P**(R,R) = $\frac{1}{4}$. The probability of getting one R out of two trials is more difficult to calculate. It is the probability of getting either the result (R,N) or the result (N,R). However, these two sequences represent exclusive possibilities. So we can use the simple addition rule:

$$\mathbf{P}(1\ \text{R}) = \mathbf{P}[(\text{R,N})\ \text{or}\ (\text{N,R})] = \mathbf{P}(\text{R,N}) + \mathbf{P}(\text{N,R}) = \tfrac{1}{4} + \tfrac{1}{4} = \tfrac{1}{2}$$

These three probabilities form a probability distribution for the relative frequency of R's in two random selections from our model population. This distribution is shown in Figure 5.20, which is our first example of a sampling distribution. You should study it carefully because most of the probability

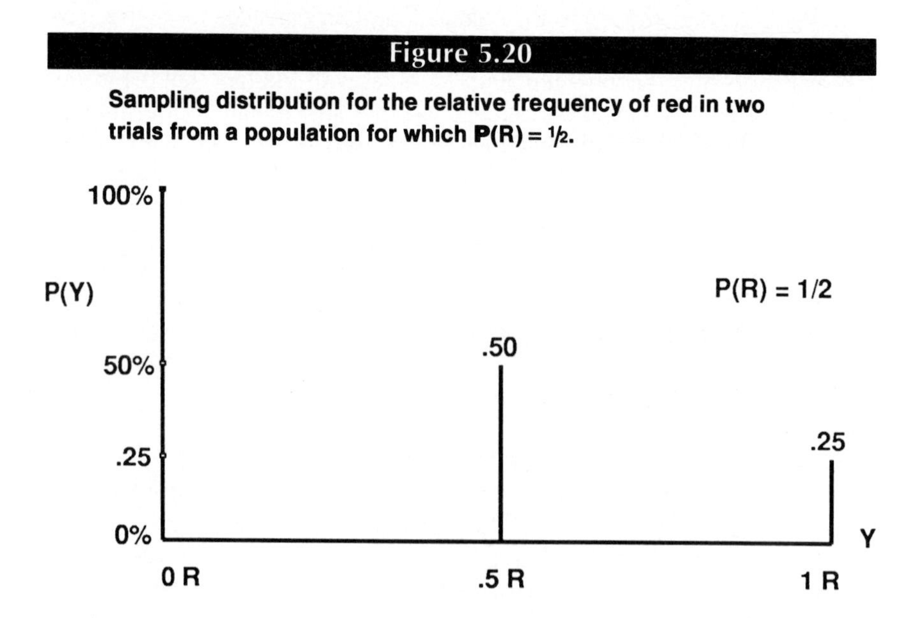

Sampling distribution for the relative frequency of red in two trials from a population for which P(R) = ½.

distributions that are important in understanding statistical reasoning are sampling distributions. If, at a later stage, you find your understanding of sampling distributions getting fuzzy, come back to Figure 5.20. You should now be able to follow every step in the derivation of this distribution. Other sampling distributions are more complicated mainly because they involve larger numbers of trials. They are not in principle any more complex.

Three, Four, and Five Trials

Let us now try three trials, with replacement, from our model population. So we are considering the sequence of random variables ($\mathbf{X}_1$, $\mathbf{X}_2$, $\mathbf{X}_3$), where each variable can take on the values R or N. In this case there are eight mutually exclusive, possible arrangements of the results of the three trials. These are:

(R,R,R) (R,R,N) (R,N,R) (N,R,R)

(R,N,N) (N,R,N) (N,N,R) (N,N,N)

As before, we imagine a large set of samplings consisting of three trials each. The assumption of random sampling implies that in this set, half the triples would exhibit an R on the first trial, half would exhibit an R on the second trial, and half would exhibit an R on the third trial. So $\mathbf{P}(\mathbf{X}_1 = \text{R}) = \mathbf{P}(\mathbf{X}_2 = \text{R}) = \mathbf{P}(\mathbf{X}_3 = \text{R}) = .50$. Random sampling also implies that the results of different trials are not correlated. To compute the probability of the result (R, R, R) we can therefore use the simple multiplication rule. That is,

$$\mathbf{P}(\text{R,R,R}) = \mathbf{P}(\mathbf{X}_1 = \text{R}) \times \mathbf{P}(\mathbf{X}_2 = \text{R}) \times \mathbf{P}(\mathbf{X}_3 = \text{R}) = \tfrac{1}{2} \times \tfrac{1}{2} \times \tfrac{1}{2} = \tfrac{1}{8} = .125$$

As $\mathbf{P}(\text{R}) = \mathbf{P}(\text{N}) = \tfrac{1}{2}$, for each random variable, each of the other seven possible outcomes of three trials will have exactly the same probability.

For a sample consisting of three trials, the random variable $\mathbf{Y}$ has four different possible values. That is, the relative frequency of R's obtained in three trials could be none out of three, one out of three, two out of three, or three out of three. We have just determined that $\mathbf{P}(3\text{ R's}) = .125$. The probability of no R's is the same.

The probability of obtaining 1 R or 2 R's is determined, as before, by using the addition rule. This time there are three, mutually exclusive sequences that have only one R. These are (R,N,N), (N,R,N), and (N,N,R). The probability of getting one or another of these three is, by the addition rule, the sum of the three probabilities—that is, $\tfrac{3}{8}$ (or .375). As you can easily verify for yourself, the probability of getting 2 R's is also $\tfrac{3}{8}$. So the sampling distribution, $\mathbf{P}(\mathbf{Y})$, for three trials is as shown in Figure 5.21.

To work out the distributions of $\mathbf{P}(\mathbf{Y})$ for four and for five trials would require looking at 16 and 32 possible sequences, respectively. That is best left as an exercise for the student. The general procedure is exactly as above. Use the multiplication rule to calculate the probabilities of the individual

Figure 5.21

Sampling distribution for the relative frequency of red in three trials from a population for which P(R) = ½.

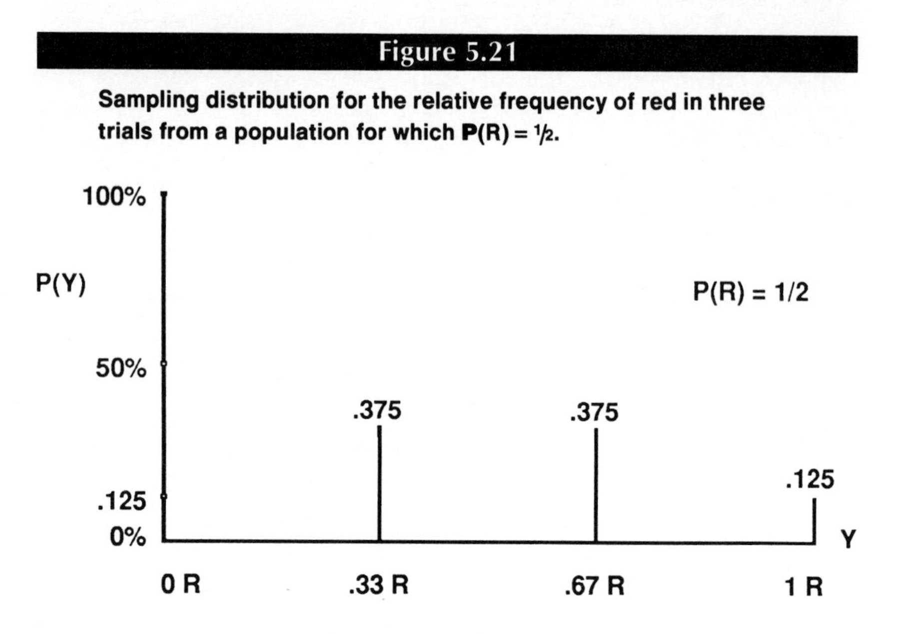

sequences, and then use the addition rule to calculate the probabilities for the various possible relative frequencies of R's. The end result of going through the calculations for both four and five trials is shown in Figure 5.22.

As you will soon learn, a sample of four or five is still much too small to evaluate any interesting hypotheses. But you should look closely at the sampling distributions in Figures 5.20 through 5.22 and make sure that you understand where they come from. Sampling distributions for more trials are very tedious to calculate by hand. For only 10 trials, there are 1024 different sequences of results to consider, and for 25 trials, over 30 million. So to work out these sampling distributions from first principles requires a good calculator or a small computer. But in principle nothing is involved except repeated applications of the simple multiplication and addition rules.

5.9 LARGE SAMPLES

Figure 5.23 exhibits the sampling distributions for the relative frequency of red marbles in 10, 25, and 50 trials on our model population. Figure 5.24 exhibits the sampling distributions for 100, 250, and 500 trials. Figure 5.25 is a blowup of Figure 5.24. That is, it exhibits the same distributions, but on an enlarged scale so that you can see the individual probabilities somewhat better.

Remember that these sampling distributions are not in principle different from the sampling distributions for smaller numbers of trials. It is just that for large numbers of trials there are many more possible relative frequencies of red marbles in the sample to be considered. So the distribution has considerably more probability values to calculate (or look up in a table).

Expected Frequency

One obvious feature of all these sampling distributions is that the most probable sample frequency for red marbles agrees with the ratio of red

Figure 5.22

Sampling distributions for the relative frequency of red in four and five trials from a population for which P(R) = ½.

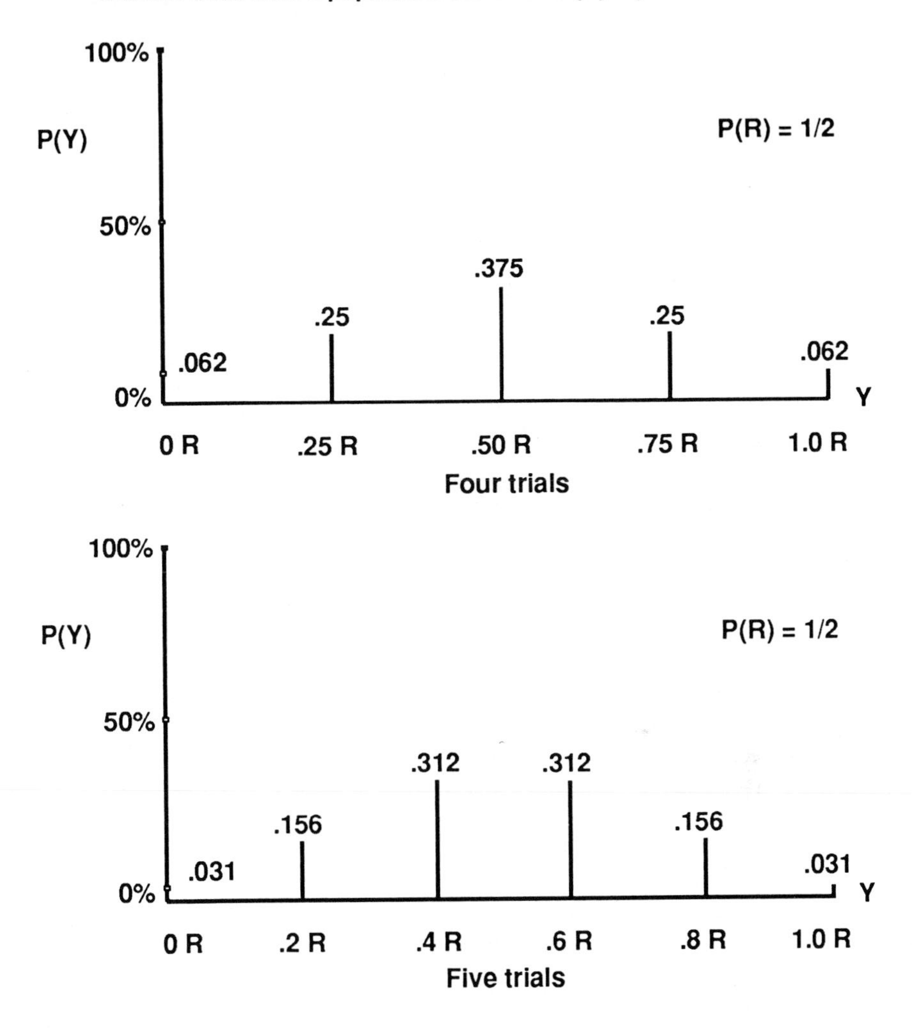

marbles in the population. In our example, the ratio of red marbles in the population is .50 and the sample frequency with the greatest probability is always .50 (or as near to .50 as possible). For most of the sampling distributions we shall encounter, the most probable frequency coincides with what, in statistics textbooks, is variously called the expected frequency, average frequency, or mean frequency.

Standard Deviation

In looking at the distributions in Figures 5.20 through 5.25, you may have noticed that the probability of the most probable frequency is lower for large samples. The probability of getting half red marbles is 50 percent for a sample

Figure 5.23

Sampling distributions for the relative frequency of red in 10, 25, and 50 trials from a population for which P(R) = ½.

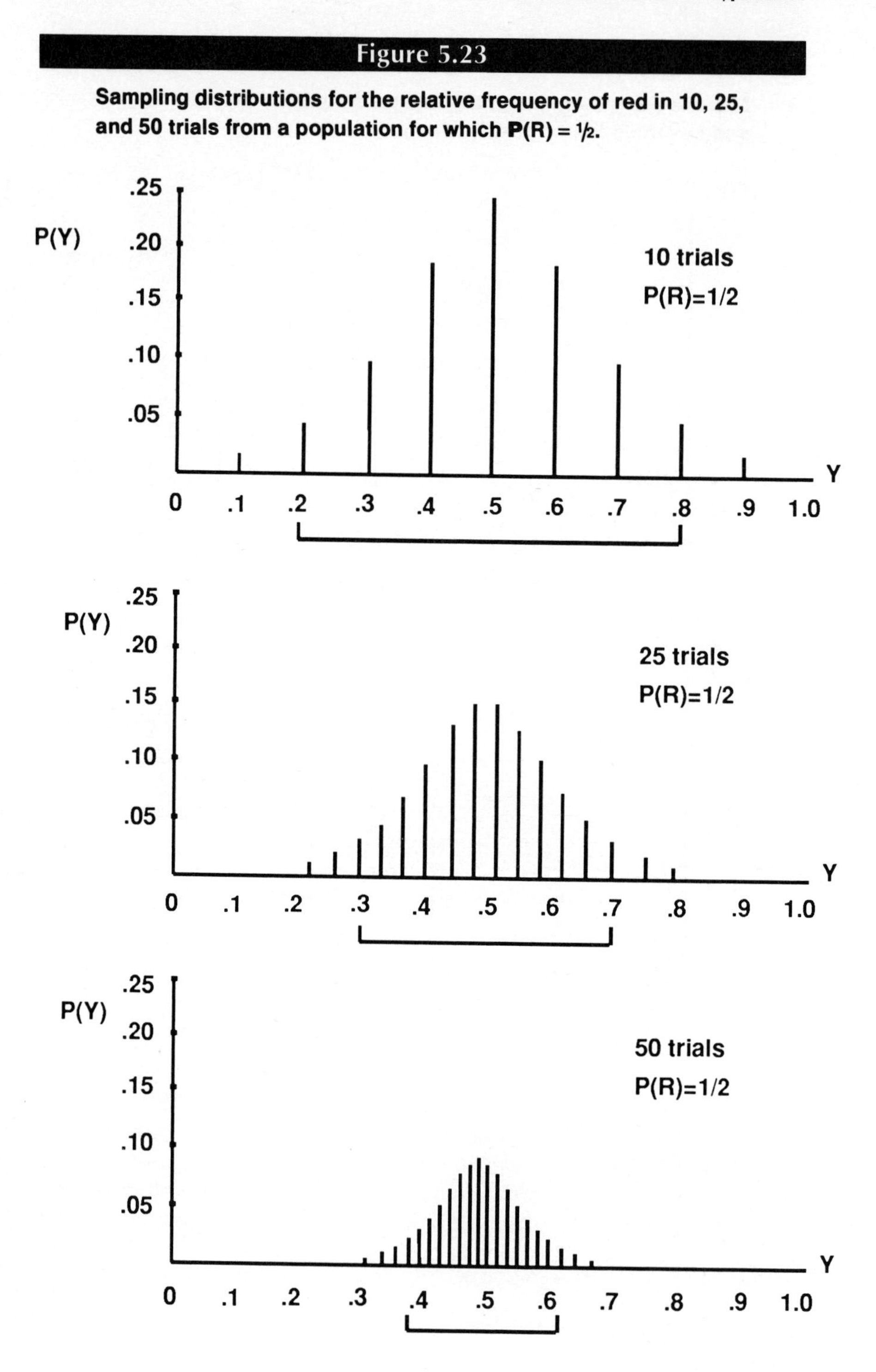

Figure 5.24

Sampling distributions for the relative frequency of red in 100, 250, and 500 trials from a population for which P(R) = ½.

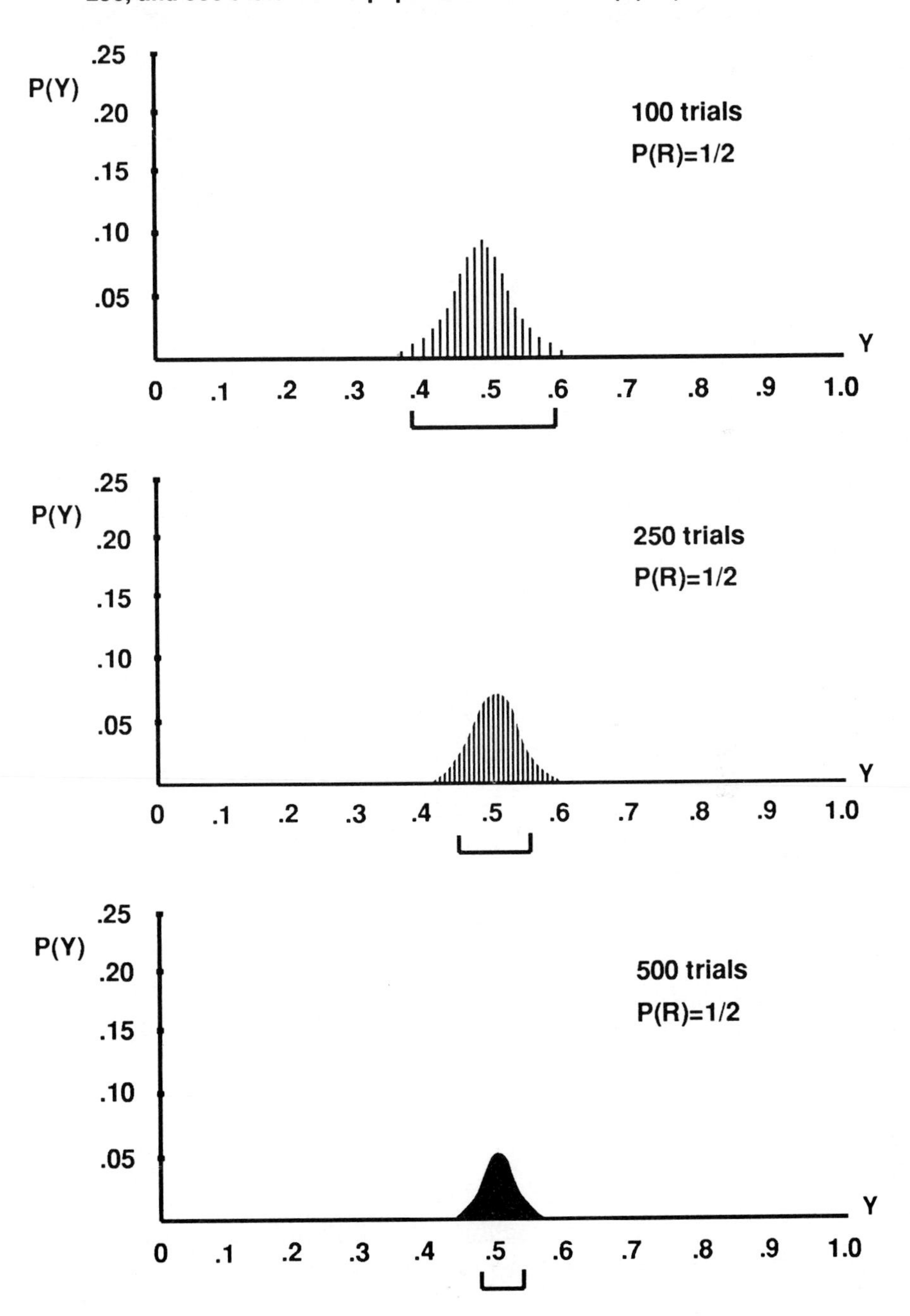

Figure 5.25

Enlarged sampling distributions for the relative frequency of red in 100, 250, and 500 trials from a population for which P(R) = ½. Note that P(Y) extends only from 0 to .10 and Y only from .40 to .60.

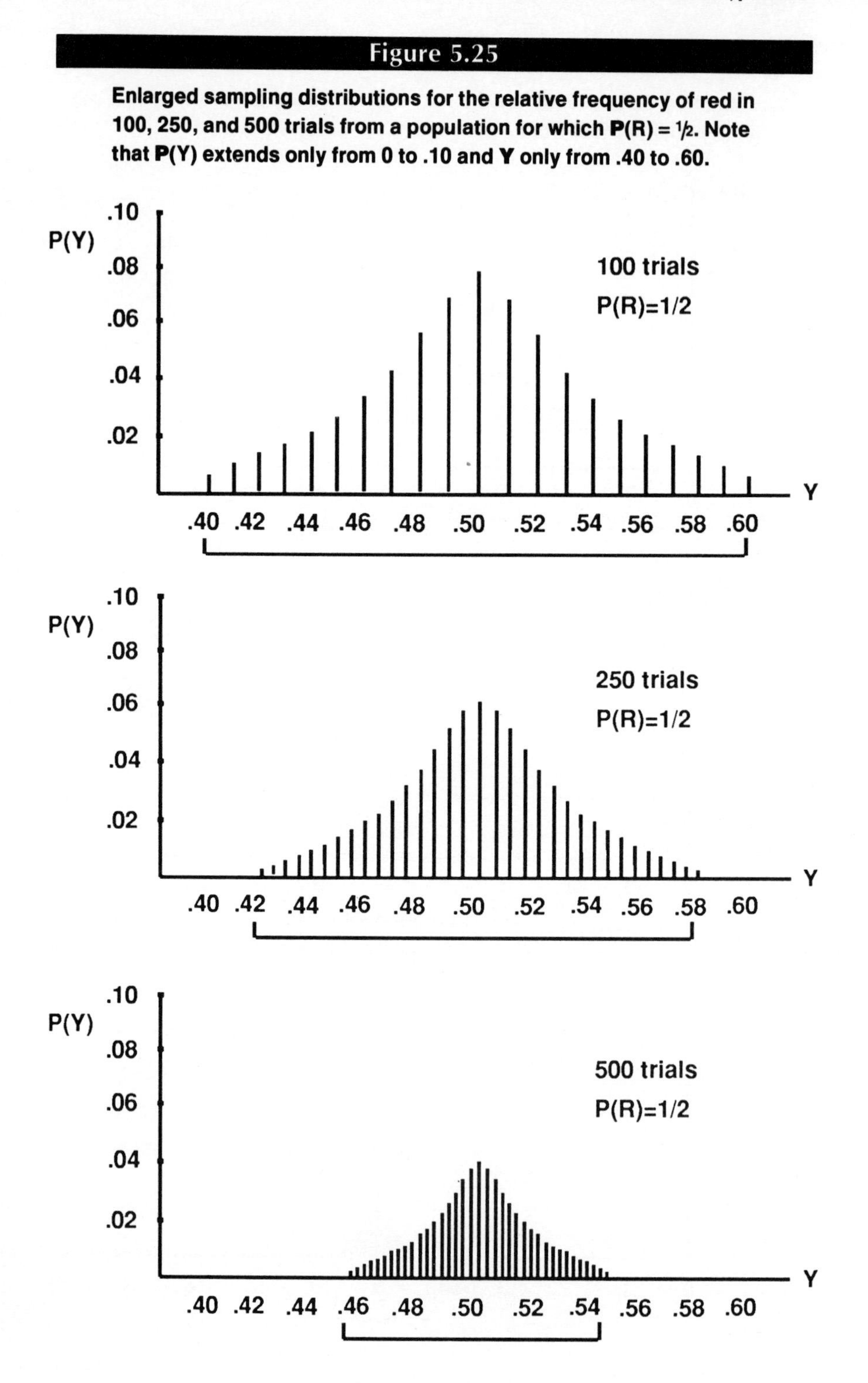

of two, 25 percent for a sample of 10, 8 percent for a sample of 100, and only 5 percent for a sample of 250. Thus, the larger the sample, the less likely you are actually to get the expected frequency. This may strike you as strange, even contradictory. But it makes sense when you think about it.

If you make only a few trials, say two or four, there is a quite good chance that half will be red marbles. (In fact, the probability is .375 for four trials.) If you do 100 trials, however, you would not really expect to get exactly 50 red marbles. You would, however, expect to get quite near to 50 red marbles. The problem is to define "nearness" within probability distributions.

Deviation from the mean is fundamentally a matter of differences in relative frequencies. A relative frequency of .40, for example, differs by .10 from a mean relative frequency of .50. However, deviations from the mean of a sampling distribution are usually measured indirectly by way of probabilities. The unit of measure for deviation is known as the standard deviation. Explaining the origin of this notion involves what for us are unnecessary complexities. The operational use of the notion, however, is quite simple.

To begin with a specific distribution, consider the sampling distribution for relative frequencies in 100 trials from a population with half its members exhibiting the property of interest. The expected number is 50, but the probability of getting exactly 50 out of 100 is only .08. However, the probability of getting between 46 and 54 is roughly ⅔, or 67 percent. All sample frequencies from .46 to .54 would, therefore, be said to be within one standard deviation of the mean. Why 67 percent defines a standard deviation is, from our standpoint, arbitrary. The real reason is mainly of historical interest.

Similarly, all sample frequencies from .40 to .60 have a combined probability of roughly 95 percent. All frequencies in this range are said to be within two standard deviations of the mean. Again, the connection between 95 percent and two standard deviations is fairly arbitrary. A similar relation holds between three standard deviations and a probability of 99 percent. The combined probability of all frequencies within three standard deviations of the mean is about 99 percent. These relationships are shown in Figure 5.26.

The reason for defining deviation, or nearness, in terms of probability is that the resulting measure is independent of the number of trials. Two standard deviations (abbreviated, 2 SD's) means the same thing for any sampling distribution, regardless of the size of the sample. It refers to all the possible sample frequencies around the mean having a combined probability of 95 percent.

The Value of Large Samples

Even with only a few trials you can begin to see why larger samples are desirable. For larger samples, the probability of observing a relative number of R's near the ratio of R's in the population (that is, 50 percent) *increases*. In Figures 5.20 through 5.22, this is best seen by looking at the ends, or "tails," of the distributions. For larger samples, the probability of selecting a relative number of R's that is not near to half decreases. Obviously, selecting either no R's, or all R's, is as far from getting one-half R's as possible. With only two trials, there is a 50 percent probability that you will select either no R's or all R's. (Use the addition rule.) For three trials, the probability of selecting no R's,

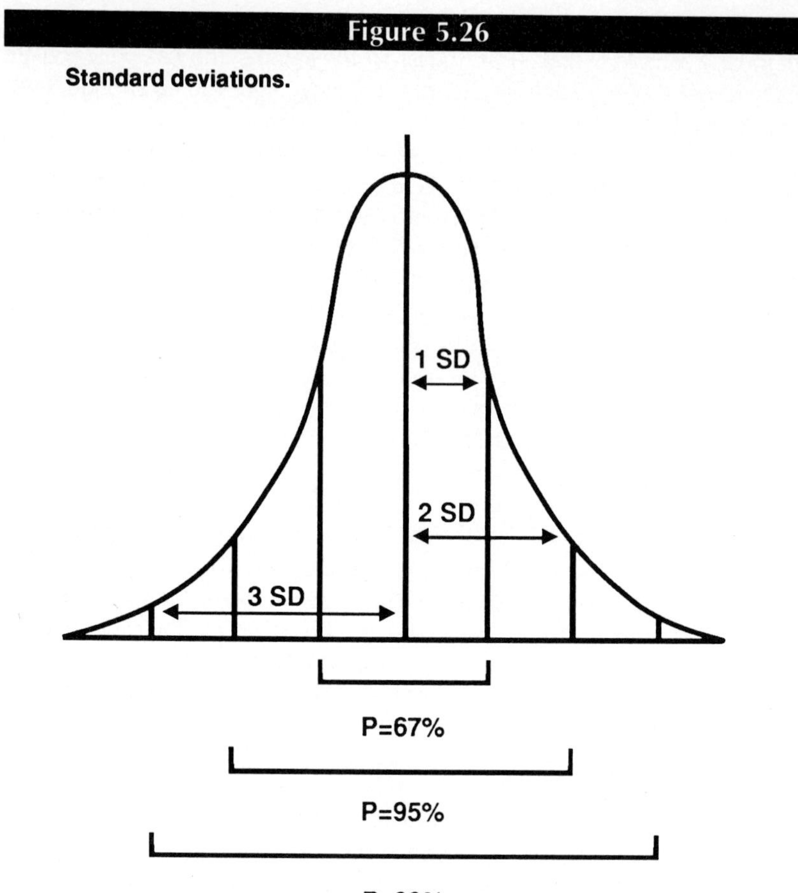

Figure 5.26

Standard deviations.

or all R's, drops to 25 percent. With four trials, it drops to about 12 percent, and with five, to about 6 percent. So if selecting near to one-half R's means merely not selecting some number as far as possible from one half, the probability of getting near to one-half R's increases from 50 percent to roughly 94 percent as the size of the sample increases from two to five. This trend becomes more pronounced as the number of trials is increased still further.

Now, look again at the sampling distributions in Figures 5.23 and 5.24. Below the axis in each graph, there is a bracket indicating all the sample frequencies corresponding to two standard deviations from the expected frequency. These frequencies have a combined probability of roughly 95 percent. In a sample of 50, for example, this set includes all possible sample frequencies from .36 to .64. As the size of the sample gets larger, the frequencies in this set get closer and closer to the expected frequency. This means that the difference between the mean and the frequencies within 2 SD's of the mean gets smaller and smaller. That is to say, the likely deviation of the sample frequency from the mean gets smaller and smaller.

We have just seen that if you pick a fixed high probability, say 95 percent, then the set of possible sample frequencies with that high a probability gets

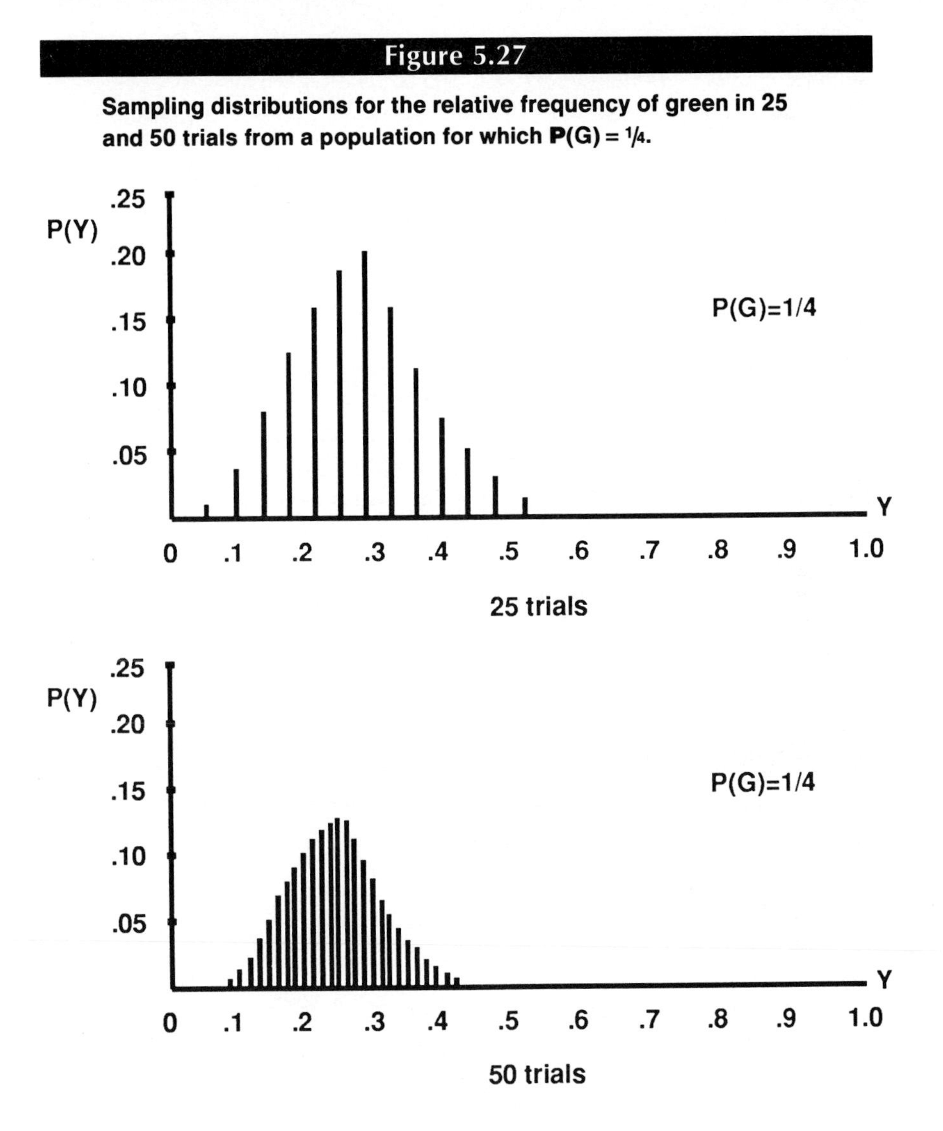

closer to the mean value for larger samples. We might turn things around, however, by fixing the range of frequencies and letting the probability change. For example, let us say that any frequency in the interval .40 to .60 is "near" to .50. Then, for a sample of ten, the probability that the sample frequency is "near" to .50 is only 65 percent. (By the addition rule, **P**(.4 or .5 or .6) = **P**(.4) + **P**(.5) + **P**(.6) = .20 + .25 + .20 = .65.) However, for a sample of 100, the probability that the sample frequency will be similarly "near" to .50 is 95 percent. Thus, the larger the sample the more probable it is that the sample frequency will be "near" to the expected frequency.

It should now be fairly clear why scientists, in general, prefer large samples to small ones. The expected frequency of any property in a random sample is the same as the proportion of that property in the population. Thus, the larger the sample, the more probable it is that the frequency observed in the sample will be close to the actual proportion in the population. So, in general, larger samples should be better indicators of the actual proportions in populations—which are, of course, what we really want to know. We will see how all this works in the next chapter.

Sampling with Unequal Probabilities

The example we have used for most of this chapter involved sampling with equal probabilities. That is, the two possible results of any single selection, red or nonred, both had a probability of one half. For most people, this case is easiest to grasp.

In order for you to see that there is no fundamental difference when sampling with unequal probabilities, Figure 5.27 shows two sampling distributions for the relative frequency of green marbles from our model population. In Model 1, **P**(G) = .25. In Figure 5.27, the number of trials is 25 and 50, respectively. You should compare these sampling distributions with those shown in Figure 5.23. The chief difference is that the distribution shifts "downward" so that its mean value is .25 rather than .50. It is also a little asymmetrical. The asymmetry, however, is less pronounced for larger samples. As before, the larger the sample, the greater the probability that the sample frequency will be near the population ratio, which in this case is .25.

Exercise 5.1

Consider a jar of marbles with the following composition:

40 Red:	*20 large,*	*20 small*
30 Green:	*20 large,*	*10 small*
20 Blue:	*15 large,*	*5 small*
10 Yellow:	*5 large,*	*5 small*

A. Diagram the statistical distribution for the color variable following the form of either Figure 5.4 or Figure 5.5.

B. Diagram the statistical distribution for the size variable following the form of either Figure 5.4 or Figure 5.5.

C. Diagram the four possible distinct correlations involving values of the color variable and the size variable.

Exercise 5.2

Use the addition rule to calculate the following probabilities for the population described in Exercise 5.1. You may check your answer by going directly to the ratios in the population.

A. **P**(*R or G*) *B.* **P**(*R or B*) *C.* **P**(*R or Y*)

D. **P**(*G or B*) *E.* **P**(*G or Y*) *F.* **P**(*B or Y*)

Exercise 5.3

Use the multiplication rule to calculate the following probabilities for the population described in Exercise 5.1. You may check your answer by going directly to the ratios in the population.

A. **P**(*R and L*) *B.* **P**(*R and S*) *C.* **P**(*G and L*)

D. **P**(*G and S*) *E.* **P**(*Y and L*) *F.* **P**(*Y and S*)

Exercise 5.4

Use the multiplication and addition rules, together with your answers to Exercises 5.2 and 5.3, to calculate the following probabilities for the population described in Exercise 5.1. You may check your answers by going directly to the ratios in the populations.

A. **P**[(*R or B*) *and L*] *B.* **P**[(*G or Y*) *and S*] *C.* **P**[(*G or B*) *and L*]

D. **P**[(*R and L*) *or G*] *E.* **P**[(*B and S*) *or Y*] *F.* **P**[(*G and S*) *or R*]

Exercise 5.5

Rank the following events in order of increasing probability, that is, number the least probable, 1, and the most probable, 3.

A. Jack and Jill went up the hill.

B. Jill went up the hill.

C. Jack or Jill went up the hill.

Exercise 5.6

In the example of the flippant judge, do all the calculations that show that the probability of the jury reaching an incorrect verdict is 1/4. (If you are mathematically inclined, you might try showing that, if the good judges have a probability, **p**, *of being correct, then the probability of the jury being correct is likewise* **p**, *no matter what* **p** *is. Thus, no matter what the reliability of the good judges, the flippant judge cancels out one of the good judges.)*

Exercise 5.7

Following the method used to determine the sampling distributions in Figures 5.20 and 5.21:

A. *Work out the details leading to the sampling distribution for four trials, as shown in Figure 5.22. You will have to consider sixteen possible results of the four trials.*

B. *Work out the details leading to the sampling distribution for five trials, as shown in Figure 5.22. You will have to consider thirty-two possible results of the five trials.*

Exercise 5.8

Following the method used to determine the sampling distribution in Figure 5.21, work out the sampling distribution for the relative frequency of green marbles on three draws from the population of Model 1, as pictured in Figure 5.13. In this case, **P**(*G*) = 1/4 *and* **P**(*N*) = 3/4.

Exercise 5.9

The following are some probabilities for the relative frequency, **f**, *of red marbles in ten draws from a jar of marbles when* **P**(*R*) = .50. **P**(**f** = 0) = .001; **P**(**f** = .1) = .01; **P**(**f** = .2) = .04; **P**(**f** = .3) = .12; **P**(**f** = .4) = .20; *and* **P**(**f** = .5) = .25. *Using these values and the fact that the distribution is perfectly symmetrical (that is,* **P**(**f** = .6) = **P**(**f** = .4), *and so on), determine the following probabilities:*

A. **P**[**f** = (0 or 1.0)]

B. **P**[**f** = (0 or .1 or .2 or .3)]

C. **P**[**f** = (.3 or .4 or .5 or .6 or .7)]

Exercise 5.10

The following are some probabilities for the relative frequency, **f**, *of red marbles in one hundred draws from a jar of marbles when* **P**(*R*) = .50:

P(**f** = .35) = .001; **P**(**f** = .40) = .011;
P(**f** = .45) = .049; **P**(**f** = .46) = .058;

P(**f** = .47) = .067; **P**(**f** = .48) = .075;
P(**f** = .49) = .078; **P**(**f** = .50) = .080.

Using these probabilities and the fact that the distribution is perfectly symmetrical, determine the following probabilities:

A. **P**[**f** = (.35 or .40 or .60 or .65)]

B. **P**[**f** = (.48 or .49 or .50 or .51 or .52)]

C. **P**[**f** = (0 through .44 or .56 through 1.00)]

Exercise 5.11

Listed below are three possible sample frequencies. For which of the following six sampling distributions are these frequencies within two standard deviations of the mean? 10 trials? 25 trials? 50 trials? 100 trials? 250 trials? 500 trials? You will have to look at Figures 5.23, 5.24, and 5.25 to answer this question.

A. **f** = .7 B. **f** = .6 C. **f** = .55

Exercise 5.12

The game of squash may be played according to two different sets of rules, one of which has 9 points to a game and the other, 15 points per game. Suppose you regularly play squash with a person whose overall squash skills are just slightly better than your own. Are your chances of winning (a) better, (b) worse, or (c) the same, with the shorter, as opposed to the longer, game? Explain.

CHAPTER 6

Evaluating Statistical Hypotheses

The purpose of learning about statistical and probabilistic models is to be able to evaluate statistical hypotheses. The relationships between models and hypotheses are exhibited in Figure 5.1. As we shall use the term, a *statistical hypothesis* is a claim about a real world population. It is a claim that the real population has the structure of a specific, statistical model. For example, the claim that a majority of all smokers in the United States are men is a claim about the population of Americans who smoke. The claim has the form of a simple *proportion*. The relevant variable is gender. The two possible values of this variable are male and female. So the statistical hypothesis claims that the real world population of American smokers fits a simple proportion model with the proportion of individuals exhibiting the value male being over 50 percent.

In principle, the truth of the statistical hypothesis could be ascertained by determining the smoking habits of every American. In practice, that is impossible. One can at most examine a sample from the population. So evaluating a statistical hypothesis is typically a matter of determining whether a given sample from the population contains data that provide evidence for the truth of the hypothesis.

Sample data consist of observed *relative frequencies* of individuals exhibiting the properties of interest. For example, we might find that a sample of 100 smokers contained fifty-five, or 55 percent, males. Is that good evidence that over 50 percent of all American smokers are male? The answer to this question is given by the statistical theory of estimation.

6.1 ESTIMATION

For any property, or any value of a random variable, call it "A," there is some probability of A in the population, **P**(A). A randomly selected sample from the population will exhibit some relative frequency of A's, **f**(A). In the typical case, the actual numerical value of **P**(A) is unknown, but the numerical value of **f**(A) is known. *Estimation* is a statistical technique for using the observed value of **f**(A) as evidence for claims about the unknown value of **P**(A).

We will first develop our understanding of the process of estimation in the context of our standard model of marbles in a jar. Later in this chapter we will begin applying our models to real world populations.

In our standard model there are two variables: *color*, with values red and nonred, and *size*, with values large and small. Earlier, we assumed that half of the marbles in the jar were red, so **P**(R) = .50. Then we determined sampling probabilities for various values of **f**(R) in samples of different sizes. With a sample of **n** = 100, for example, the probability is roughly 8 percent that exactly 50/100 red marbles would be selected. That is, the sampling probability that **f**(R) would turn out to be exactly .50 is only 8 percent. Ordinarily one would not regard 8 percent as a high enough probability to claim that one had good evidence for thinking that the process of sampling 100 marbles would yield exactly 50 red ones. This leaves us with two questions. How high a probability is required to have good evidence for a claim? What claim about **f**(R) could one make that had a suitably high probability?

In answer to the first question, statisticians typically adopt the convention that a probability of 95 percent is sufficient for good evidence. This is a reasonable convention. In typical circumstances, most people are willing to regard a 95 percent chance of being right as sufficient assurance. That is the standard we will use from now on.

What, then, can we say about **f**(R) that has a 95 percent probability of being correct? Well, we know that for any randomly selected sample from a jar with **P**(R) = .50, the probability that **f**(R) will be within two standard deviations of .50 is 95 percent. That holds for any sample size. For a sample of 100, two standard deviations includes all possible values of **f**(R) from .40 through .60. So we can say that there is good evidence that random selection of 100 marbles from a population with **P**(R) = .50 will produce a sample in which **f**(R) lies between .40 and .60.

So far, we have been reviewing the situation when the proportion of red marbles in the population, **P**(R), is *known.* We have been exploring what one could expect regarding the relative frequency of red marbles, **f**(R), in a randomly selected sample. Our typical real situation, however, is roughly the reverse of this. We learn the relative frequency, **f**(A), of some property in a randomly selected sample, and wish to say something about the *unknown* proportion, **P**(A), of that property in the population.

Although beyond the technical level of this text, there is an elementary result in statistical theory that allows us to extend our analysis of the case in which **P**(R) is known, to the case in which **P**(R) is unknown. Suppose, this time, that we randomly select **n** marbles from a jar in which the proportion of red marbles, **P**(R), is *unknown.* Our sample, of course, exhibits some definite relative frequency, **f**(R), of red marbles. We now construct an interval of values for the unknown value of **P**(R) by adding and subtracting two standard deviations from the observed value of **f**(R). That is, we construct the interval **P**(R) = **f**(R) ± 2SD's. Statistical theory tells us that there is a 95 percent probability that any interval, so constructed, will include the true value of **P**(R).

For example, with a sample size of 100, two standard deviations amount to a difference of about .10. If we observe 50 red marbles in a random sample of 100 marbles from a jar whose proportion of red marbles is unknown, we construct the interval, **P**(R) = .50 ± .10. That interval either will or will not include the unknown value of **P**(R). Statistical theory assures us, however, that 95 percent of all intervals constructed by this general method will in fact include the true value of **P**(R).

In statistical terms, the interval, .40 to .60, would be called the *95 percent Confidence Interval* for the numerical value of **P**(R). So, in general, whenever a randomly selected sample turns up a frequency **f**(A) of some property, there is good evidence that the confidence interval corresponding to the size of the sample includes the proportion, **P**(A), in the population. *Estimation,* then, turns out to be a matter of constructing the appropriate confidence interval.

Margin of Error

In the context of deriving sampling probabilities from a population with a known proportion of red marbles, we introduced the notion of a standard deviation. For any sample size, the probability was 95 percent that the observed numerical value, **f**(R), would fall within two standard deviations of the numerical value of **P**(R). The corresponding notion for estimation is the notion of a *Margin of Error* (**ME**). For an observed relative frequency, **f**(R), the 95 percent interval estimate for the numerical value of **P**(R) can be constructed by taking the numerical value of **f**(R) plus or minus the margin of error. Thus, upon observing 50 red out of 100 randomly selected marbles, one constructs the 95 percent confidence interval for **P**(R) as being equal to **f**(R) plus or minus .10. That is: estimated value of **P**(R) = **f**(R) ± **ME** = .50 ± .10.

It cannot be emphasized too strongly that every estimate of a population probability from a sample frequency must carry a margin of error. So every estimate must be an interval. One of the greatest mistakes you can make in statistical reasoning is simply to equate the unknown population probability, **P**(A), with the observed sample frequency, **f**(A). If you did that as a regular policy, you would be wrong most of the time. As we noted above, if **P**(R) in our model population were, in fact, .50, the chance that a random sample of 100 would yield 50 red marbles is only about 8 percent. Thus, in general, if you went around equating **P**(A) with **f**(A) for samples of 100, you would be right only about 8 percent of the time. Adding the appropriate margin of error means that you will be right about 95 percent of the time.

Sample Size and Margin of Error

The concept of a margin of error, like that of standard deviation, is defined so as to be independent of sample size. However, the particular proportions of **P**(A) included within the margin of error depend crucially on the sample size. This should already be evident from Figures 5.23 to 5.25, which show the standard deviation encompassing an ever-narrower range of sample frequencies as the sample size increases. The same is true for the margin of error. Table 6.1 shows the margin of error for a range of sample sizes from **n** = 10 to **n** = 10000. As you will note, the corresponding margin of error ranges from .30 to .01.

Table 6.2 provides a shorter version of Table 6.1 in which the margins of error are rounded off so as to make them easier to memorize. Thus, for example, Table 6.1 gives the correct margin of error for **n** = 500 as being close to .04. In Table 6.2, the margin of error for **n** = 500 is given as .05. This is much easier to remember. Think of Table 6.2 as providing convenient "rules of

Table 6.1

Approximate 95% Confidence Intervals for an Observed Sample Frequency of .50 in Samples of Various Sizes

Sample Size	Confidence Interval	Margin of Error
10	.20 to .80	±.30
25	.28 to .72	±.22
50	.36 to .64	±.14
100	.40 to .60	±.10
250	.44 to .56	±.06
500	.46 to .54	±.04
1000	.47 to .53	±.03
2000	.48 to .52	±.02
10000	.49 to .51	±.01

Table 6.2

"Rule of Thumb" Margins of Error for Selected Sample Sizes

Sample Size	Margin of Error
25	±.25
100	±.10
500	±.05
2000	±.02
10000	±.01

thumb" for margins of error. You should have these numbers on the tip of your tongue so that you can give a quick evaluation of statistical data as you come across it every day. For sample sizes in between those covered by these rules of thumb, simply interpolate. For example, knowing the margins of error for **n** = 500 to be .05 and for **n** = 2000 to be .02, you can easily interpolate that the margin of error for **n** = 1000 must be about .03.

Figure 6.1 summarizes the process of estimating proportions in a population on the basis of sample frequencies. First note the sample frequency, **f**(A). Then use the sample size, **n,** together with the rules of thumb provided by Table 6.2, to determine the appropriate margin of error, **ME.** Estimate **P**(A) to be in the interval **f**(A) ± **ME**.

Note that the *sample size* is an absolutely crucial piece of information in estimating proportions in populations. Without some knowledge of the sample size, knowledge of the sample frequency is quite useless. What if **n** were

Figure 6.1

Estimating a proportion in a population using the observed relative frequency in a random sample.

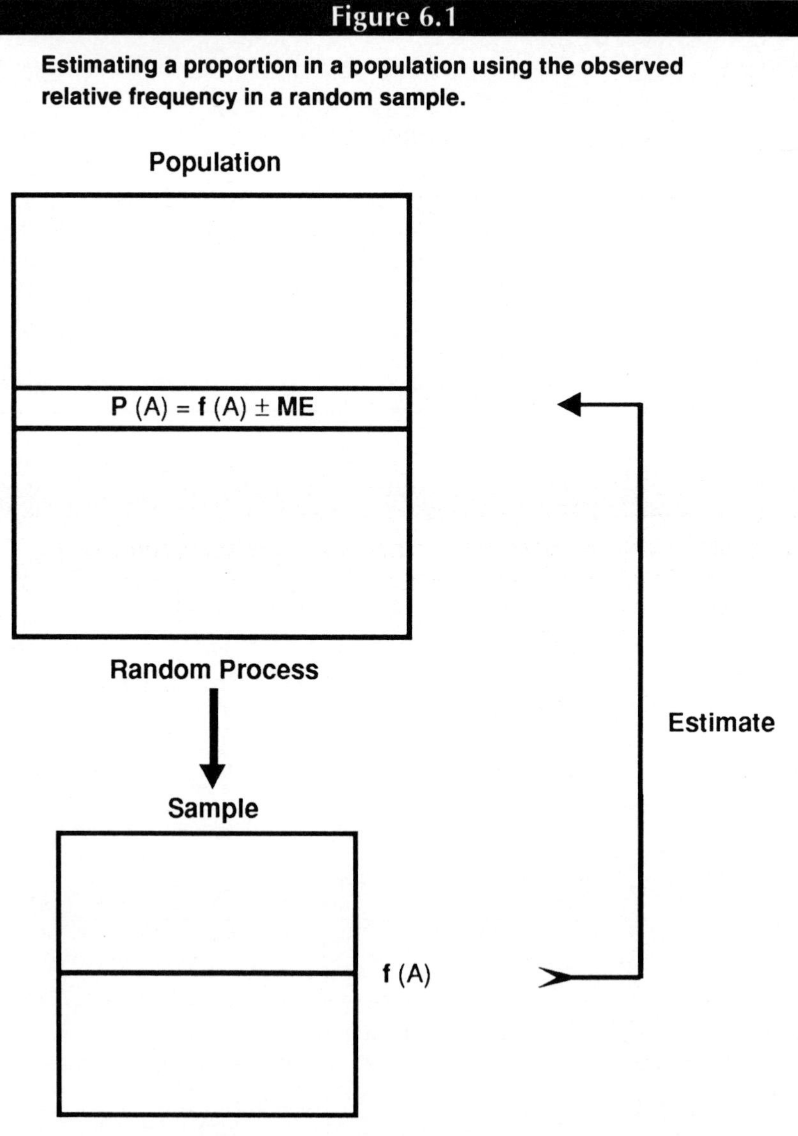

only 10? That would mean a margin of error of .30 and an interval more than half of the whole range from 0 to 1. That is better than no information at all, but not very informative.

Confidence Level and Margin of Error

We have adopted a 95 percent *confidence level* in constructing our margins of error. This means that the probability of the range **f**(A) ± **ME**, including the true value of **P**(A), is 95 percent. Although we will generally stick to the convention of using a 95 percent confidence level, it is instructive to consider the consequences of adopting a different convention.

Let us for the moment stick with a sample size of 100. Because margin of error is an extension of deviation in sample frequencies, let us look first at sample deviations. From Chapter 5 we know that the probability of getting exactly 50/100 red marbles in a random sample of 100 from a population with 1/2 red marbles is about .08. Likewise, the probability of getting 49/100 is .078. Because this sampling distribution is symmetrical, the probability of getting 51/100 is likewise .078. So the probability of getting either 49 or 51 out of 100 is .156. Similarly, the probability of getting 48 or 52 out of 100 is .150. Because these are all exclusive possibilities, the probability of getting between 48 and 52 (inclusive) out of 100 is the sum, or .386. Continuing this process until the probability sums to .95, we have to include all possible sample frequencies out to 40/100 and 60/100, which individually have probabilities of only .011.

If we wish to *raise* the probability of including the actual sample frequency to 99 percent, it is clear what we have to do. We have to include some more possible sample frequencies until the sum of all their probabilities equals .99. In fact, we need only include four more: 39/100, 61/100, 38/100, and 62/100. So *raising* the probability of inclusion *increases* the corresponding deviation.

Returning to *margins of error,* raising the confidence level increases the margin of error. Thus, for a sample of 100, *raising* the confidence level from 95 percent to 99 percent *increases* the margin of error from ±.10 to ±.12. By similar reasoning, *lowering* the confidence level from 95 percent to 90 percent *decreases* the margin of error from ±.10 to ±.08.

In general, then, there is an *inverse* relationship between confidence level and margin of error. Raising one lowers the other, and vice versa. The only way to decrease the margin of error without lowering the confidence level is to *increase* the *sample size,* as we have already seen.

Sample Frequency and Margin of Error

The margins of error given in Tables 6.1 and 6.2 are most accurate for sample frequencies around .50. The further from .50 the sample frequency is, the less accurate the given margins of error. In general, as **f**(A) moves toward 0 or 1, the **ME** becomes smaller and less symmetrical. It becomes shorter on the side toward 0 or 1, depending on whether **f**(A) is less than or greater than .50. These distortions are not too serious so long as **f**(A) is between .10 and .90. For smaller and larger values the distortions can become substantial. Apart from the above general comments, however, there are no simple rules that can serve as useful rules of thumb. You must just be aware that your rules of thumb are misleading for very small and very large values of **f**(A). The intervals in these cases are narrower and less symmetrical than your simple rules of thumb would indicate.

A Formula for Margin of Error

For those who can appreciate the mathematics, there is a fairly simple formula for calculating an approximate value for the margin of error, **ME,** as a function of both the sample size, **n,** and the observed sample frequency, **f.** The formula, which gives the value of one standard deviation, is:

$$1\ SD = [(\mathbf{f})\,(1 - \mathbf{f})/\mathbf{n}]^{1/2}.$$

To obtain our standard margin of error, which produces a 95 percent confidence interval, simply multiply the result by two. For a 99 percent confidence interval, multiply by three. A few minutes spent with a hand calculator will give you a good idea of how the margin of error varies when **f** has values near one or zero. However, because this formula is too complex to compute in one's head, we shall not typically use it in our evaluations.

6.2 EVALUATING DISTRIBUTIONS AND CORRELATIONS

What we have just learned about estimating proportions from sample frequencies can be applied to the evaluation of both distributions and correlations.

Evaluating Distributions

Figure 6.2 shows how estimation works for a distribution in which there are more than two values of a single variable. The model shows a population of red, green, and blue marbles in unknown proportions. A randomly selected sample with **n** = 250 yields **f**(R) = .56, **f**(G) = .24, and **f**(B) = .20. The margin of error is .06. So the estimates are: **P**(R) = .56 ± .06; **P**(G) = .24 ± .06; and **P**(B) = .20 ± .06. The actual population probabilities are, of course, constrained by the fact that the proportions of red, green, and blue marbles must add up to 100 percent.

Evaluating Correlations

When evaluating correlations, there are two kinds of situations. In one, the sample frequencies provide evidence for the existence of a correlation. In the other, the sample frequencies fail to provide evidence for the existence of a correlation. The sample frequencies cannot provide evidence that there is no correlation because estimates always include a margin of error. The most one could find is evidence for a correlation that has very low, if any, strength.

Evidence for a Correlation

Return to the model of marbles in a jar with two variables, color and size, with values, red or nonred and large or small, respectively. The proportions, however, are *unknown.* Imagine a randomly selected sample with **n** = 600, of which 500 turn out to be red and 100 nonred. Among the red marbles in the sample, 325/500, turn out to be large. Thus, adopting the notation for conditional probabilities to sample frequencies, **f**(L/R) = 65%. Among the nonred marbles in the sample, 35/100 are large, so **f**(L/N) = 35%.

Figure 6.2

Estimating a distribution in a population using the observed relative frequencies in a random sample.

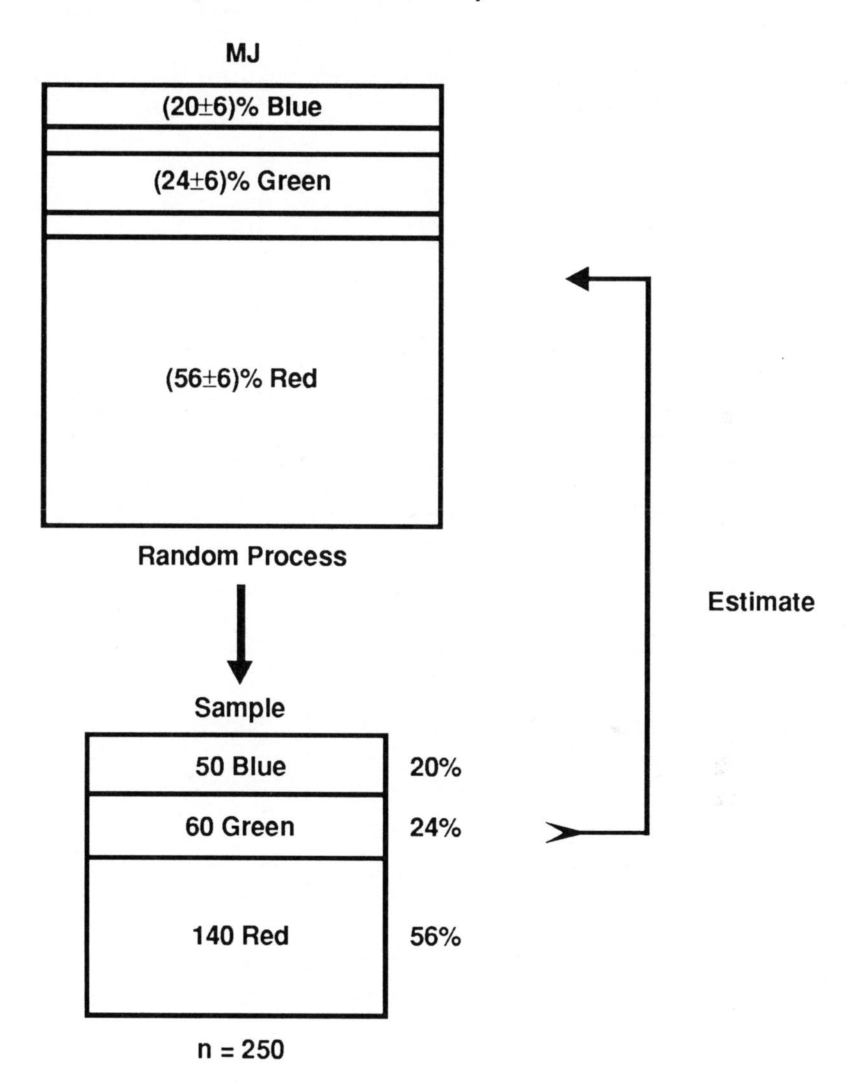

There is a clear difference in the *sample* between **f**(L/R) and **f**(L/N). But is this difference in sample frequencies good evidence for the existence of a correlation in the *population?* To find out, one must construct the relevant interval estimates. To estimate **P**(L/R) the appropriate sample size is not the total number of marbles selected, but only those that turned out to be red, which in our example is 500. By our rules of thumb, **ME** = .05. So we estimate **P**(L/R) = **f**(L/R) ± **ME** = (65 ± 5)%. Similarly, for **n** = 100, **ME** = .10. So we estimate **P**(L/N) = **f**(L/N) ± **ME** = (35 ± 10)%. These conclusions are illustrated in Figure 6.3.

Figure 6.3

An example of evidence for the existence of a correlation.

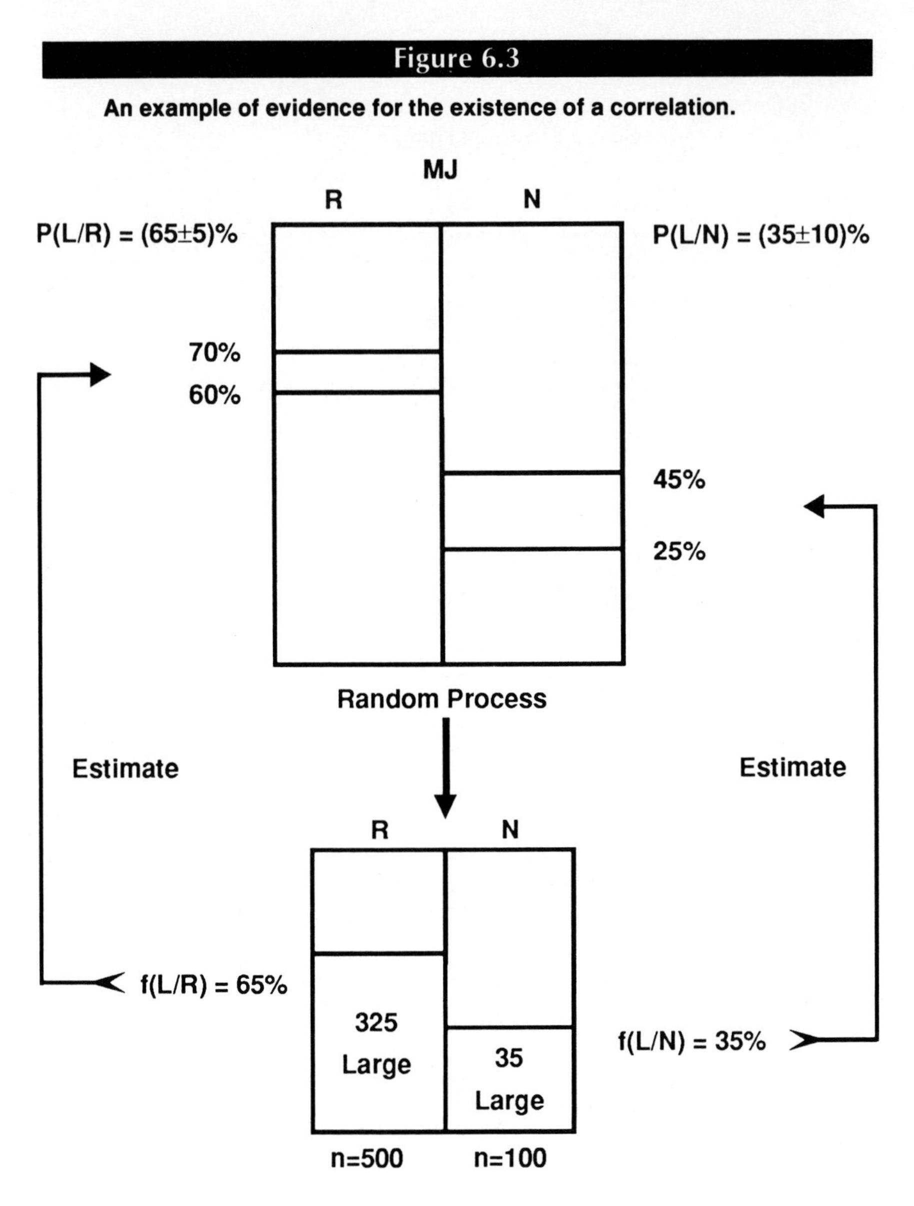

Inspecting Figure 6.3 we see that the two interval estimates *do not overlap*. That is, the interval estimate for **P**(L/R) is wholly above the interval estimate for **P**(L/N). That is good evidence for concluding that **P**(L/R) really is larger than **P**(L/N) in the population. This is just to say that color and size are correlated in the population, and, indeed, that being large is positively correlated with being red.

Lack of Evidence for a Correlation

To illustrate this case, we need only make a small change in the above example. This time suppose that the observed sample frequencies of large marbles among the red and nonred are 270/500 (54 percent) and 46/100 (46 percent), respectively. We estimate **P**(L/R) = **f**(L/R) ± **ME** = (54 ± 5)%. Similarly, we estimate that **P**(L/N) = **f**(L/N) ± **ME** = (46 ± 10)%. This new example is shown in Figure 6.4.

In this case, the two interval estimates *overlap.* This means that as far as the evidence of the sample is concerned, the proportion of large marbles might be 50 percent among both the red and nonred marbles. Thus, for all we know from examining the sample, there might be no correlation at all between

Figure 6.4

An example of failure to have evidence for the existence of a correlation.

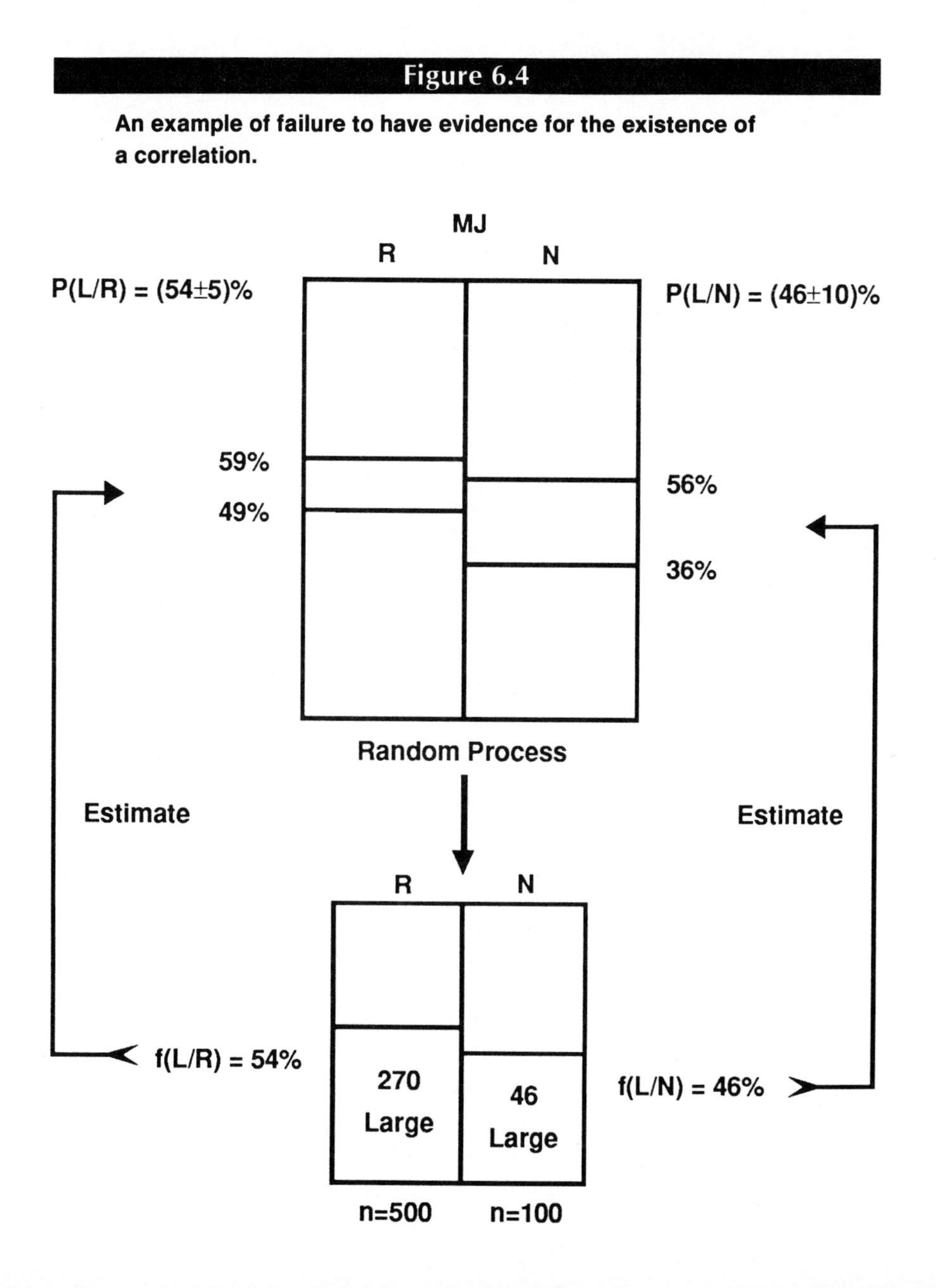

color and size in the population. However, we do not have good evidence that this is, in fact, the case. There might also be a correlation. For all we can tell, it could be the case that **P**(L/R) = 57% and **P**(L/N) = 37%. We do not have good evidence that this is not the case. In general, whenever the interval estimates overlap, the appropriate conclusion is that there is no good evidence for there being a correlation between size and color in this population.

Estimating the Strength of a Correlation

Whether or not the sample frequencies provide good evidence for the existence of a correlation in the population, there is one more thing one can do by way of evaluation. One can estimate the *strength of the correlation.* In Chapter 5 we defined the strength of a correlation as the difference between two proportions, or conditional probabilities, in the population. Here we do not know the exact proportions. We have only interval estimates of the proportions. But we can take the difference between these intervals. How does one do that?

The difference between two intervals must itself be an interval that runs from the maximum to the minimum difference allowed by the two intervals. In our first example, shown in Figure 6.3, the maximum difference in proportions is .70 − .25 = .45. The minimum difference is .60 − .45 = .15. So the estimated strength of the correlation is the interval (.15, .45). In the second example, shown in Figure 6.4, the maximum possible difference in proportions is .59 − .36 = .23. The minimum difference is .49 − .56 = −.07. So the estimated strength of the correlation is the interval (.23, −.07). Note that when the initial intervals overlap, the minimum strength of the correlation must be a negative number. This indicates that the interval includes zero strength as a possibility, and thus that it is possible that in the population there is a correlation going in either direction.

Summary

When faced with sample data indicating the possibility of a correlation among variables in the population, your strategy is clear. Note the sample frequencies and the corresponding sample sizes. Determine the relevant margins of error and construct the appropriate interval estimates for the population proportions. See whether the intervals so constructed overlap. If they do not, conclude that there is good evidence for there being a correlation between the variables and note the direction of the correlation. If the intervals do overlap, conclude that there is no good evidence for the existence of a correlation. In either case, estimate the strength of the correlation. Narrow, overlapping intervals resulting from large samples provide good evidence for little or no correlation between the variables.

6.3 STATISTICAL SIGNIFICANCE

Statisticians often talk about the "statistical significance" of data, and this terminology sometimes turns up in popular reports of statistical studies. It is important, therefore, to be able correctly to interpret claims about statistical significance, or the lack thereof.

Statistically Significant Differences

Statistical significance applies primarily to *sample frequencies.* Mostly, it applies to *differences* between sample frequencies. There are fundamentally two factors that contribute to differences between sample frequencies. First, there may be a real difference between the corresponding proportions in the population from which the sample has been drawn. Second, the sampling process itself introduces random variations in sample frequencies that exist whether or not there is a real difference in population proportions. Statistically significant differences are differences in sample frequencies that are unlikely to have been produced by the sampling process itself. They are, therefore, differences that are likely to indicate real differences among proportions in the population.

We have already learned enough to make rough determinations of when differences in sample frequencies are statistically significant. In brief, differences in sample frequencies are *statistically significant* whenever the corresponding interval estimates *do not overlap.* If the corresponding interval estimates *do overlap,* then the differences in sample frequencies are *not statistically significant.* For example, the sample differences shown in Figure 6.3 are statistically significant. Those shown in Figure 6.4 are not statistically significant.

A Formula for Statistical Significance

Once again, there is a relatively simple formula for determining whether a difference in observed sample frequencies is or is not statistically significant. Here we designate the two observed sample frequencies as $\mathbf{f_1}$ and $\mathbf{f_2}$ respectively. The corresponding sample sizes we designate as $\mathbf{n_1}$ and $\mathbf{n_2}$. The formula, which yields the value of one standard deviation in the *difference* between the two sample frequencies, is as follows:

$$1\ \mathrm{SD} = [\mathbf{f}_1 (1 - \mathbf{f}_1)/\mathbf{n}_1 + \mathbf{f}_2 (1 - \mathbf{f}_2)/\mathbf{n}_2]^{1/2}.$$

Although clearly too complex to compute in one's head, it is worth taking a little time to examine this formula because it reveals a systematic bias in our simple method for determining statistical significance.

Suppose we observe samples for which both $\mathbf{n_1}$ and $\mathbf{n_2}$ are 100, and in which $\mathbf{f_1}$ has the value .60 while $\mathbf{f_2}$ has the value .40. In our standard model, we might be supposing that 60 out of 100 large marbles are red while 40 out of 100 small marbles are red. In this case, the formula yields the value (rounded off) of .07. That is one standard deviation. Two standard deviations would be .14. What this means is that the 95 percent confidence interval for the *difference* between $\mathbf{P_1}$ and $\mathbf{P_2}$ would be $.20 \pm .14$. In particular, the lower end of this interval is .06, which is clearly above zero. So, according to the formula, the difference between $\mathbf{f_1}$ and $\mathbf{f_2}$ is clearly statistically significant. According to our simple rules for determining statistical significance, however, this difference (.60 – .40) is right on the margin for being statistically significant. Thus, our rules will sometimes lead us to judge differences as not statistically significant when in fact they are. In other words, our method for determining statistical significance is biased in the direction of requiring a greater difference in sample frequencies than is necessary to have good evidence of a real difference in the population.

One way of understanding this bias is that, by using the margins of error for simple proportions when evaluating correlations, we are tacitly raising our confidence level. In the example given above, we have tacitly raised our confidence level from 95 percent to 99 percent. Three standard deviations would be .21 ($3 \times .07$), which is greater than the difference between the sample frequencies. The confidence interval for the difference between $\mathbf{P}_1$ and $\mathbf{P}_2$ includes the value zero. So, using a confidence level of 99 percent, the difference between .60 and .40 (with samples of 100) is, indeed, not statistically significant (at the .01 level).

For the aims of this text, having to remember only one set of margins of error for various sample sizes, and avoiding explicit calculations, is well worth introducing the bias explored above. One must just remember that, when evaluating *correlations,* a small overlap in intervals is compatible with the difference in sample frequencies in fact being statistically significant at the .05 level. If one really wants to know what counts as a "small" overlap in a particular case, one must consult the above formula.

Statistical Significance and Strength of Correlations

All too often one comes across reports of statistical studies in which all that is reported is whether differences in sample frequencies are statistically significant. In this case, one has much less information to go on than if the actual sample frequencies are given. Some evaluation, however, is still possible, depending on whether differences are reported as statistically significant.

Differences Reported as Statistically Significant. Imagine a report about a sample from a jar of marbles that are either red or nonred, and either large or small. Apart from the sample sizes, all that is reported is that the difference between the sample frequencies of large marbles among the red and among the nonred is statistically significant, with a greater percentage of large among the red. What can you say about the strength of the correlation between large and red marbles in the population?

In fact, all you know is that the relevant interval estimates *do not overlap.* So the strength of the correlation between large and red must be greater than zero. How much greater might it be? You have no way of knowing. The gap between the intervals might be small or large. For all you know, the strength of the correlation might be close to one! Thus, you have good evidence for concluding that there is a positive correlation between large and red, but no evidence at all regarding the strength of the correlation.

Note that knowing the size of the sample is of little help. Small samples yield wide intervals and large samples yield narrow intervals. However, not knowing by how much the intervals are separated, you are left in the dark about the strength of the correlation. All you can say is that it is greater than zero.

Differences Reported as Not Statistically Significant. Now imagine another sample from the same model population. This time it is reported that the difference between the sample frequencies of large marbles among

Table 6.3

Summary of Relationships among Sample Size, Reported Statistically Significant Differences, and Estimates of Effectiveness

Sample Size	SSD	Estimate of Effectiveness
Large	No	Low effectiveness
Large	Yes	Can't tell
Small	No	Can't tell
Small	Yes	Can't tell

the red and among the nonred is not statistically significant. Here one's evaluation of the situation does depend on what one knows about the sample size. Of course, if no information about sample size is given, then all you can say is that there is no evidence for the existence of any correlation between large and red marbles.

If the sample size is known to be relatively small, then a simple report of no statistically significant difference in the frequency of large marbles is quite uninformative. You know that the relevant intervals overlap, but you do not know by how much. You do know the intervals are quite wide. If in fact they overlap only slightly, then the appropriate estimate of the strength of the correlation would be an interval from zero to the sum of the individual margins of error each doubled. For example, if there were 100 red and 25 nonred marbles, a report of no statistically significant difference could still be evidence of a strength of correlation as great as $(2 \times .10) + (2 \times .25) = .70$, which is considerable.

The only case in which a simple report concerning statistical significance without sample frequencies is really informative is if it claims no statistically significant difference with very large samples. Very narrow intervals, even if only slightly overlapping, still yield a low estimate for the strength of correlation. For example, with 2000 each of red and nonred marbles sampled, the maximum estimated strength of correlation compatible with overlapping intervals is .08, which is quite small.

These possibilities are summarized in Table 6.3. The general lesson is that it is better to be told the actual sample frequencies than simply that their difference is or is not statistically significant.

6.4 AN OPINION POLL

The following report appeared in *The New York Times*, Sunday, August 20, 1989. A number of passages have been slightly edited and several paragraphs have been eliminated, but mostly the report appears as written. Some of the data accompanying the article are reproduced as Tables 6.4 and 6.5. As you read this report, keep in mind that this is a very good poll and is reported in an expert and complete fashion. Most polls you learn about through the popular media are not nearly so well-executed nor so well-reported.

Bars to Equality of Sexes Seen as Eroding Slowly

By Lisa Belkin

A quarter-century after the start of the women's rights movement, American women say that despite their gains, it is still a man's world.

Though they have seen a closing of the gap between men and women, both in the workplace and at home, a *New York Times* poll shows that most women say the goals of the women's movement have not been fully realized, and many say the gains have come at too high a price.

Men, while generally expressing support for women's pursuit of equality, said there had been more changes than women saw, with less cost to women than women reported. They suggested they had overcome sexism more thoroughly than women acknowledged, and they saw less need for further changes than women did.

The basic goal of the women's movement was to eliminate the barriers that kept women from achieving as much as men. But the poll found that 56 percent of women say American society has not changed enough to allow women to compete with men on an equal basis—a view held by only 49 percent of men.

Moreover, the poll found sharp differences between women of different ages and races, and between men and women of the same age—reflecting a dissonance and potential tension that can rub ordinary life raw.

"Sometimes I think my husband is talking another language," said Alison Mellin. "He thinks he's being sympathetic and all that, but he just doesn't *get* it." Ms. Mellin is 36 years old and works in a Los Angeles flower shop.

Some men, like Jim Colmer, a 32-year-old cleaning supplies salesman from Detroit, say women fail to keep problems in perspective. "I have some women friends who have children, and that's all they talk about," said Mr. Colmer, who is single and childless. "The women aren't trying to so-called balance anything that men haven't balanced for years and years. They just complain more."

Particularly confused and frustrated are the women who came of age at the height of the women's movement—the baby boomers between the ages of 30 and 44. This is a generation of women who became adults against the background of debates on such questions as economic equity and reproductive freedom, and those debates shaped their thinking. Sixty-two percent of the women in that age group agreed with the statement: "Most men are willing to let women get ahead, but only if women do all the housework at home."

The *Times* poll, conducted June 20 to 25, shows that women generally identify their problems in the same general terms as they have at least since

Betty Friedan's 1963 book, *The Feminine Mystique,* helped to start the women's movement.

Problems relating to equality on the job were often cited by women as the most important ones they faced, with 23 percent of women mentioning them. Eight percent of women called abortion the most important issue before them.

But the poll also showed nearly as large a set of grittier and more specific concerns about how to balance family life and work. Nineteen percent cited these problems as most important. These concerns cut across racial and economic lines far more clearly than some of the grander philosophical issues of the 1960s and 1970s. Today's questions are: "Who takes care of the children?" "Why do working women *still* do more of the housework?" "How do you deal with sexism that is subtle rather than overt?"

As Seen by Society, As Seen by Themselves

Here are some of the other main findings of the *Times* poll of 1025 women and 472 men. It was intended to measure how society views women and how women view themselves. The margin of sampling error was plus or minus three percentage points for women and five for men.

Thirty-three percent of the women, and 37 percent of the men, said children's needs are slighted the most when a woman combines a job with marriage and motherhood. Twenty-six percent of women, and 20 percent of men, said the marriage was slighted the most. But only 6 percent of women and men said the woman's job suffered the most. Many others refused to choose, and only 17 percent of women and 18 percent of men said nothing suffered.

When asked whether women have "given up too much" in exchange for gains in the workplace, almost half of all women say yes, as do a third of the men. Most women say that what they gave up is time with their children and quality of their family life.

Women still do most of the work at home. But perhaps reflecting vestiges of traditional attitudes, 61 percent of them still say that what their husbands do is their "fair share." Even among women with full-time jobs and children at home, 40 percent say their husbands do their fair share, although 42 percent say the husbands do less. Substantial majorities of married women say they do most of the food shopping, cooking, house cleaning, bill paying, and child care, whether they work outside the home or not. Men agree that their spouses do more around the house, and about half concede that women do "more than their fair share" of household chores.

While three-fifths of women say men's attitudes toward women have improved in the last two decades, the improvement perceived is limited. Fifty-three percent of women say most of the men they know think they are better than women. Men are even more likely to say men's views of women have improved, but 48 percent of them say most men they know consider themselves superior to women.

Women working full time are less likely than men to call their work a "career" as opposed to a job. But among 18-to-29-year-olds, women are at least as likely as men to call their work a career.

The optimism of women 18 to 29 was shown frequently. Only half of them agreed that "men still run almost everything and usually don't include women when important decisions are made," a view held by solid majorities of older women. And about a quarter of the younger women, far more than in older groups, said nothing is slighted when a woman combines work, marriage, and children. But this group still said it supported a "strong women's movement to push for changes that benefit women." Seventy-one percent of them said the United States needed one.

Areas of Agreement, Areas of Annoyance

The portrait that emerges from the poll includes a good amount of irritation felt by women toward men. Women say men's attitudes have not changed nearly enough.

"They're proud of themselves because they treat women as equal," said 22-year-old Tracey Clements, a military graphic designer from Edgewood, Maryland. "But when they're driving along and the person ahead makes some sort of mistake they say, 'That must be a woman driver.' They think that's a joke, but it's not a joke, and it's not equal."

There were some important areas of agreement between men and women. For example, big majorities of both said the women's movement had helped make all sorts of relationships between men and women more honest. Perhaps surprisingly, 70 percent of women with full-time jobs said that where they worked, women had an equal or better than equal chance of promotion. Fifty-nine percent of men with full-time jobs said the same.

The poll reflected one example after another of sharp differences between men and women in the same age group. Asked whether there are more advantages of all kinds to being a woman or a man, slightly more than half of women between the ages of 30 and 44 say men have more advantages, as compared to almost one-third of the men in same age group.

Chasms in Perceptions

On the question of whether men still run everything and exclude women from important decisions, only 34 percent of men in the 45-to-64 year age group said they did, but 59 percent of women that age said so—a vast difference in one age group.

And when asked to name the most important problem faced by women today, the conflict between work and family or child care was named by 17 percent of women between 18 and 29, but by just 9 percent of their male peers.

"Sometimes it seems like that's all I think about," said Beth McGuire, a 28-year-old administrative nurse in Baltimore who juggles her job and caring for her 1-year-old twins.

Great Expectations, Great Frustrations

Almost as striking as the differing perspectives of men and women are the varying views of women of different age groups. To generalize roughly, women 18 to 29—those least likely to have faced all the concerns of marriage, child care, and the work force and more likely to have mothers who worked—have the fewest complaints, despite examples to the contrary like Mrs. McGuire.

Women 30 to 44—who are now at an age that includes marriage, work, and children, and who are the first generation raised to expect to have all of those—are most confused and frustrated.

Thirty-year-old Deborah Lenz, who left a career in fashion merchandising for a seasonal job with the Internal Revenue Service so she could spend more time with her two toddler sons, has views typical of her age group.

"There are so many hard questions," said Ms. Lenz, who is married and lives in Indianapolis. "When there was no choice, maybe it was easier. Now there's a choice, which is good, but it's hard to make that choice. There are times that I'd really like to be out there doing the career I more or less gave up. But I couldn't stand losing the time with my children."

And women 45 and older—whose children are generally older, whose jobs are probably more stable and whose memory of the days before the women's movement are clearer—are somewhat more sanguine.

Hope Luder, a 48-year-old high school anthropology teacher from Billerica, Massachusetts, who has developed stress-related health problems that have forced her to cut her work hours, is typical of her age group.

"What a lot of women try to do is get as many ulcers as men," she said. "You see a lot of women that try to do too much. I think when I was younger I bought into the idea somewhat that women had to do everything, but I don't think so anymore."

How the Poll Was Taken

The *New York Times* poll on women's issues is based on telephone interviews conducted June 20 through 25 with 1497 adults around the United States, excluding Alaska and Hawaii.

The sample of telephone exchanges called was selected by a computer from a complete list of exchanges in the country. The exchanges were chosen so as to assure that each region of the country was represented in proportion to its population. For each exchange, the telephone numbers were formed by random digits, thus permitting access to both listed and unlisted numbers. The numbers were then screened to limit calls to residences.

Women were sampled at a higher rate than men so that there would be enough women interviewed to provide statistically reliable comparisons among various subgroups of women. The results of the interviews with 1025 women and 472 men were then weighted to their correct proportions in the population.

Results were also weighted to take account of household size and number of residential telephone lines and to adjust for variations in the sample relating to region, race, age, and education.

In theory, in nineteen cases out of twenty the results based on such samples will differ by no more than three percentage points in either direction from what would have been obtained by seeking out all American adults.

The potential sampling error for smaller subgroups is larger. For example, for men it is plus or minus five percentage points. For women, it is plus or minus three percentage points. For women aged 18 to 29, it is plus or minus six percentage points.

In addition to sampling error, the practical difficulties of conducting any survey of public opinion may introduce other sources of error into the poll.

6.5 EVALUATING STATISTICAL HYPOTHESES

We will now proceed to develop a program for evaluating statistical hypotheses similar to that developed in Chapter 2 for evaluating theoretical hypotheses. We will first develop the program informally using an example from the *New York Times* poll. Then we will set it out in a form that can be easily remembered for future use.

The Real World Population of Interest

Every statistical study is done with the intent of learning something about a specific *real world population.* The first task in evaluating any statistical hypothesis is to identify that population. For the *New York Times* poll, the task is easy. The population is American adults, or, more specifically, American men and women 18 years of age and older. The description of the poll at the end of the report explicitly refers to adults in the United States. The lowest age category referred to in the report is men and women aged 18 to 29. The highest age category referred to is 45 and above.

Not all reports are this clear. Sometimes you will have to "read between the lines" to determine the population of interest. In some cases, it will be impossible to tell from the report exactly what the population of interest might be. All you can do is note this failing in your evaluation.

Sample Data

Like our *New York Times* report, many reports of statistical studies provide a great deal of information about the *sample.* It is generally impossible to evaluate it all at once. You have to pick out some particular piece of data on which to focus your evaluation. When you have finished evaluating that bit of data, you can go on to another. With experience, you can learn to go through much data relatively quickly, but even so you will have to take it one bit at a time.

The first numerical data presented in the *Times* report appears in the fourth paragraph in the sentence: "The poll found that 56 percent of women say American society has not changed enough to allow women to compete with men on an equal basis—a view held by only 49 percent of men." Let us focus our analysis on this data. It is important to be clear that the percentages reported refer only to the *sample,* not to the population of interest. The whole point of evaluating the reported data is to determine whether the reported data from the sample provide good evidence for any statistical hypothesis referring to the *population* of interest.

Statistical Model

Formulating a *statistical hypothesis* about the real world population requires constructing a *statistical model.* What model? That is best determined by looking at the data and deciding for what statistical model the data might be relevant. This is not as difficult as it might seem because we have only three possible models from which to choose: proportions, distributions, and correlations.

The most efficient way to proceed is to determine what *variables,* and what *values* of these variables, might be introduced to represent the data. Recall that a proportion has one variable with two values; a distribution has one variable with three or more values; a correlation has two variables with two values each.

The statement of the data we are evaluating suggests one obvious variable, *gender,* with values, *men* and *women.* But there is also indication of another, less obvious, variable. It might be described as a person's *attitude* toward the statement: "American society has not changed enough to allow women to compete with men on an equal basis." The two values of this variable are *agree* and *disagree*.

There might have been a third value, no opinion, but that is not reported. In many polls, the subjects are not given the option of refusing to express a definite opinion. If they are, you can lump together the no-opinion value with one of the other two, for example, disagree or no opinion. This tactic allows you to produce a simple correlation model with only two values for each variable. No such tactics are necessary in dealing with the example at hand.

Our data, therefore, can be represented as having the form of a *correlation* with 56 percent of the women and 49 percent of the men agreeing with the offered statement. The obvious *statistical hypothesis* to consider is that there is a similar correlation in the real world population. In particular, the hypothesis would be that there is a positive correlation in the population between being a woman and agreeing to the statement that American society has not changed enough to allow women to compete with men on an equal basis. Our task now is to determine whether the data provide good evidence for this hypothesis.

Can We Assume a Random Sampling Model?

Before we can evaluate the relevance of the data for the hypothesis, we must first determine whether our model for *random sampling* fits the actual process by which the sample was selected from the real world population.

That is, does the model pictured on the right-hand side of Figure 5.1 fit the real world situation pictured on the left? All our calculations involving sample sizes and confidence intervals were developed in the context of a random sampling model. If that model does not fit the real world situation, none of our rules about overlapping or nonoverlapping intervals have any relevance.

Recall from Chapter 5 that a randomly selected sample is one in which (1) all members of the population who might exhibit a property of interest have an equal chance of being selected, and (2) there is no correlation between the outcome of one selection and any other. Polls like the *Times* poll are exceptional in that strenuous efforts are taken to insure randomness in sampling. For example, telephone numbers are selected by random-number generators and the total sample is proportioned to the known population in various areas of the country.

Note that the *Times* poll follows the convention of using confidence intervals with a 95 percent level of confidence. This comes out clearly in the final description of the poll in which they say: "In theory, in nineteen cases out of twenty the results based on such samples will differ by no more than three percentage points in either direction from what would have been obtained by seeking out all American adults." Here they are referring to the whole sample of 1497 men and women. Since $^{19}/_{20}$ is strictly equivalent to $^{95}/_{100}$, they are saying that, with a 95 percent confidence level, the margin of error for the sample is plus or minus .03. That agrees with our rules of thumb.

The only major failing of the sampling procedure used in the *Times* poll is inherent in the use of telephones in polling. People without telephones, such as the homeless, have no chance at all of being selected. Nevertheless, at the present time in the United States the percentage of people without telephones is very small. So missing people without telephones cannot make much difference to the final results. Our conclusion, therefore, is that the actual sampling process in this poll is quite well represented by a random sampling model.

Evaluating the Hypothesis

Assuming, then, that the actual sampling process can be represented by a model of randomly selecting marbles from a jar, what does the data tell us about the population? In particular, is there evidence of a positive correlation in the population between being a woman and agreeing to the statement that American society has not changed enough to allow women to compete with men on an equal basis?

To answer this question one must consider the relevant margins of error. For the 1025 women sampled, the margin of error is $\pm.03$. For the 472 men, the margin of error is $\pm.05$. Thus, the estimated value of **P**(A/W) = **f**(A/W) $\pm$ **ME** = (56 $\pm$ 3)%, or an interval from 53 percent to 59 percent. Similarly, the estimated value of **P**(A/M) = **f**(A/M) $\pm$ **ME** = (49 $\pm$ 5)%, or an interval from 44 percent to 54 percent. These relationships are diagramed in Figure 6.5.

The confidence intervals overlap. By our usual method for determining statistical significance, we would judge that the difference in sample frequencies is not statistically significant and, therefore, that there is little

Figure 6.5

A diagram of the *Times* study evaluated in the text.

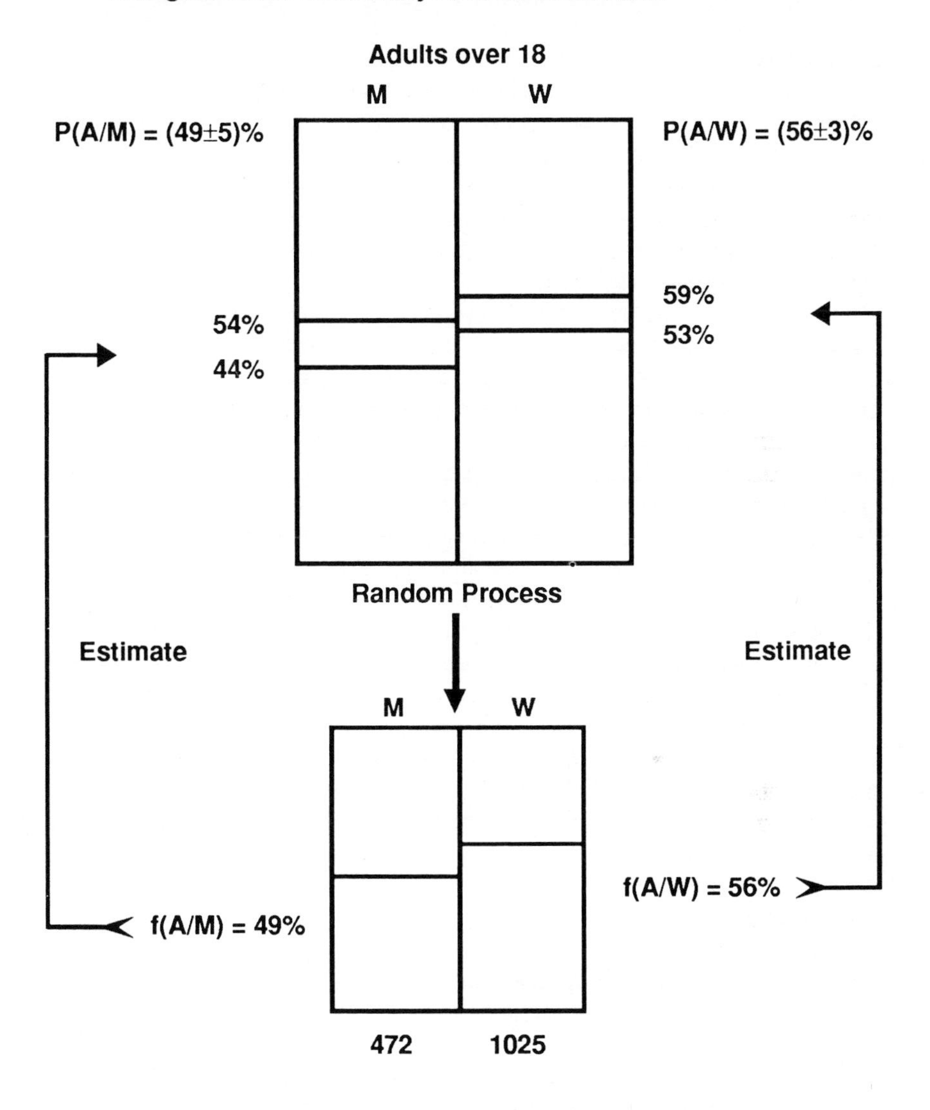

evidence that the suggested correlation exists. The overlap, however, is only one percentage point, and we know that our method tends to require a greater difference in sample frequencies than is necessary for statistical significance (at the .05 level). So here one should probably go along with the *Times'* suggestion and assume that the difference is statistically significant, though perhaps only just barely so.

In fact, if one uses the formula given earlier, one standard deviation for the difference in sample frequencies turns out to be .03. So the (95 percent) margin of error for this difference is .06, which is less than the observed difference in sample frequencies, .07 (.56 – .49). So the observed difference in sample frequencies is indeed statistically significant at the .05 level (though not at the .01 level).

Summing Up

At this point in the evaluation of any statistical hypothesis, one should review what has been learned in order to extract the "take home lesson" from the information presented. Two points are particularly important to review.

One is how well the actual sampling process can be modeled by a random sampling model. Here you can make only rough qualitative judgments. Except for the fact that it was a telephone poll, the *Times* poll scores exceptionally high in this regard.

Second is the actual conclusion you are to draw. By our usual method of determining statistical significance, the 95 percent confidence intervals overlap. The overlap, however, is sufficiently small that, even without doing any calculations, we could be fairly confident that the exact 95 percent confidence interval for the difference in population ratios would not include zero. So a considered judgment would be that there is "reasonable" evidence that women in general are more likely than men to agree with the statement. That is to say, there is reasonable evidence that the suggested correlation exists. There is, however, no evidence that it is a very strong correlation.

This is a somewhat "wishy-washy" conclusion, but it is the best we can do. Interpreting statistical data cannot be done completely mechanically. It requires some judgment as well. As you evaluate more cases, you will learn the kinds of judgments that one can reasonably make.

6.6 A PROGRAM FOR EVALUATING STATISTICAL HYPOTHESES

We will now set out our program for evaluating statistical hypotheses and then apply it to some of the other data from the *Times* poll.

The Program

Step 1. The Real World Population. Identify the real world population that is the intended object of the study.

Step 2. The Sample Data. Identify the real world sample and the particular data from that sample whose relevance for hypotheses about the population you wish to evaluate.

Step 3. The Statistical Model. Identify the statistical model of the population that is appropriate for evaluation given the data already identified. Identify the relevant *variables* and the *values* of these variables. If the data have the form of a correlation, give a clear statement of the statistical hypothesis asserting the existence of the corresponding correlation in the real world population.

Step 4. Random Sampling. How well does a random sampling model represent the actual process by which the sample was selected from the population?

Answer: a) very well; b) moderately well; c) somewhat well; or d) not very well. Explain the factors relevant to your answer.

Step 5. Evaluating the Hypothesis. Diagram the study as it applies to the real world population, the sample, and the data you have identified. Label all relevant parts of the diagram. Assuming a random sampling model is applicable, what can you conclude about the real world population? If a correlation is possible, is there good evidence for the hypothesis stated in Step 3? Evaluate the *strength* of the correlation.

Step 6. Summary. Review your evaluation, particularly your analyses in Steps 4 and 5. Give a summary statement of how well the statistical data support the resulting statistical hypothesis about the population.

A Woman's Place

Now let us employ the above program to evaluate several more statistical hypotheses suggested by the *Times* poll. In particular, let us begin with the data on the first line of Table 6.4.

Table 6.4

A Woman's Place: Views of Two Generations

		Women		Men	
		Age			
All Adults	Percentage of people who say:	18–44	45+	18–44	45+
60%	For themselves, they prefer to combine marriage, children, and a career.	55%	40%	76%	69%
52	Society has not changed enough to allow women to compete with men on an even basis.	56	55	45	54
50	Most men they know think they are better than women.	56	47	53	41
48	Men are willing to let women get ahead, but only if women still do all the housework at home.	57	54	35	43
37	All things considered, there are more advantages in being a man in America today.	44	41	31	30
63	Men's attitudes toward women have changed for the better in the past 20 years.	70	46	77	54
59	The United States continues to need a strong women's movement to push for changes that benefit women.	71	60	55	45

Based on interviews with 1025 women and 472 men nationwide, conducted by telephone June 20–25, 1989.

Step 1. The Real World Population. American adults over 18 years old.

Step 2. The Sample Data. Sixty percent of 1497 adults sampled agreed with the statement that, for themselves, they prefer to combine marriage, and children, and a career.

Note that this data is adjusted to correspond with what would have been expected if roughly equal numbers of women and men had been sampled. You can see that this is so by averaging the percentages for women below and above 44 years old, averaging the percentages for men below and above 44 years old, and then averaging the two results. Rounding off to whole percentages yields 60 percent. This calculation assumes roughly equal numbers of both women and men below and above 44. However, that is in accord with the reported figures, evaluated above, that 56 percent of women and 49 percent of men agree that American society has not changed enough to allow women to compete with men on an equal basis.

Step 3. The Statistical Model. Here there is only one *variable* to consider, namely, attitude toward the statement that, for themselves, they prefer to combine marriage, and children, and a career. The two values of this variable are *agree* (A) and *disagree* (D). So the relevant statistical model is that of a simple proportion.

Step 4. Random Sampling. As discussed earlier, the actual sampling process is well represented by a random sampling model.

Step 5. Evaluating the Hypothesis. For $\mathbf{n} = 1497$, $\mathbf{ME} = .03$. So the interval estimate with a 95 percent confidence level is that $\mathbf{P}(A) = \mathbf{f}(A) \pm \mathbf{ME} = (60 \pm 3)\%$. The relevant diagram is shown in Figure 6.6.

Step 6. Summary. The sampling process fits a random sampling model very well (Step 4) and determining the 95 percent confidence interval is straightforward (Step 5). There is good evidence for the statistical hypothesis that $(60 \pm 3)\%$ of all adults prefer to combine marriage, children, and a career.

This example illustrates an important point. Knowing the overall proportion of members of a population exhibiting a given property can be very misleading if the population contains subgroups that exhibit very different percentages of that same property. In other words, the overall proportion *averages* overall subgroups, and the average may be very misleading for some subgroups. In the present example, the fact that a clear majority of all American adults express a preference for combining marriage, children, and a career hides the fact that *less* than a majority of women express this preference. In the overall proportion, women's preferences are masked by the large majority of men in favor of combining marriage, children, and career.

Figure 6.6

A diagram of the *Times* study regarding the proportion of adults favoring combining marriage, children, and a career.

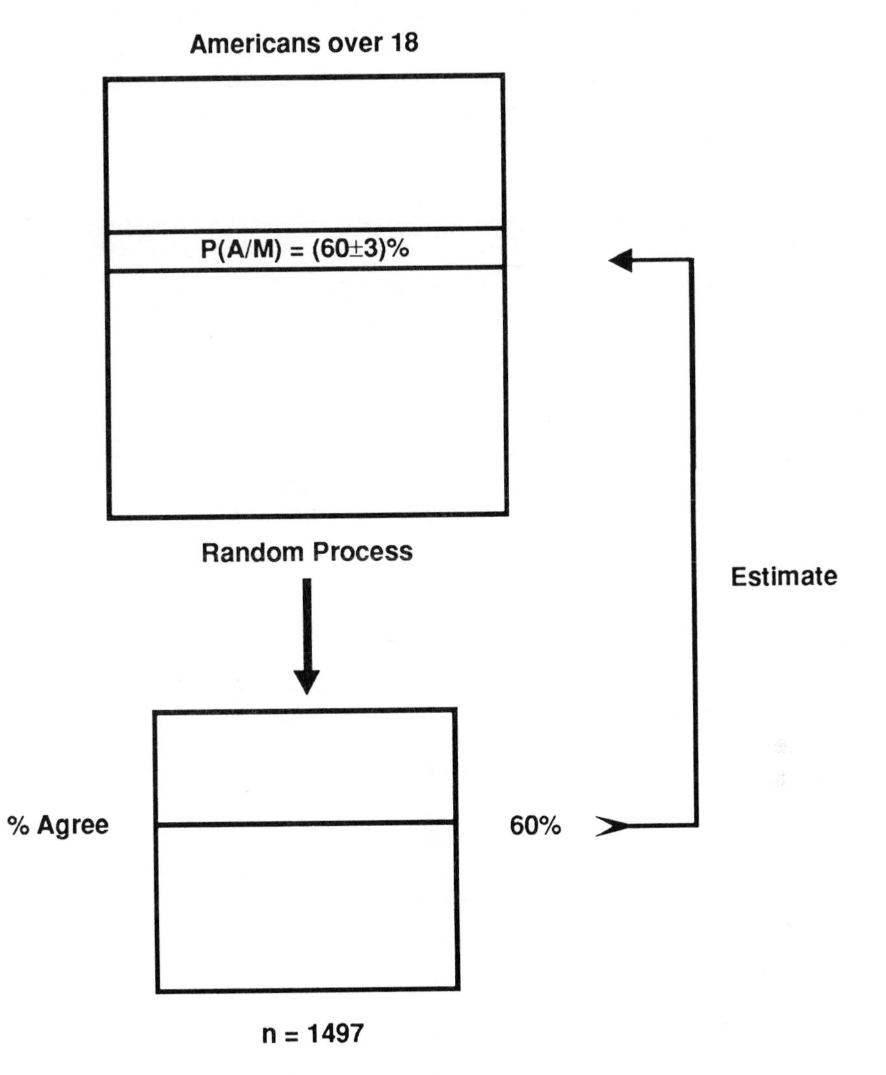

Averaging the percentages for women below and above 44 years old, and then for men below and above 44 years old (and rounding off to whole percentages) yields the result that roughly only 47 percent of all the women sampled, but 72 percent of the men sampled, express agreement with the idea of combining marriage, children, and a career. The corresponding interval estimates are $(47 \pm 3)\%$ for women and $(72 \pm 5)\%$ for men. There is a large gap between these intervals. There is, therefore, good evidence for a *negative correlation* between being a woman and expressing a preference in favor of combining marriage, children, and career. The evaluation of this correlation is shown in Figure 6.7. The estimated strength of this correlation is in the interval (.33, .17).

We can take the evaluation one step further. Within the category of women, 55 percent of women 18 to 44 years of age, but only 40 percent of women 45 or older, express a preference in favor of combining marriage, children, and career. Here the population of interest is adult American women. One variable is age, with values 18 to 44 and 45 or over. The other variable is expressed agreement or disagreement with the statement. The sample sizes

Figure 6.7

A diagram of the *Times* study evaluating the negative correlation between being a woman and favoring combining marriage, children, and a career.

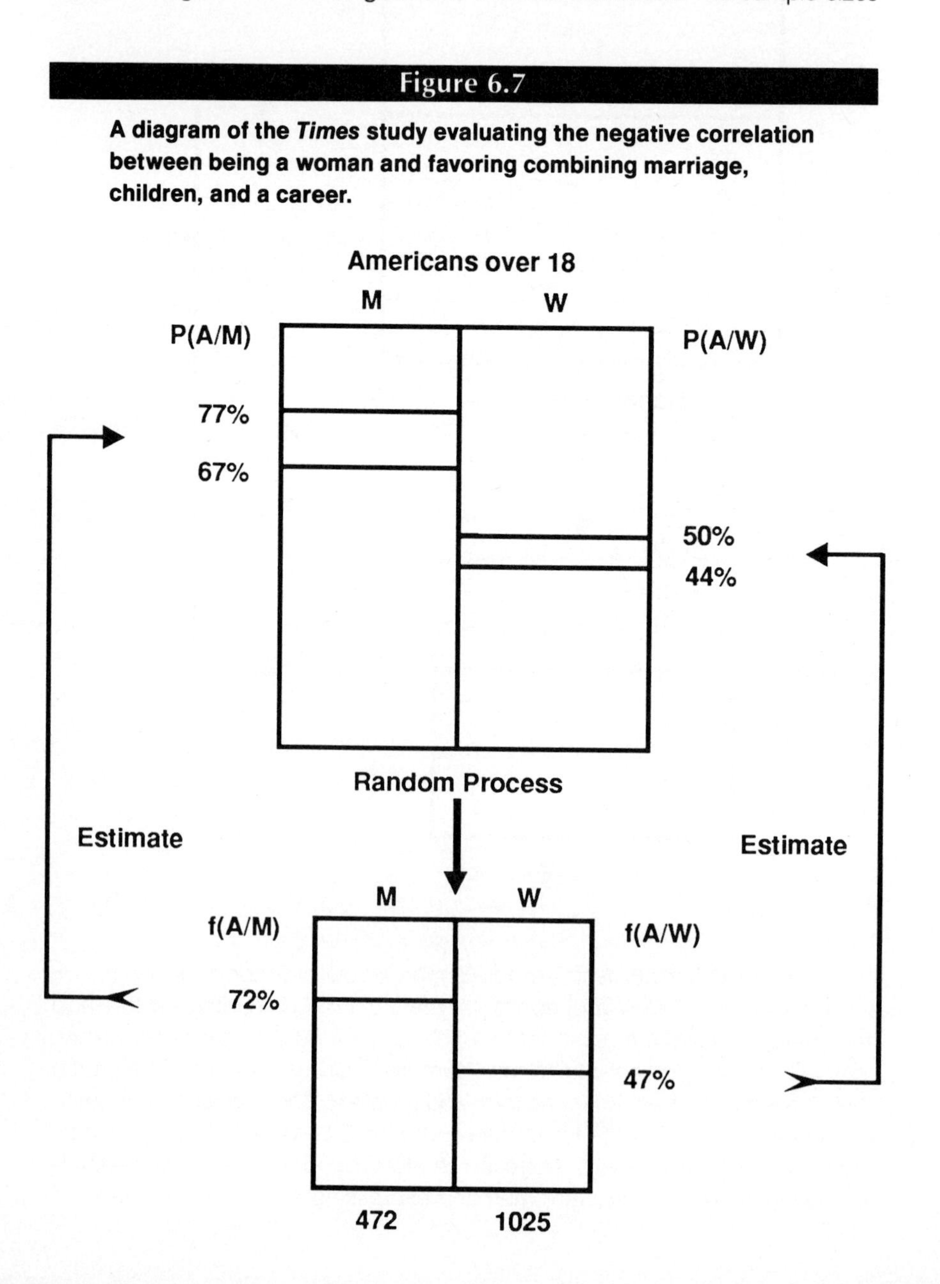

for different age groups are not explicitly given. But, as noted above, there is internal evidence in the report that there were roughly equal numbers in the two age categories represented in the table. That would mean roughly 500 women in each category, which implies a margin of error of .05. So the estimated percentage of women agreeing with the statement in the younger age group is $(55 \pm 5)\%$. The estimated percentage in the older age group is $(40 \pm 5)\%$. These intervals clearly do not overlap. So there is good evidence of a negative correlation among women between being in the older group and expressing a preference in favor of combining marriage, children, and career. The estimated strength of the correlation is in the interval (.25, .05).

Note that even if we are terribly mistaken in thinking there are equal numbers of women in the two age categories, there is still strong evidence for the correlation in the population. If the sample size for one age group were only 100, and therefore 900 for the other, the respective margins of error would be .10 and .03, and the intervals still would not overlap. Or, in other terms, the observed difference in the sample is statistically significant.

In these past few paragraphs we have been evaluating several statistical hypotheses without actually going through the steps in detail. That is because once one becomes familiar with a particular example, some elements of the evaluation remain the same. Also, one learns quickly to spot statistically significant differences, so a detailed evaluation is not necessary. For the purposes of homework exercises and examinations, however, following the steps is necessary to insure sufficient uniformity of answers to allow efficient and fair grading. Therefore, you have to learn the steps.

Work Attitudes

We will perform one more analysis on the data in the *Times* poll. Here again, we will not follow the program in detail, but proceed more informally, as you would if you were just reading this poll for your own information.

Look at the data on work attitudes compiled in Table 6.5. From the comment at the bottom, it is clear that the original sample of 1025 women and 472 men turned up 483 women and 350 men reporting holding full-time jobs. The table compiles the percentages of women and men in these two groups agreeing with eight different statements. So there are eight possible correlations to investigate. This can be done quite efficiently once one determines the respective margins of error.

For a sample of 483 women, a margin of error of .05 is appropriate. For a sample of 350 men, .06 is a reasonable interpolation. That means that in order for differences in agreement between working women and men to be statistically significant, they must differ by at least eleven percentage points.

Looking down the list of questions, there are only three out of the eight questions for which the difference in response is clearly statistically significant. There is, therefore, good evidence for three correlations: a negative correlation between being a woman and reporting about supervising other employees at work; a positive correlation between being a woman and saying that at work most men do not take women seriously; and a positive correlation between being a woman and saying that women have had to give up too much to get ahead. In addition, there are two cases in which there clearly is

Table 6.5

Work Attitudes: A Comparison

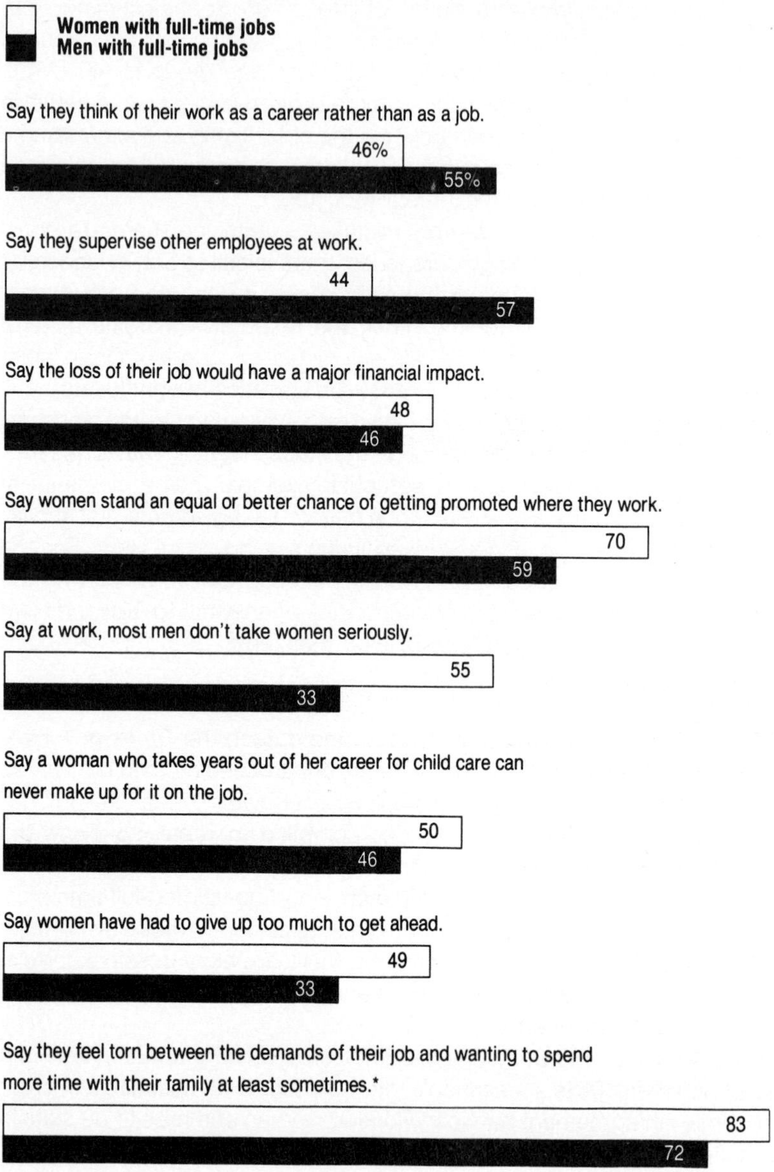

Women with full-time jobs
Men with full-time jobs

	Women with full-time jobs	Men with full-time jobs
Say they think of their work as a career rather than as a job.	46%	55%
Say they supervise other employees at work.	44	57
Say the loss of their job would have a major financial impact.	48	46
Say women stand an equal or better chance of getting promoted where they work.	70	59
Say at work, most men don't take women seriously.	55	33
Say a woman who takes years out of her career for child care can never make up for it on the job.	50	46
Say women have had to give up too much to get ahead.	49	33
Say they feel torn between the demands of their job and wanting to spend more time with their family at least sometimes.*	83	72

* Respondents with children under 18.

Based on interviews with 483 working women and 350 working men conducted by phone June 20–25, 1989.

no good evidence of a correlation, two cases that are borderline, and one that is between borderline and clearly failing to be evidence of a correlation.

This analysis illustrates that, once you get the hang of it, you can evaluate a great deal of statistical information quite easily and efficiently.

6.7 PROBLEMS WITH SURVEY SAMPLING

Interval estimation is a procedure that can be used whenever a randomly selected sample is found to exhibit some property with a definite relative frequency. Survey sampling falls into this category but presents special problems because it is used to assess the attitudes of people in large and diverse populations. These special problems can be grouped into two general categories: problems about the *reliability* of the *data* and problems concerning the *randomness* of the *selection process.* You should learn to identify these problems in reports of surveys.

Unreliable Information

One reason why information obtained in an opinion survey may be unreliable is that people do not always give true answers. The following example is somewhat extreme, but nevertheless illuminating.

In 1972, the staunchly segregationist governor of Alabama, George Wallace, ran in the presidential primaries in a number of northern states, including Indiana, Wisconsin, and Massachusetts. In all of these states, Wallace got a much higher percentage of the vote than indicated by the polls prior to primary day. In Indiana, for example, the polls said Wallace would get about 20 percent of the vote, but on election day he got over 40 percent. The generally accepted explanation for this wide discrepancy is that, because of his segregationist reputation, people in the north were reluctant to admit to an interviewer that they preferred Wallace to the other candidates. So they gave an "acceptable" reply to the interviewer, but voted for Wallace in the privacy of the voting booth.

Even when people are not deliberately lying, they may often be mistaken. Consider, for example, investigating the relationship between smoking and health problems by surveying smokers. This is typically done by asking people about their smoking habits (such as when they started smoking, how much they smoke in a day, and whether they inhale) and about their general health. In such a situation, you may not get very accurate answers. People forget when they began smoking, probably thinking it was later than it actually was. People in poor health might see smoking as a convenient scapegoat and overestimate the amount they smoke. People in good health, but worried about possible bad effects, might well underestimate how much they smoke, not wanting to admit even to themselves that it is not only a pack a day, but well over a pack and a half. Similar answers result from questions about one's state of health. One might simply forget having had a mild case of pneumonia or strep throat. People often do not realize how much they actually cough, and so on. It is because of this type of problem that most serious studies of health issues are not based solely on interviews but employ actual medical examinations.

In reading reports of polls, then, you must be on the alert for ways in which the data may be unreliable or misleading. In particular, be aware that what people *say* to an interviewer may differ from what they really believe, or from what is in fact the case.

Nonrandom Sampling

In survey sampling, especially public-opinion polling, the population of interest is usually a fairly large subgroup of all Americans—for example, all adults of voting age, all registered voters, all women. It would be very difficult and costly to set up a system of sampling that gave every person in the whole population an equal chance of being selected, which is what a truly random sampling system would have to do. Thus, the population of people actually sampled, that is, those who do have an equal chance of being selected, is often only a subpopulation of the original population of interest. So there is always the danger that the subgroup actually sampled is not sufficiently representative of the population of interest. If it is not, then no matter how large the sample, the conclusion will not be correct for the original population of interest.

The classic example of such a mistake was the political poll conducted by the *Literary Digest* in 1936. Ten million sample ballots were mailed to people randomly selected from telephone directories around the country. More than 2 million were returned and showed a clear majority for Republican Alfred M. Landon over Democrat Franklin Roosevelt. In fact, Roosevelt got 60 percent of the vote on election day. Stop and try to guess the mistake before reading any further.

The mistake was that in 1936 the country was in the middle of the Great Depression; people with telephones tended to be people in the upper economic brackets. People in the lower brackets, especially people out of work, tended not to have telephones. However, these poorer people were strongly for Roosevelt. So because of strong correlations involving telephone service, economic status, and party preference, the poll showed a majority for Landon when in fact the majority of voters preferred Roosevelt. Largely because of this fiasco, the *Literary Digest* ceased publication in 1937.

The moral of this story is that you should pay particular attention to the method of sampling used—if you are fortunate enough to be given this information. Try for yourself to think of ways the sampled population might differ from the population of interest, and then consider whether these differences might be relevant to the conclusions reached. If you are not told anything about the sampling method, you should retain a healthy degree of skepticism about the results.

Of the various methods of conducting surveys, sending questionnaires by mail is particularly bad because the randomness of the sampling process is strongly influenced by the *return rate*—the percentage of questionnaires returned. Usually, this rate is fairly low simply because most people are not willing to take the time to fill out a questionnaire and send it back. Note that the response rate in the *Literary Digest* poll was only 20 percent. Moreover, those who are willing to take the time tend to have strong opinions on the questions at issue. Finally, if the sponsor of the poll is known to support a particular view, the people *agreeing* with this view are generally more likely to respond.

To see how this works, consider the example of a former congresswoman who sent 200,000 questionnaires to her constituents asking their opinions on the issue of school busing for purposes of integration. The congresswoman was widely known for her opposition to busing. Her poll showed 80 percent of the respondents also against busing. She claimed that this figure was representative of her district. It turns out, however, that only about 23,000, or roughly 12 percent, of the questionnaires were returned. What about the other 177,000 people? We do not know. Perhaps the 18,400 people who sent back supportive responses constituted the majority of her hard-core support. In any case, because supporters are more likely to respond than opponents, her support among the remaining 177,000 was almost certainly less than among the 23,000 respondents. Whatever the actual facts, there was little justification for her claim of 80 percent support throughout her district.

The moral of this story is always to look for the *return rate* when you read the results of a poll done by handing out questionnaires by mail or otherwise. If the return rate is low, ask yourself whether there is likely to be any strong correlation between having the motivation to return the questionnaire and answering the questions one way or the other. If so, the percentages in the returned questionnaires are bound to misrepresent the whole population. Even if you cannot think of any obvious correlations between answers and motivation to respond, there probably are some. You should remain somewhat skeptical.

In fairness to the major polling organizations, it must be said that they devote much effort to getting random samples, and generally use personal or telephone interviews to get their information. As in the *Times* poll reviewed above, the typical sample in standard polls is between 1200 and 1500. Moreover, professional polling organizations generally *stratify* the whole population into a number of groups known from previous polls to be relatively uniform. The groupings are done by economic status, occupation, place of residence, region of country, race, sex, age, and so forth. Steps are then taken to make sure that the sample reflects the correct percentages of these groups. For example, if half of all adults live in cities with populations over 100,000, then half the sample is selected from this subpopulation. When such care is taken, one can be reasonably confident that something close to the theoretical margin of error will be realized.

6.8 A FINAL EXAMPLE

The following is a lightly edited version of a type of short report that appears frequently in newspapers and magazines. An evaluation of this report will serve as a summary to our study of statistical hypotheses.

Eight Percent Say They Were Abused in Childhood

(A) Fifteen percent of American adults claim to know children they suspect have been physically or sexually abused, and 8 percent say they themselves were abused as children.

Knowledge of victims is no higher now than it was in 1981, but awareness of the scope of the problem recently has been enhanced by highly publicized cases, such as the beating death of 6-year-old Lisa Steinberg by her adoptive father in New York City. The weight of public opinion, in fact, attributes the larger number of reported child-abuse cases to greater public awareness of the problem than to an increase in actual victimization.

The National Committee for Prevention of Child Abuse said in its annual report that child-abuse deaths in the fifty states rose 5 percent in 1988 from the year before. It estimated, based on data from forty-one states and the District of Columbia, that there were 3 percent more child-abuse reports filed last year than in 1987.

More women than men claim to have been abused as children. Reported victimization also is higher among 30- to 49-year-olds than among younger or older adults.

(B) Social and economic factors, such as education and income, apparently have little bearing on child abuse, with adults from widely varying walks of life reporting about the same incidence of childhood maltreatment.

(C) Knowledge of child-abuse cases and personal victimization are strongly interrelated. Four in ten of those who say they were child-abuse victims claim to know children who have been abused. In sharp contrast, only one in ten of those who were not abused are aware of children they suspect are victims.

These findings are based on nationwide telephone interviews with 1000 adults 18 and over, conducted by the Gallup organization between February 28 and March 2, 1989. In addition to sampling error, question wording and practical difficulties in conducting surveys can introduce error or bias into the findings of opinion polls.

We will focus on the information presented in the paragraphs labeled (A), (B), and (C). Rather than performing three separate evaluations, which would be repetitious, we will follow the strategy of evaluating all three pieces of information at once. This requires only one pass through the entire program.

Step 1. The Real World Population. The population of interest is clearly described as "adults 18 and over" in the United States.

Step 2. The Sample Data. (A) Fifteen percent of the sample of 1000 claim to know children they suspect have been physically or sexually abused. Eight percent say they themselves were abused as children. (B) Sample members with different levels of income and education report about the same incidence of childhood mistreatment. (C) Four in ten of those who say they were child-abuse victims claim to know children who have been abused. Only one in ten of those who were not abused are aware of children they suspect are victims.

Step 3. The Statistical Model. (A) Here we have data in the form of two separate *proportions*. In the first, the single *variable* is knowledge of children they suspect to have been physically or sexually abused. The two *values* of this variable are "have such knowledge" and "do not have such knowledge." For the second proportion, the *variable* is "report being themselves abused as children." The two *values* of this variable are "report abuse" and "do not report abuse." (B) The data here have the form of *correlations.* One variable is level of education, with implied values being "high" and "low." The potentially correlated variable is report of childhood maltreatment, with values "yes" and "no." There is also the possibility of a correlation between the reporting variable and level of income. (C) Here again, the data have the form of a *correlation.* One variable is abuse as a child, with values "yes" and "no." The potentially correlated variable is awareness of children suspected to be victims, with values "have such awareness" and "do not have such awareness."

Step 4. Random Sampling. The fit with a random sampling model must be fairly good simply because this poll was done by the Gallup organization. It is, however, subject to the worries that attend any telephone poll.

Step 5. Evaluating the Hypothesis. (A) For **n** = 1000, **ME** = ±.03. The estimate is that between 12 and 18 percent of all American adults would claim to know children they suspect have been physically or sexually abused. Similarly, the estimate is that between 5 and 11 percent of all American adults would admit that they themselves were abused as children.

(B) The sample apparently shows no statistically significant difference in percentages of subjects reporting abuse as children among subjects with a high, as opposed to a low, level of education. There is therefore no evidence of a correlation between educational level and reported childhood abuse. There is no indication of the number of subjects with high or low educational levels. If we were to assume 500 each, the margin of error would be .05, and the strength of any correlation between educational level and reported abuse would be estimated as less than .10. The evaluation for income level, and any other reported social and economic factors would be similar.

(C) From our evaluation of the data in (A), we know that roughly 10 percent of all adults report having been abused as a child. So in a sample of 1000, there would be roughly 100 reporting abuse and 900 not so reporting. The data are that 40 percent of those who say they were child-abuse victims claim to know of children who have been abused. With **ME** = .10, we estimate that between 30 percent and 50 percent of all adults who say they were child-abuse victims would also claim to know of children who have been abused. The data are that only 10 percent of those who were not abused are aware of children they suspect are victims, for **n** = 900, **ME** = .03. So we estimate that between 7 percent and 13 percent of all adults who were not abused would be aware of children they suspect are victims. There is a wide gap between these two intervals. So there is good evidence of a positive correlation between having been abused as a child and claiming to know children who have been abused. We estimate the strength of the correlation as being between .43 and .17.

Step 6. Summary. We have determined that the actual sampling process pretty well followed a random sampling model. There is, therefore, quite good evidence for the proportions and correlations, or lack thereof, stated in Step 5.

EXERCISES

Analyze these reports following the six-point program for evaluating statistical hypotheses developed in the text. Number and label your steps. Be as clear and concise as you can, keeping in mind that you must say enough to demonstrate that you do know what you are talking about. A simple "yes" or "no" is never a sufficient answer. Many of these reports are taken directly from recent magazine or newspaper articles and are presented here with only minor editing.

Exercise 6.1

Study: People Tell Lies to Have Sex

Asking about a sexual partner's health or sexual history is not a reliable means of avoiding AIDS, because many people lie to have sex, a researcher says. One-third of sexually experienced male college students and 10 percent of women admitted in a study that they had lied to get someone to have sex with them, the researcher said.

"Asking partners about risk factors is probably not a very good strategy for reducing the risk of AIDS," said Susan Cochran, associate professor in the Psychology Department of California State University at Northridge. She described results of her survey at the annual meeting of the American Psychological Association.

Cochran said she began the study after noticing that many women were reassuring themselves about having sex by asking sex partners about past behavior that carries a risk of AIDS infection. The virus that causes acquired immune deficiency syndrome is spread by having sex with an infected person or using an infected needle or contaminated blood products. Questioning partners about their drug use and sexual history is not an effective substitute for taking precautions, such as using condoms, she said.

The study focused on 422 unmarried, sexually experienced students at two colleges in southern California. Ages ranged from 18 to 25 with an average of 19. Forty-six percent were men.

Thirty-five percent of men admitted they had lied to have sex. In addition, 47 percent of men and 60 percent of women reported that somebody had lied to them in order to have sex. The study did not specify what the respondents had lied about.

"If you ask somebody if they have risk factors and they say yes, you can be reasonably assured they're not lying," Cochran said. "But if they say no, you're not really any further along than not asking."

Exercise 6.2

Study Disputes View of Dyslexia, Finding Girls as Afflicted as Boys

Despite a widespread belief to the contrary, girls are just as likely as boys to have the reading impairment dyslexia, three studies say.

But teachers are far more likely to diagnose the condition in boys, researchers said, to the point that too many boys are in special education classes and too many girls with reading problems are struggling in silence.

Dyslexia afflicts 10 to 15 percent of the population.

One study is being published today in the Journal of the American Medical Association. *Another is being prepared for publication, and the third has been published by Dr. Sally Shaywitz, the co-director of the Center for the Study of Learning and Attention Disorders at Yale University and other researchers. Their study involved 445 Connecticut children who were studied from the time they entered kindergarten until they reached third grade. The researchers found that the schools identified more than four times as many second-grade boys than girls as being dyslexic, and more than twice as many third-grade boys than girls were said to have reading problems.*

When the researchers independently tested the students, however, they found equal numbers of boys and girls in both grades with reading difficulties. Also, they found fewer than half the children referred for reading problems actually had them. . . .

Most research on dyslexia has . . . worked with children identified by school systems as having dyslexia, although the new study did not, and dealt with an unbiased group of subjects. Most school systems rely on the recommendations of teachers before testing students for the impairment, rather than testing all students for it. . . .

Dr. Shaywitz said she hoped the findings would lead to a fundamental change in the way reading disabilities were viewed. "Boys are really being penalized by being mislabeled, and girls, who really need help, are failing to be noticed," she said.

She and others said they suspect the frequent diagnosis of dyslexia in boys and the failure to diagnose it in girls is caused partly by teachers' and parents' expectations that boys, not girls, will have problems. "It becomes a self-fulfilling prophecy," Dr. Shaywitz said.

Another factor may be the exuberant behavior of many boys, while many girls are calmer and quieter. Teachers pay more attention to noisy boys and are more likely to refer them for help. "Girls, who may have problems with achievement but are sitting with their hands folded nicely may be overlooked," Dr. Shaywitz said.

Exercise 6.3

Poll Shows Men Sinking in the Opinion of Women

An increasing number of American women believe that most men are mean, manipulative, oversexed, self-centered and lazy, according to a survey, and they are getting annoyed.

A poll by the Roper organization found growing numbers of women expressing sensitivity to sexism and unhappiness with men on many issues. It compared data from identical questions asked twenty years ago.

Some of the changes were sizable. In 1970, two-thirds of women agreed that "most men are basically kind, gentle, and thoughtful." In the new poll, only half of the 3000 women who were surveyed agreed.

The reason? "Women's growing dissatisfaction with men is undoubtedly derived from their own rising expectations," the survey's authors said. "The more independent women of today expect more from men." Those expectations apparently are going unfulfilled. Most women rated men negatively on their egos, libidos, and domesticity. Sizable minorities went further: 42 percent called men "basically selfish and self-centered."

Prurience took a particular pounding. Fifty-four percent of the women who were surveyed agreed that "most men look at a woman and immediately think how it would be to go to bed with her." In 1970, 41 percent had agreed.

The survey was conducted by in-person interviews with a random sample of women across the country. It had a margin of error of plus or minus two percentage points.

Respondents overwhelmingly said women have made progress in obtaining job opportunities, equitable salaries, and acceptance as political leaders, but they also expressed greater awareness of continuing discrimination.

Moreover, six in ten working women said juggling jobs and families put them under "a lot of stress," and nearly as many felt guilty about time they spend at work and away from their families.

The second-greatest cause of resentment in survey respondents' lives was their mates' failure to help with household chores, cited by 52 percent. Only money was a greater cause of problems.

"As women contribute more to family income, they expect a more equitable division of the household responsibilities in return," the survey said. While many men acknowledge responsibility for household work, it added, "Women indicate that men are failing to live up to this ideal, and that their failure is a major source of irritation."

But diapers and dishes are by no means the sole problem: Many women in the survey expressed negative views of men's attitudes, as well as their behavior. Among the findings:

- *Fifty-eight percent of women agreed that "most men think only their own opinions about the world are important." That was up from 50 percent in 1970.*
- *Given the statement, "Most men find it necessary for their egos to keep women down," 55 percent agreed, up from 49 percent 20 years ago.*
- *Fifty-three percent agreed that "most men are interested in their work and life outside the home and don't pay much attention to things going on at home." That was up from 39 percent in 1970.*

Exercise 6.4

Heavy TV Viewer Sees Scary World

Violence on television leads viewers to perceive the world as more dangerous than it really is. That is the conclusion of a recent four-year study of the effects of television on adults. A large number of adults—men and women, old and young, well educated and not—were interviewed. Subjects were classified as "light" viewers if they watched two hours or less of TV per day and as "heavy" viewers if they watched four or more hours per day.

When asked, "Can most people be trusted?" the heavy viewers were 35 percent more likely to check "Can't be too careful." (Forty-eight percent of light viewers and 65 percent of heavy viewers checked this response.)

When viewers were asked to estimate their own chances of being involved in some type of violence during any given week, they provided further evidence that television can induce fear. The heavy viewers were 33 percent more likely than the light viewers to pick such fearful estimates as 50-50, or 1 in 10, instead of a more plausible 1 in 100. (Thirty-nine percent of light viewers and 52 percent of heavy viewers picked the higher estimates.)

Exercise 6.5

Study Links Heavy Drinking and High Blood Pressure

Regular consumption of three or more alcoholic drinks each day has been linked to the development of high blood pressure, according to results of a large study in California. This link was reported by epidemiologists at the Kaiser-Permanente Medical Center in Oakland on the basis of their statistical analysis of responses to health-checkup questionnaires and the medical records of 83,947 men and women of three races.

The findings showed a solid statistical association between consumption of alcohol and high blood pressure. The study found that the blood pressures of men taking two or fewer drinks each day was similar to those of nondrinkers. Women who took two or fewer drinks each day had slightly lower blood pressures. Men and women who took three or more drinks each day had higher blood pressures.

Exercise 6.6

Poll Finds More Liberal Beliefs on Marriage and Sex Roles

Americans are more likely to believe that marriages in which the partners share the tasks of breadwinner and homemaker are a more "satisfying way of life" than they are to prefer the traditional marriage in which the husband is exclusively a provider and the wife exclusively a homemaker and mother.

Of those interviewed, 53 percent said they preferred the idea of shared-marriage roles and 47 percent said they preferred the "traditional" marriage. But among those under 40 years of age, 61 percent preferred shared-marriage roles, while 59 percent of those over 40 preferred the traditional marriage.

The survey also detected sharp differences of opinion as to whether working women make better mothers than women who do not work. Of women who work, 43 percent said working women make better mothers, whereas only 14 percent of women who do not work shared this opinion.

The survey was conducted by telephone interviews with 1603 adult Americans from all parts of the nation and representing different races, religions, ages, and occupations.

Exercise 6.7

Homicide Is Top Cause of Death From On-Job Injury for Women

The leading killer of American women in the workplace is homicide. The Centers for Disease Control reported today that 42 percent of the deaths of American women resulting from on-the-job injury from 1980 to 1985 were homicides.

The second most-frequent cause of death was vehicle accidents, at about 39 percent, said Catherine Bell, an epidemiologist with the centers' National Institute for Occupational Safety and Health. "If a woman's going to die from an injury at work, she's probably going to be murdered," Ms. Bell said.

Nationwide, for both sexes, about 13 percent of on-the-job injury deaths in the six years studied were homicides, the centers reported. Homicides accounted for 12 percent of men's occupational fatalities, as against the 42 percent for women. Findings regarding killings of men in the workplace are not complete. But most on-the-job deaths involve male workers, who traditionally dominate more-dangerous occupations, and most of those deaths are accidental.

Investigators for the safety institute identified 950 such killings over the six years of the study. Of that number, 389 of the women—or 41 percent—worked in the retail field. Researchers are not certain, but it appears that strangers in the store—not co-workers—are responsible for most workplace killings of women, Ms. Bell said. "They appear to be robbery-associated crimes in the retail trade," she said. "It quite likely is stranger violence."

The killings peaked in the winter. Sixty-four percent of the killings were shootings; 19 percent were stabbings.

The workplace homicide rate for black women was nearly double that for white women, but that was not necessarily unexpected. Nationwide, black women are nearly four times as likely to be murdered as are white women, the centers said, citing a 1986 study.

Exercise 6.8

What Men Are Really Like

The typical American man believes that a faithful marriage is the ideal relationship. He wants sincerity, affection, and companionship in women above everything else. That is the conclusion of the recently published book Beyond the Male Myth.

The book is based on the results of a 40-item multiple-choice questionnaire designed to answer the questions women most wanted to know about men. The questionnaire was distributed by a public-opinion polling organization to men in communities around the country. The men, who were approached in such places as shopping malls and office buildings, were asked to fill out the questionnaire and to deposit it into a sealed box. Four thousand sixty-six men responded.

More than half of the men surveyed regarded a faithful marriage as their ideal. And almost three-quarters of the men said that, besides love, their primary reason for getting married is the desire for companionship and a home life.

Exercise 6.9

Aggression in Men: Hormone Levels Are a Key

When men are domineering and intensely competitive, it may be just another case of raging hormones, specifically the male sex hormone testosterone, researchers say.

Scientists have long linked abnormally high levels of testosterone to an unusually early involvement with sex and drug use and to certain violent crimes, such as particularly vicious rapes. But now a series of new findings shows that high testosterone levels play a role in the normal urge for the upper hand.

Those men who are most likely to try to dominate in a social situation, be it in a prison yard or a boardroom, are likely to have higher testosterone levels than their peers, new data show.

Scientists caution against placing too much stock in the importance of hormones like testosterone in human affairs, since so many other factors, from childhood experiences to social status, shape the expression of a given behavior, like dominance or competitiveness.

"This doesn't mean that people can't be dominant or aggressive without high testosterone," said Dr. Robert Rose, a psychiatrist at the University of Minnesota Medical School. "A woman, of course, can be as competitive as a man, even though her testosterone levels are much lower. . . ."

The newest and strongest evidence about the effect of testosterone comes from a study of 1706 men in the Boston area, aged 39 to 70. The men were selected at random as part of a larger study on aging. Those who agreed to participate were interviewed in their homes and given psychological tests, and blood samples were taken.

Those who had higher levels of testosterone and related hormones had a personality profile that researchers described as "dominant with some aggressive behavior."

"The picture we get is of a man who attempts to influence and control other people, who expresses his opinions forcefully and his anger freely, and who dominates social interactions," said John B. McKinlay, a psychologist at the New England Research Institute in Watertown, Massachusetts, who was on the research team.

The study, which will be published in Psychosomatic Medicine, *is considered significant because it is the first to examine so many randomly chosen men. Most studies of testosterone in humans have been on small numbers in select populations, like prison inmates. . . .*

Exercise 6.10

Tying Abortion to the Death Penalty

Do women who have had an abortion have less regard for human life than those who haven't? Some psychologists think the answer may be "yes," based on a survey conducted on attitudes toward capital punishment.

Shortly after a highly publicized execution in Utah, researchers asked more than 300 residents of Pasadena, California, whether they were totally against the death penalty, favored it for heinous crimes, or thought it should be used more frequently.

Overall, about 75 percent of the respondents in this conservative community favored capital punishment in some form. About half as many men as women were opposed to the death penalty under any circumstances.

Among women who had undergone abortion, only 6 percent were unqualifiedly against the death penalty; 11 percent thought it should be used more frequently; and the rest favored it for some crimes. Twenty-six percent of the women who had not had an abortion were against execution under any circumstances; 20 percent favored more frequent use; and the rest wanted it for heinous crimes.

Exercise 6.11

AIDS Plague Hits Teen Heterosexuals

The AIDS epidemic clearly has reached teen-age heterosexuals, growing evidence shows. . . . The most ominous signs are in New York City, where the epidemic is most advanced, says Dr. Karen Hein, director of a program that serves and studies teens with AIDS.

"The very heterosexual epidemic everyone says has not occurred in the adult population has occurred among teen-age

girls in New York City," says Hein, who . . . pointed to danger signs picked up by several recent New York state and city studies:

One in 100 19-year-old women giving birth statewide is infected with the AIDS virus, as are 1 in 1000 15-year-olds.

Half of the city's teen-age girls with AIDS say they got infected by having sex with males.

State researchers found about 6 percent of homeless boys and girls at one city shelter were infected. . . .

However, teen-age mothers, runaways, and clinic patients are not representative of all teens, says Lloyd Novick, of the New York State Health Department, Albany. Also, he says, studies of new mothers show teens are less likely to be infected than older mothers—meaning there is "substantial room for prevention," especially among young teens.

Hein adds that teens who do not have sex or share needles may worry too much about AIDS. But, she says, studies show that more than half of all teens do have sex by age 19. . . . Teens who are considering sex "need to learn you can't look into the eyes of someone you love and tell if they have the virus."

Exercise 6.12

Scientists Seek Secrets of Happiness

Happiness, a favored subject of philosophers, writers, and poets since the beginning of history, has only recently come under scientific scrutiny. The results so far suggest that many traditional beliefs about happiness are incorrect.

The results of many recent studies are contained in the book, Happy People, *by Jonathon Freedman, professor of psychology at Columbia University. In one study, Professor Freedman and associates published questionnaires in* Psychology Today *and* Good Housekeeping. *The questionnaires were returned by more than 100,000 people around the country. Among the conclusions were:*

1. *Sixty-five percent of all Americans are either moderately or very happy.*

2. *The percentage of happy people is greater among those who are married than those who are single.*

3. *The percentage of happy people does not differ among those with strong religious feelings from those without such feelings.*

4. *Considering just those people with high incomes, there are fewer happy people among the college graduates than among those who did not go to college.*

In the book, Professor Freedman suggests that some people may have a genetic disposition toward being happy—a disposition that may be connected with the presence of certain recently discovered chemicals in the brain.

Exercise 6.13

Sixty-Four Percent of Americans Are Regular Drinkers

Sixty-four percent of Americans acknowledge some level of alcoholic consumption. That is the overall conclusion of a recent survey involving 500 in-home interviews in twenty cities around the country. Asked to characterize their own drinking habits, 35 percent said they were nondrinkers, 46 percent are light drinkers, 17 percent are moderate drinkers, and only 1 percent are heavy drinkers. One percent refused to answer the question.

Among those with a college education, 75 percent classified themselves as either light or moderate drinkers, but only 49 percent of those with a high school education gave these responses. Fifty-seven percent of blue-collar workers gave the light or moderate response. The corresponding percentage for white-collar workers was 71 percent.

There were also variations by age and income. Sixty-two percent of those under 35 admitted to being light or moderate drinkers, compared with 69 percent for those 35 to 54 and 54 percent of those 55 and over.

Variation by income was similar, with 59 percent of those earning less than $15,000 a year classifying themselves as light or moderate drinkers. In the $15,000 to $25,000 category, the percentage was 64 percent and rose to 76 percent for those earning more than $25,000.

Exercise 6.14

Majority of Women Perceive Job Bias

Fifty-four percent of women responding to a recent Gallup poll said they do not think that women have equal employment opportunities with men. Forty-one percent thought they did have equal opportunities, while 5 percent had no opinion.

Perception of bias depends strongly on whether one has actually been in the labor force. Among those women who have been working, 65 percent thought women's opportunities were not equal to men's but this opinion was shared by only 32 percent of women who had never been formally employed.

Education is also a factor. Among college-educated women, 68 percent thought there is bias against women, whereas only

49 percent of those with a high school education or less agreed with this view.

When asked if a woman of equal ability has an equal chance of becoming chief executive officer of a company, 71 percent of the college graduates said she did not. That opinion was shared by 50 percent of women with less education, and by 56 percent of all women in the survey.

The survey, conducted by interview, has a margin of error of plus or minus 5 percent.

Exercise 6.15

Project

Find a report of a statistical study. You may find an example in a newspaper or newsmagazine. The Sunday supplement to your local newspaper is a good bet. When you have found something that you find interesting and substantial enough to work on, analyze the report following the standard program for statistical hypotheses.

CHAPTER 7

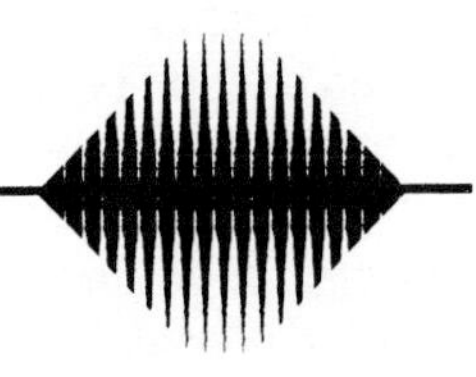

Causal Models

Many interesting and useful scientific findings concern *causal* relationships. Does a high level of cholesterol cause heart attacks? Does taking vitamin C prevent colds? We will begin our study of causal relationships by distinguishing *causation* from *correlation.* Then we will develop several models of causation that will help us to understand and evaluate *causal hypotheses*.

7.1 CORRELATION AND CAUSATION

One of the most common mistakes in statistical reasoning is inferring the existence of a causal connection from a known correlation. This is true in spite of the fact that standard textbooks are full of examples illustrating the dangers of such reasoning. A few of these examples are worth repeating.

Correlation Without Causation

Comparing lung cancer patients with people without lung cancer, one finds that a history of using ashtrays is positively correlated with having lung cancer. That is, there is a much higher percentage of ashtray users among the lung cancer patients than among others. Clearly, using ashtrays does not cause lung cancer. Rather, cigarette smoking causes lung cancer, most lung cancer patients are former cigarette smokers, and most people who smoke use ashtrays.

Again, among small children there is a positive correlation between exhibiting red spots associated with measles and having a fever. But the spots do not cause the fever. Nor does the fever cause the spots. Both are produced by the measles virus.

More interestingly, anthropologists studying a tribe in the South Seas found the natives believing that body lice promote good health. It turns out that this was not just superstition. Almost every healthy person had some lice, but many sick people did not. So the incidence of lice among healthy people was clearly greater than the incidence of lice among sick people. Thus, there was a clear, positive correlation between having lice and being healthy.

The reason for the correlation, however, was not that having lice made you healthy. It was that being sick caused you not to have lice. Lice are not stupid. They prefer healthy bodies to sick ones, particularly to feverish ones. When a person's temperature gets much higher than normal, the lice start looking for cooler surroundings.

In a more serious vein, researchers at a hospital connected with a major state university began comparing the recovery rates of patients. They discovered, among other things, that among all patients treated in the university hospital, those patients living within 50 miles of the university hospital had a much higher recovery rate than patients from farther away. That is, among the university hospital's patients, there was a positive correlation between recovering and living within 50 miles of the hospital. There were many different hypotheses suggested in an attempt to explain why living near the university hospital causally promoted recovery.

The correct explanation for the correlation was that the university hospital had the best facilities in the region. Thus, seriously ill patients from all over the state were brought to this hospital. Less seriously ill patients, of course, were not. They were treated at their local hospitals. So, of course, the recovery rate at the university hospital for patients living farther away was lower. These patients were, on the average, much more seriously ill to start with.

Causation Is Not Symmetric

These examples are instructive, but they fail to reveal the underlying nature of the difference between correlations and causal connections. A more enlightening difference is revealed by recalling that correlation is a *symmetrical* relationship. If A is positively correlated with B, then B will be positively correlated with A, and vice versa.

Now whatever else may be true of causation, it is certainly not a symmetrical relationship. Speeding causes accidents, but obviously accidents do not cause speeding. Taking poison may cause death, but death certainly cannot cause the taking of poison. In general, if being an A causes you to be a B, it does not follow that being a B would cause you to be an A, although there are some cases of mutual causation. So causation and positive correlation are fundamentally different kinds of relationships.

What is the source of the asymmetry of most causal relationships? Here philosophers and statisticians disagree. Some emphasize *temporal* order. Causes, it is widely believed, operate only forward in time. You cannot cause something to happen in the past. Temporal order, however, cannot be the sole source of the asymmetry of causation. After all, using ashtrays typically precedes the onset of lung cancer.

Could one generate the required asymmetry using temporal order plus a more complex pattern of correlations involving other variables besides the two in the initial, simple correlation? Some philosophers and statisticians think so, but we shall not pursue this idea here.

Causal Production

In our discussion of causal hypotheses we shall employ a primitive notion of *causal production*. Causes, we shall say, *produce* their effects, or at least contribute to their production. The asymmetry of causation is part of the asymmetry of production. Exercise produces fatigue, but fatigue does not produce exercise. Production operates forward in time.

Talk about causal production does not go very far toward explaining what causation is, but it does at least give us a convenient way of talking about causal relationships. In the end, the *evaluation* of causal hypotheses will not depend very much on deep, metaphysical issues about the nature of causation. So we may as well use a simple, intuitive notion of causal production, even if some find it metaphysically suspect.

7.2 CAUSAL MODELS FOR INDIVIDUALS

We will begin by developing several simple causal models for individuals that we will later use to represent such diverse individuals as people and laboratory rats. These models are only a few of many similar, but more complex, models one could develop. Nevertheless, they are applicable to a wide variety of actual cases one finds reported in the popular media.

A Deterministic Model

For purposes of the model, an individual will be characterized by a *set of variables.* For the most part we will consider only simple qualitative variables such as "is red" or "is not red," which we will use to represent such characteristics as "smokes cigarettes" or "does not smoke cigarettes." Among the variables characterizing an individual, we will pick out one that will represent a single characteristic that is under consideration as being a *causal factor* related to another single characteristic, a possible *effect.* We will designate the causal variable by the bold-faced letter, **C**, which has possible values C and Not-C. Similarly, we will designate the possible effect by the bold-faced letter **E**, which has possible values E and Not-E. The rest of the variables $\mathbf{S}_1$, $\mathbf{S}_2$, $\mathbf{S}_3$, . . . , $\mathbf{S}_N$, characterize the *residual state* of the system. We will use the bold-faced letter, **S**, to designate the variable representing the 2^N possible residual states including such particular states as S_1, Not-S_2, S_3, . . . , Not-S_N, which we will call simply S. Figure 7.1 pictures this model of an individual causal system.

Using this simple model we can now give a more precise characterization of the notion of a *causal factor.*

> C is a positive causal factor (deterministic) for E in an individual I, characterized by residual state S, if C produces E, and Not-C produces Not-E.

Figure 7.1

A model for causation in an individual.

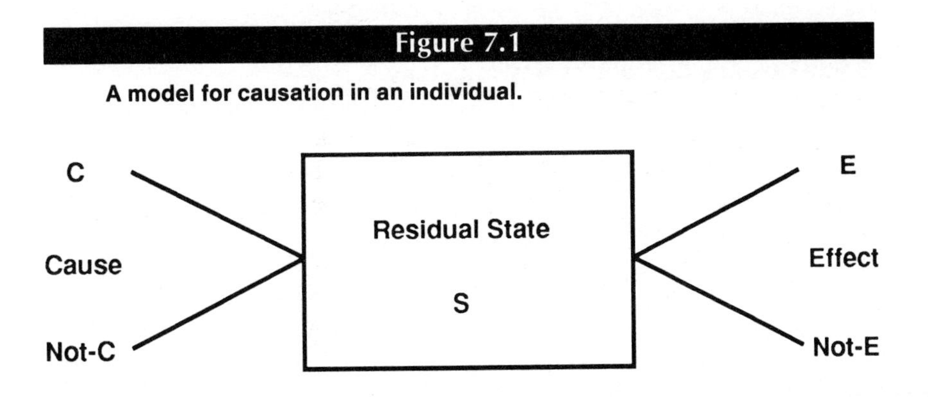

> C is a negative causal factor (deterministic) for E in an individual I, characterized by residual state S, if C produces Not-E, and Not-C produces E.

If C is either a positive or negative causal factor for E in I with S, then we say that the variables **C** and **E** are *causally related.* If C is neither a positive nor a negative causal factor for E in I given S, then we say that the variables **C** and **E** are *causally irrelevant* in I given S.

The reason this is called a *deterministic* model is that, given the residual state S, the presence or absence of C completely determines the presence or absence of E in the individual I. Or, to put it the other way around, the presence or absence of E is completely determined by the presence or absence of C.

Several implicit features of this model are worth making explicit. First, to say that C is a positive causal factor for E in I is to say much more than that both C and E are present in I. Being a positive causal factor implies that if C were not present, then E would also not be present. But the hypothetical situation in which C is not present in I may never exist. Once C is present it may remain for the whole time I exists.

Second, the specification of the *residual state* is crucial. Another individual characterized by the same total set of variables might exhibit a different specific residual state. For example, its residual state might include Not-S_1, rather than S_1. For such an individual, C might not be a positive causal factor for E. It could be causally irrelevant, or even negative. This implies that some of the variables making up the residual state of I may themselves be causally related to E.

To relieve the sense of abstraction involved in building up a causal model, let us apply the model to a real case that will be discussed in the following chapter. There it will be claimed that ingesting saccharin is a positive causal factor for bladder cancer in laboratory mice. Consider an individual laboratory mouse, Izzy. Among the residual states of Izzy are that he is male, is white, is healthy, and so on. To say that saccharin is a positive causal factor for bladder cancer in Izzy, given his residual state, is to say that if he is given saccharin he will get bladder cancer, and if he is not given saccharin he will not get bladder cancer. As far as bladder cancer is concerned, for Izzy, given his residual state, saccharin makes all the difference.

Could Humans Be Deterministic Systems?

There are many people who would question whether such a simple, deterministic model could possibly be used to represent such a complex system as a human body. Before proceeding we should pause to consider this point of view. The fact is that most people intuitively apply such a model and much of modern medical research presumes that such a model is applicable.

Imagine two people as much alike as possible—say, identical twins. Suppose that both come down with the same fairly serious disease. Not having any other good options, the doctor in charge decides to try treating the disease with a new drug. Both twins get the same dose. However, one twin recovers almost immediately but the second remains ill for a long time. Now, if you were the doctor, would you not wonder why the drug worked on the one twin and not on the other? And would you not immediately assume that there must be *some difference* in the two cases that explains this difference in results? If this is your reaction, then you are assuming a deterministic model. That is, you are assuming that there is some combination of treatment and residual state that leads deterministically to recovery and some other combination that leads to failure to recover.

The alternative is to say, "No, there is no difference whatsoever in the two cases, except in the result. Although the internal makeup and residual states were identical, the final states differed." Most people find this response almost incomprehensible. So most people assume a deterministic model, perhaps in spite of themselves. Certainly most doctors and medical researchers proceed as if deterministic models do apply. It is difficult to imagine a medical researcher saying, "Too bad. Some people get cancer and others don't. There is no other difference. There is nothing to research. That's just the way it is." Few, if any, medical researchers would give such a response. Moreover, any medical researcher who thought this way would probably not be very successful.

A Stochastic Model

There is, however, another approach. This is to assume that human bodies and other complex systems are to be represented not by deterministic, but by *stochastic,* or probabilistic, models. In spite of much intuitive resistance to such an assumption, it is sometimes very useful and thus worth developing in further detail.

Let us continue thinking about the model pictured in Figure 7.1. This time, however, there is no longer a unique connection between the value of the causal variable and the value of the effect variable. For each value of the causal variable, either value of the effect variable might be realized. What the value of the causal variable does is change the *probability* of the value of the effect variable. More precisely,

> C is a *positive causal factor* (stochastic) for E in an individual I, characterized by residual state S, if the probability of E given C is *greater than* the probability of E given Not-C.
>
> C is a *negative causal factor* (stochastic) for E in an individual I, characterized by residual state S, if the probability of E given C is *less than* the probability of E given Not-C.

If the probability of E is the *same* for both values of the causal variable we would say that C is *causally irrelevant* to E in I, given residual state S. As in the case of a deterministic model, specification of the residual state is crucial. For a different residual state the probabilities could be different.

There is some difficulty with the understanding of probability in the above characterization of a stochastic model. With a single individual, there is no obvious population to consider. What can we be talking about when we refer to the probability of E given C?

The suggestion most in keeping with our previous understanding of probability is to imagine the same individual, with the same residual state, being subjected time after time to alternating values of the causal variable. The relative frequency of E's when the causal variable has value C would be the probability of E given C. Similarly, the relative frequency of E's when the causal variable has value Not-C would be the probability of E given Not-C. In actual applications, however, these probabilities are often purely hypothetical. There may never be more than one "trial" for a given individual.

Another suggestion is to suppose that causal production can vary in strength. That is, both C and Not-C can produce E, but the causal tendency of C to produce E may be stronger than the causal tendency of Not-C to produce E. Probability, on this account, is a measure of the strength of these causal tendencies.

These are all difficult notions. In what follows, we will use both deterministic and stochastic models for talking about individual systems. Most of the time it will not really make much difference which model of individuals is assumed. This is because the methods used for evaluating causal hypotheses in *populations* are essentially the same no matter which model of *individuals* is employed.

7.3 CAUSAL MODELS FOR POPULATIONS

For many causal relationships, particularly those studied in the biomedical sciences and those of interest to public health officials, it is impossible to investigate the causal relationship by studying just individuals. Not enough is known about the chemistry and physiology of individuals. The only way to get at the causal relationship is to study large groups of individuals. So we need a model of causality that can be applied to populations. Also, it must be a model that will help us evaluate causal hypotheses discussed in the popular media.

To develop such a model, we will begin with the *deterministic* model for individuals and extend it to populations of individuals. Later, we will briefly consider how the model would differ if we started with a stochastic model of individuals.

A Comparative Model for Causation in Populations

Our model for causation in a *population* will consist of a set of individuals, each of which is modeled by a deterministic model of causation in individuals. The basic idea is this: We will say that the variables **C** and **E** are causally related *in the population* U, if there are any individuals in U for whom **C** and **E** are causally related. This basic idea, however, needs to be further developed because, if there is no way to pick out individuals for whom **C** and **E** are causally related, there will be no way to determine whether the relationship holds in the population.

The key to a more useful model is to remember that whether the variables **C** and **E** are causally related in an individual is independent of which values of each variable that individual, in fact, happens to exhibit at the moment. The important question for causation is whether a particular value of **C** *produces* a particular value of **E.** So what we want to know is whether there are any individuals in the population for which C produces E, and Not-C produces Not-E, or vice versa.

One way to get at this question is to consider what would happen in the population if every individual exhibited the value C of the variable **C**. In particular, would any individuals now exhibiting values Not-C and Not-E exhibit value E for the effect variable if the value of their cause variable were changed from Not-C to C? Similarly, would any individuals now exhibiting values C and E exhibit value Not-E for their effect variable if the value of their cause variable were changed from C to Not-C? If there are any such individuals, then, by the above definition, **C** is causally related to **E** *in the population.*

In the original population U, there will be some percentage of members exhibiting the effect E. This percentage is the probability of E in the population U, which we will symbolize as $\mathbf{P}_U(E)$. Now imagine that every member of U that exhibits the value Not-C of the variable **C** is changed so as to exhibit the value C. This change results in a new population, which we will call X. X is just like U except that every individual member of X exhibits the value C of the causal variable **C.** In the hypothetical population X, there will be some percentage of members exhibiting the effect E. This percentage is the probability of E in X, which we will symbolize as $\mathbf{P}_X(E)$. It follows that if there are any individuals in U for which C is a positive causal factor for E, but which in U exhibit Not-C and Not-E, then changing those individuals from Not-C to C will also change them from Not-E to E in X. Thus, $\mathbf{P}_X(E)$ will be *greater than* $\mathbf{P}_U(E)$.

Similarly, imagine that every member of U that exhibits the value C of the variable **C** is changed so as to exhibit the value Not-C. This change results in a new population, which we will call K. K is just like U except that every individual member of K exhibits the value Not-C of the causal variable **C.** In the hypothetical population K, there will be some percentage of members exhibiting the effect E. This percentage is the probability of E in K, which we will symbolize as $\mathbf{P}_K(E)$. It follows that if there are any individuals in U for which C is a positive causal factor for E, and which in U exhibit C and E, then changing those individuals from C to Not-C will also change them from E to Not-E in K. Thus, $\mathbf{P}_K(E)$ will be *less than* $\mathbf{P}_U(E)$.

Putting these two results together, if there are any individuals in U for which C is a positive causal factor for E, no matter whether these individuals exhibit C and E, or Not-C and Not-E, in U, it must turn out that $\mathbf{P}_X(E)$ is *greater than* $\mathbf{P}_K(E)$. Thus,

> C is a *positive causal factor* for E in the population U whenever $\mathbf{P}_X(E)$ *is greater than* $\mathbf{P}_K(E)$.

A parallel chain of reasoning applies for the case in which C is a negative causal factor for E in some individuals in the initial population. In that case, it will turn out that $\mathbf{P}_X(E)$ *is less than* $\mathbf{P}_K(E)$. Thus,

> C is a *negative causal factor* for E in the population U whenever $\mathbf{P}_X(E)$ *is less than* $\mathbf{P}_K(E)$.

Finally,

> **C** is *causally irrelevant* for **E** in the population U whenever $\mathbf{P}_X(E)$ *is equal to* $\mathbf{P}_K(E)$.

These relationships are all pictured in Figure 7.2.

One interesting fact about these models is that it could turn out that **C** is causally irrelevant for **E** in the population U even though **C** is not causally irrelevant for **E** in all individuals in U. This could happen if there were some individuals in U for which C is a positive causal factor for E and an equal number for which C is a negative causal factor for E. This is possible so long as different individuals in U have different residual states. In this case, $\mathbf{P}_X(E)$ would turn out to be equal to $\mathbf{P}_K(E)$. So long as we deal with individuals that

Figure 7.2

A comparative model for causation in populations.

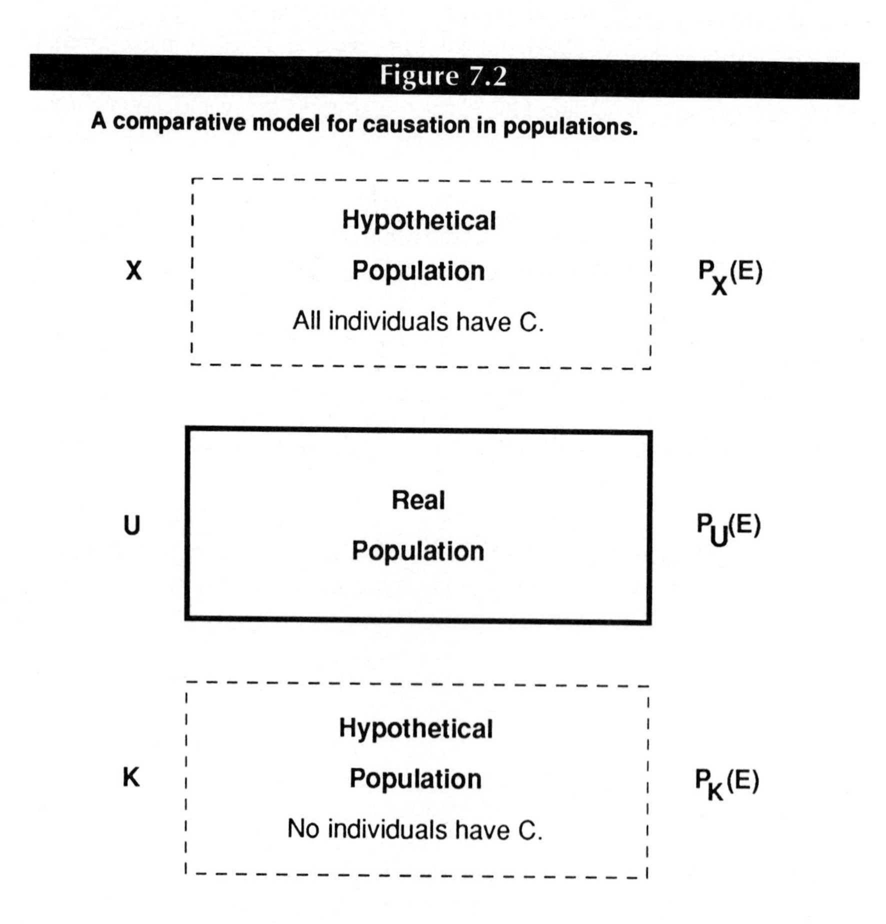

differ in causally relevant ways, there will always be this sort of theoretical gap between models for individuals and models for populations. Population models always "average" over individuals and, thus, ignore what may be important differences among individuals.

What if the Individuals Are Stochastic?

The above model of causal factors in populations assumed that the *individuals* making up the population fit a simple, deterministic model for causal factors in individuals. What if the individuals are stochastic? The only difference is that $\mathbf{P}_X(E)$ and $\mathbf{P}_K(E)$ are no longer definite numbers, but only averages over probabilities. Suppose, for example, that there is exactly one individual in U for which C is a positive causal factor for E, but which in fact exhibits Not-C and Not-E. On a deterministic model of that one individual, the number of cases of E in population X will definitely be greater by one than the number in population U. On a stochastic model of that one individual all we can say is that there is some *probability* that the number of cases of E in X will be greater than the number in U. Here it is perhaps easier to understand probabilities as causal tendencies than as relative frequencies. In any case, it turns out that the evaluation process is much the same whether we assume a deterministic or stochastic model for individuals. In most of our further work we will assume the deterministic model.

7.4 EFFECTIVENESS OF CAUSAL FACTORS

For some purposes, such as making a decision, it is often less important to know that something *is* a causal factor than to know *how much* of a factor it is. In the case of medical treatments, for example, it is more important to know how many people a given treatment is likely to help than simply to know that there would be some. In the case of treatments, the term we would use to talk about relative numbers would be *effectiveness.* We want to know how effective a given treatment would be if it were applied to a given population. Let us generalize this notion and use the word effectiveness to talk about any causal factor, whether it is good, such as vaccinations, or bad, such as air pollution.

Effectiveness in Individuals

On a *deterministic* model of individuals, there are only three grades of effectiveness. A positive causal factor in an individual is maximally effective in bringing about the effect. C produces E, and Not-C prevents E. A negative causal factor is maximally effective in the negative direction. C prevents E, and Not-C produces E. The intermediate case is causal irrelevance. The presence of C is irrelevant to the presence or absence of E.

On a *stochastic* model of individuals, there is a full range of degrees of effectiveness. Here C being a positive causal factor for E means that E is more probable given C, than given Not-C. In the extreme case, $\mathbf{P}(E/\text{Not}-C) = 0$ and $\mathbf{P}(E/C) = 1$. This coincides with C being a deterministic, positive causal factor for E. C being a negative causal factor means that E is less probable given C, than given Not-C. Here the extreme case is when $\mathbf{P}(E/\text{Not}-C) = 1$ and

P(E/C) = 0. This coincides with C being a deterministic, negative causal factor for E. The simplest definition of the effectiveness of C in producing E, **Ef**(C,E), in an individual I, is the simple difference between **P**(E/C) and **P**(E/Not–C) for I:

$$\mathbf{Ef}(C,E) = \mathbf{P}(E/C) - \mathbf{P}(E/\text{Not}-C).$$

This measure has maximum value +1 and minimum value −1, with zero effectiveness corresponding to *causal irrelevance* between the variables **C** and **E**. That is, causal irrelevance means that **P**(E/C) = **P**(E/Not–C).

Effectiveness in Populations

For a population model assuming a deterministic model for individuals, the simplest measure of the effectiveness of a causal factor in population U is the difference between $\mathbf{P}_X(E)$ and $\mathbf{P}_K(E)$. That is,

$$\mathbf{Ef}(C,E) = \mathbf{P}_X(E) - \mathbf{P}_K(E).$$

This measure again ranges from −1 to +1, with zero effectiveness corresponding to causal irrelevance. This measure applies not to individuals but to populations. However, just as a given causal agent might be quite effective for one individual and not for others, a causal agent might be quite effective in one population and completely ineffective in another.

If we assume *stochastic* individuals, the corresponding measure of the effectiveness of a causal agent in a population would be the difference in the expected values of $\mathbf{P}_X(E)$ and $\mathbf{P}_K(E)$ in the two hypothetical populations.

7.5 SUMMARY

There is an obvious parallel between our models of positive *correlation* and of positive *causal factors.* Similarly, there is a parallel between our definitions of the *strength* of a correlation and of the *effectiveness* of a causal factor. Causation is not the same as correlation, however, and now we can see more clearly why this is so.

To make this summary more concrete, let us apply our models to the example of smoking, the cause, and lung cancer, the effect. No one questions whether there exists a *positive correlation* between smoking and lung cancer. There are clearly more cases of lung cancer among smokers than among nonsmokers. Yet it is still regularly questioned whether there is good evidence for the *causal hypothesis* that smoking is a positive causal factor for lung cancer. There is a basis for this difference in opinion in the great difference between the two sorts of hypotheses.

A *correlation* is a relationship between properties that exist in some *actual* population. Thus, the correlation between smoking and lung cancer is defined by the relative numbers of lung cancer patients among smokers and nonsmokers in the population of adult Americans as it now exists.

Causal factors, on our model, are defined by relationships between two *hypothetical* versions of the real population. To say that smoking is a positive causal factor for lung cancer is not to say merely that there are, in

Figure 7.3

The difference between a correlation and a causal hypothesis.

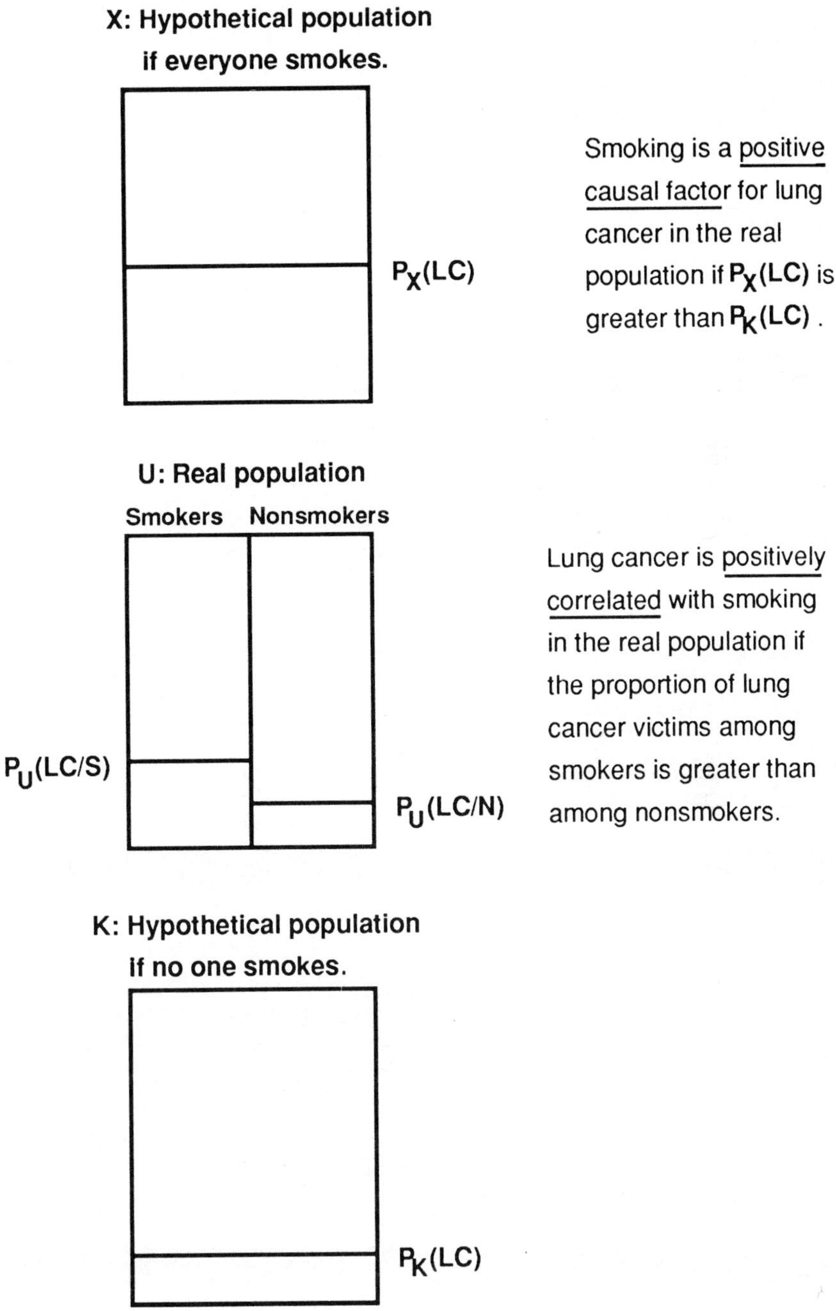

X: Hypothetical population if everyone smokes.

$P_X(LC)$

Smoking is a <u>positive causal factor</u> for lung cancer in the real population if $P_X(LC)$ is greater than $P_K(LC)$.

U: Real population

Smokers **Nonsmokers**

$P_U(LC/S)$

$P_U(LC/N)$

Lung cancer is <u>positively correlated</u> with smoking in the real population if the proportion of lung cancer victims among smokers is greater than among nonsmokers.

K: Hypothetical population if no one smokes.

$P_K(LC)$

fact, more cases of lung cancer among smokers than among nonsmokers in the existing population. That is only a correlation. It is to say there *would be* (or probably would be) more cases of lung cancer if everyone smoked than if no one smoked—everything else being the same. That is a very different claim. These differences between correlations and causal relationships are pictured in Figure 7.3.

You should make sure you have a good grasp of the difference between correlations and causal relationships in populations. This difference will be crucial when we come to consider evaluating causal hypotheses. It takes much more to have good evidence for a causal hypothesis than for a claim of correlation.

EXERCISES

Exercise 7.1

Diagram the following hypotheses discussed in Section 7.1 of the text:

(A) Among the islanders, having lice is positively correlated with being healthy.

(B) Among the islanders, having lice promotes good health.

(C) Among the islanders, being healthy is positively correlated with having lice.

(D) Among the islanders, good health causes one to have lice.

Exercise 7.2

It is well established that the death rate from heart attacks among widows is greater than the rate among married women in general. This fact has been cited as evidence for the hypothesis that being married prevents heart attacks. Diagram the established correlation between being married and suffering a heart attack. Then diagram the suggested causal hypothesis. Explain in general terms why the established correlation by itself provides no evidence for the causal hypothesis. Can you think of an alternative explanation for the correlation?

Exercise 7.3

During World War I, the U.S. Navy argued in its recruiting literature that being in the Navy during the Spanish-American

War was safer than living in New York City. The death rate in New York City during that war, they pointed out, was higher than that among sailors in the Navy. Diagram the correlation appealed to in the Navy's literature. Then diagram the suggested causal hypothesis. Explain in general terms why the established correlation by itself provides no evidence for the causal hypothesis. Can you think of an alternative explanation for the correlation?

Exercise 7.4

A careful study of college students came up with the conclusion that there is a positive correlation between having a low grade-point average and smoking cigarettes. Opponents of cigarette smoking concluded that smoking causes students to get lower grades. Others concluded that getting low grades causes people to smoke. Diagram the correlation revealed by the study. Then diagram both of the suggested causal hypotheses. Is there any reason to think that the correlation provides better evidence for one causal hypothesis than for the other? Explain briefly.

Exercise 7.5

Imagine three individuals designated by their respective internal states as I_1, I_2, and I_3. Suppose that, given their differing internal states, C is a positive causal factor for E in I_1, a negative causal factor for E in I_2, and causally irrelevant for E in I_3. In the actual world, I_1 exhibits C and E, I_2 exhibits Not-C and E, and I_3 exhibits C and Not-E. Now consider the population *consisting of just these three individuals. In this population, is C a positive causal factor for E, a negative causal factor for E, or causally irrelevant for E?*

CHAPTER 8

Evaluating Causal Hypotheses

Does saccharin cause bladder cancer? Does smoking cause heart attacks? Does "the pill" cause fatal blood clots? In this chapter we shall examine these three questions in order to illustrate three basic methods that may provide evidence for *causal hypotheses*. The first method provides the best evidence, and we shall use it as the standard by which to judge the shortcomings of the other two methods. The reason for taking the other methods seriously is that there are many questions that, for scientific, practical, or moral reasons, cannot be investigated using the scientifically best method.

The three studies we will examine are interesting in themselves. You should be careful, however, not to get so involved in the specifics of a particular example that you fail to grasp the essential features of the method the example illustrates. Your main goal is to learn to recognize the type of study at issue. So treat these cases as *analog models* to be used as guides when you come across new cases to evaluate. Popular reports of scientific studies do not usually tell you exactly what type of study was involved. You have to figure that out for yourself. Because the quality of the evidence provided by these three methods is quite different, it is important to learn to recognize the differences.

8.1 SACCHARIN AND CANCER

In March of 1977, officials of the U.S. Food and Drug Administration (FDA) learned of a Canadian study that showed that saccharin, a then-popular artificial sweetener, causes cancer in laboratory rats. Two days after learning of this study, the FDA issued a preliminary ban on the sale of substances containing saccharin. The legal basis for the ban was a rule (the Delaney Clause) in a 1958 amendment to the Food, Drug, and Cosmetic Act stating that any additive found to induce cancer in humans or *animals* shall be deemed unsafe. The proposed ban was strongly opposed by the soft-drink and diet-food industries and by large segments of the general public. Bills to revoke the Delaney Clause were introduced in Congress. Hearings were held and the Congressional Office of Technology Assessment (OTA) was requested to review the evidence. The final OTA report upholding the Canadian findings was issued in October. In November, a bill was passed by the U. S. Congress temporarily suspending the FDA ban in favor of a warning label on foods containing saccharin.

Our concern at the moment is not with the public-policy or decision-making aspects of this episode, but with the scientific question whether saccharin does indeed cause cancer in laboratory rats. As a matter of fact, few people questioned this scientific conclusion. Why, then, did the Canadian study provide such good evidence for this causal hypothesis?

The following are excerpts from the 1977 OTA report, *Cancer Testing Technology and Saccharin,* which reviewed and evaluated the Canadian study. You should study this report carefully because it is a "textbook" example of *randomized experimental design.*

Saccharin Risks

The best evidence to date for concluding that saccharin is a potential human carcinogen comes from two-generation rat-feeding experiments. These tests demonstrated that, over a long period, diets high in saccharin produced bladder tumors in rats. Evidence for carcinogenicity by other routes of administration and in other species of laboratory animals, while not convincing by itself, supports the conclusions from the two-generation rat experiments.

Animal Studies

The route of administration in animal tests is crucial. Accepted experimental protocols require that if human exposure is by ingestion, then the experimental animals must be fed the substance being tested. Forced feeding, although it allows a more precise dose, is not accepted. The induction of cancer in test animals following ingestion of saccharin shows that saccharin can potentially cause cancer when ingested by humans.

A situation that as nearly as possible duplicates human exposure to saccharin would be desirable. But for reasons mentioned earlier, using small animals is a necessary compromise. These practical modifications lead to other problems in relating the results of animal experiments back to human experience. Nonetheless, the weight of scientific evidence shows that these methods are valid for predicting carcinogenic effects in humans.

The positive feeding experiments were conducted over two generations. Rats of the first generation were placed on diets containing saccharin at the time of weaning. These animals were bred while on this diet, and the resulting offspring were fed saccharin throughout their lives. Members of the second generation were thus exposed to saccharin from the moment of conception until the termination of the experiment. Each animal of the second generation was examined for cancers at its death or at its sacrifice after two years on the experiment.

Each experiment had appropriate control groups that did not ingest saccharin. Compared to control animals, the saccharin-fed animals showed an excess of bladder tumors. These differences were sufficiently convincing to lead to the conclusion that saccharin caused cancer in rats.

A statistical analysis of the results of the two-generation rat-feeding experiment is shown in (Table 8.1). Cancers were observed only at the highest dose levels of saccharin, 5 percent of the diet. Results are presented as a fraction; the numerator is the number of bladder cancers, and the denominator is the number of animals (at least 18 months old) examined. The number of cancers found in the saccharin-fed animals exceeded the number found in the controls in each experiment.

Statistical methods are available to calculate the significance of the differences in cancer incidence observed between the experimental and control animals. Such a calculation was made by a standard procedure at the National Cancer Institute. The value "p" when multiplied by 100 gives the percent probability that the observed differences would arise by chance alone, if there actually was no difference between the experimental and control animals. In the first experiment in the table, "p" is equal to .075. Multiplying by 100 gives 7.5 percent. Thus, in that experiment, 7.5 percent is the probability that chance alone would produce at least this large a difference in cancer incidence even though there was no difference between the experimental and control groups. Any "p" value equal to or less than .05 is considered statistically significant. The results for the offspring generation in the experiments were significant by this criterion.

Bladder cancers were more frequent in male animals than in females. The explanation for this difference is unknown. In human populations, bladder cancers occur about twice as frequently among males, but that observation is ascribed to differences in exposure to carcinogens in the workplace and to smoking.

To conclude, the two-generation experiment showed that saccharin caused an increase in bladder cancer in second-generation animals, especially among males. In the experiment in which the first generation was examined, the increase fell just short of the standard test of significance. No cancer of any other site has been convincingly associated with saccharin.

We will now proceed to develop a standard model for evaluating studies, such as these Canadian experiments.

8.2 RANDOMIZED EXPERIMENTAL DESIGNS

The Canadian study is a perfect example of a *randomized experimental design* (*RED*). We shall first examine the structure of such designs in order to understand how they can, in appropriate circumstances, produce good evidence in favor of causal hypotheses. At the end of this section, we will briefly take up some of the broader issues concerning the use of animal experiments to study human diseases.

The Real World Population and the Causal Hypothesis

One might naturally think that the *real world population of interest* is human beings. After all, who really cares whether laboratory rats get cancer from too much saccharin? There is nothing inherently mistaken about this point of view. For our purposes, however, we cannot allow so loose a relationship between the population of interest and the population actually sampled. So we will insist that the sample actually studied consist entirely of members of the population of interest. There were no humans in the samples involved in this study—only rats. For purposes of our evaluation, then, the real world population of interest consists of laboratory rats. The possible relationship between what happens to rats and what might happen to humans is a separate issue.

The *cause variable* in this study is saccharin, or, rather, the ingestion of saccharin. The *effect variable* is the occurrence, or absence, of bladder cancer. The *causal hypothesis* of interest is that ingesting saccharin is a positive causal factor for the development of bladder cancer in laboratory rats.

According to the model developed in the previous chapter, the causal hypothesis asserts that there would be more cases of bladder cancer in the population of laboratory rats if all members of this population were exposed to saccharin than if none were. Figure 8.1 shows the populations, both real and hypothetical, that model this hypothesis.

Figure 8.1

A model of the hypothesis that ingesting saccharin is causally related to developing bladder cancer.

Hypothetical population (X) **Hypothetical population of lab rats in which all are exposed to saccharin.**	$P_X(E)$	(i.e., the proportion that would develop bladder cancer in population X)
Real population (U) **The real population of lab rats.**	$P_U(E)$	(i.e., the proportion that do develop bladder cancer in the real population U)
Hypothetical population (K) **Hypothetical population of lab rats in which none is exposed to saccharin.**	$P_K(E)$	(i.e., the proportion that would develop bladder cancer in population K)

The Sample Data

The sample data for the experiment are presented in Table 8.1. Actually there are several sets of data. For the moment we will focus on the totals (male and female) in the parental generation. Here the data are that among seventy-four rats given no saccharin, only one developed bladder cancer, while among seventy-eight ingesting a diet of 5 percent saccharin, seven developed bladder cancer.

The Experimental Design

As we already know, it is rarely feasible to examine the whole of any population. Scientists usually settle for a sample and then infer from the sample back to the population. But it is obviously impossible to sample a *hypothetical* population, let alone two hypothetical populations with incompatible characteristics. So how does one test the causal hypothesis?

The answer is that scientists create real samples to play the role of samples from the two hypothetical populations. This is done by randomly selecting a sample from the real population and then randomly dividing it into two groups. All the individuals in one group are experimentally manipulated so that they have the condition that defines the hypothetical population we have called X (for "experimental"). The individuals in the other group are manipulated so that they have the condition that defines the hypothetical population that we call K (for "kontrol"). These two groups, then, are just as if they had been sampled from the two hypothetical populations. These groups are called the *experimental group* (x group) and the *control group* (k group), respectively.

To convince yourself that this procedure really is a satisfactory substitute for sampling the hypothetical populations, consider a single rat in a real population having N members. Now consider the following two procedures: (1) Depending on the flip of a coin, every rat in the real population either is, or is not, put on the saccharin diet. A rat is then selected at random. (2) A rat is randomly selected from the real population and then given a saccharin diet, or not, depending on the toss of a coin. The probability of our chosen rat ending up in the experimental group is the same either way, that is, N/2. This would also be true for any other rat. Thus, since our conclusion is to be based on what happens to rats in the experimental and control groups, we would reach

Table 8.1

Results and Statistical Analysis of Rat-Feeding Experiments

Generation	Dose	Cancers (M)	Cancers (F)	Total	Significance
Parental	0%	1/36	0/38	1/74	p = .075
	5%	7/38	0/40	7/78	
Offspring	0%	0/42	0/47	0/89	p = .003
	5%	12/45	2/49	14/94	

the same conclusion either way. The second procedure has the advantage that it can actually be carried out in normal circumstances.

In the saccharin experiment, the researchers began with a group of laboratory rats, that, given the careful conditions under which these rats are bred, may be regarded as a random sample of all such rats. We can safely assume that designation of individual rats as experimentals or controls was done by some random process. The rats in the one group (the x group) were fed a 5 percent saccharin diet; the rats in the other group (the k group) received the same diet minus the saccharin.

In the end, of course, there was an observed relative frequency of rats in each group that developed bladder cancer. We shall designate these two frequencies as $\mathbf{f}_x(E)$ and $\mathbf{f}_k(E)$, respectively. A model of the experiment concerning the parental generation is pictured in Figure 8.2. The two hypothetical populations are represented by broken lines indicating that they do not really exist. In general, our diagrams representing experimental studies will not include these two hypothetical populations. The first time through, however, it is convenient to have them handy for easy reference.

Random Sampling

In randomized experimental designs, random sampling occurs in two places. First, the whole sample of subjects (in this case laboratory rats) should be randomly selected from the whole population of interest. For example, one would not want a batch of rats that had accidentally been exposed to a rare chemical that sensitized them to saccharin. In such circumstances, the results of the experiment could be totally misleading regarding the population as a whole.

Second, the division of the initial sample into experimental and control groups must be done randomly. For example, one would not want the control group to contain an excess of sickly rats who were already prone to bladder cancer. In that case, one might not get a statistically significant difference in the number of bladder cancers in the experimental and control groups, even if saccharin did cause bladder cancer in some members of the experimental group.

There is little explicit mention of random sampling in the OTA report on the Canadian saccharin studies. That must be because it is taken for granted in highly professional studies of this sort. If there had been even the slightest suspicion that the Canadian researchers had not followed the most careful procedures to insure random sampling, one can be sure that the OTA would never have come out with such a politically unpopular review of their work.

Evaluating the Causal Hypothesis

Figure 8.3 shows a diagram of the data from the first part of the saccharin experiment. You will notice that this looks very much like a diagram of data that would be used to evaluate the existence of a correlation between two variables. In this case, one variable is the *cause* variable, the presence or absence of saccharin in the diet, that characterizes the x group and the k group respectively. The second variable is the *effect* variable, the occurrence (or absence) of

Figure 8.2

A model of the rat-feeding experiment for the parental generation.

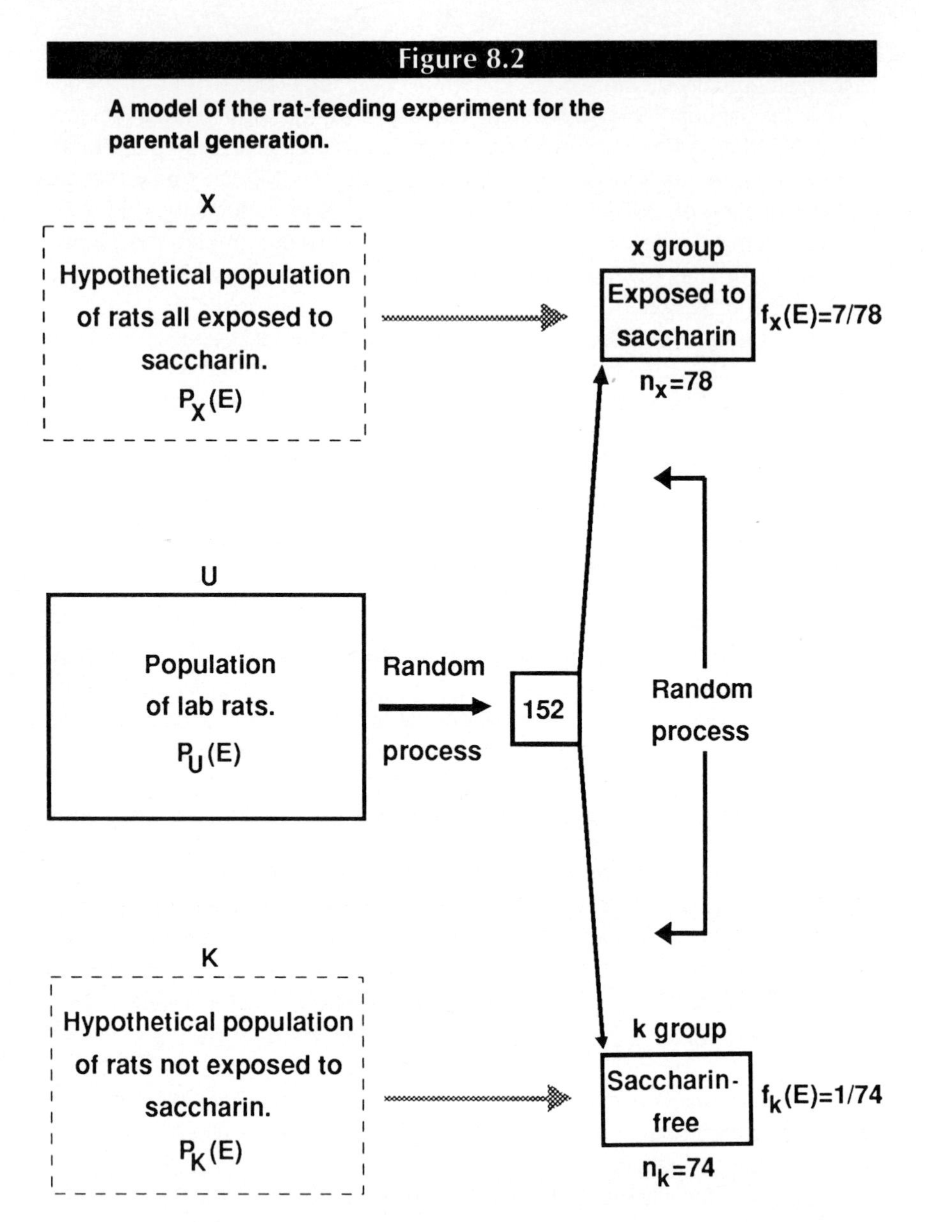

bladder cancer. The quantity $\mathbf{f}_x(E)$ represents the relative frequency of bladder cancer in the x group; $\mathbf{f}_k(E)$ represents the relative frequency of bladder cancer in the k group. Evaluating the claim of a causal relationship between the cause and effect variables in the real population is thus similar to evaluating whether there is a correlation in the real population.

The relevant question is whether there is a statistically significant difference between $\mathbf{f}_x(E)$ and $\mathbf{f}_k(E)$. This is ascertained by using the values of $\mathbf{f}_x(E)$ and $\mathbf{f}_k(E)$ to estimate the values of $\mathbf{P}_X(E)$ and $\mathbf{P}_K(E)$. As before, we construct the relevant intervals, $\mathbf{P}_X(E) = \mathbf{f}_x(E) \pm \mathbf{ME}$ and $\mathbf{P}_K(E) = \mathbf{f}_k(E) \pm \mathbf{ME}$.

Figure 8.3

A diagram of the data from the parental generation in the rat-feeding experiment.

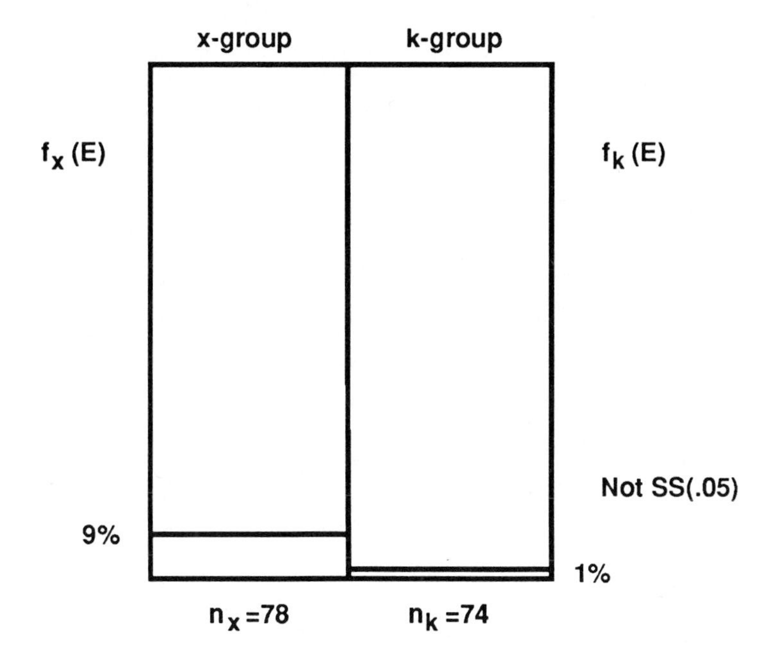

If the intervals do *not* overlap, the difference between $\mathbf{f}_x(E)$ and $\mathbf{f}_k(E)$ *is* statistically significant. That means that there is good evidence for a real difference between $\mathbf{P}_X(E)$ and $\mathbf{P}_K(E)$, which in turn means that there is good evidence for a causal relationship between the cause and effect variables.

If the intervals *do* overlap, the difference between $\mathbf{f}_x(E)$ and $\mathbf{f}_k(E)$ is *not* statistically significant. That means that there fails to be good evidence for a real difference between $\mathbf{P}_X(E)$ and $\mathbf{P}_K(E)$, which in turn means that there is not good evidence for the existence of a causal relationship between the cause and effect variables.

Note that in this latter case there is no basis for claiming good evidence for a complete lack of any causal connection between the cause and effect variables. Only if there were an overlap of intervals with very large numbers of subjects in both the x group and the k group would one have good evidence for a very low level of *effectiveness* of the cause.

Estimating the *effectiveness* of the causal factor is similar to estimating the *strength* of a correlation. One simply calculates the difference between the two nearest and two farthest ends of the estimated intervals for $\mathbf{P}_X(E)$ and $\mathbf{P}_K(E)$. The resulting interval is the estimate of the effectiveness of the causal factor in the population. We need not repeat the details here.

In the OTA report, the difference in the incidence of bladder cancer in the experimental and control groups for the parental generation was reported as *not* statistically significant. That means that the first part of the experiment failed to produce good evidence in favor of the causal hypothesis that ingesting saccharin is a positive causal factor for bladder cancer in laboratory rats.

Actually, this particular report provides more information than is usual in more popular reports of statistical studies. From the information provided, we can easily infer that the difference between $\mathbf{f}_X(E)$ and $\mathbf{f}_K(E)$ for the parental generation would have been statistically significant if we used a confidence level of 92.5 percent rather than the standard 95 percent. That is, with a confidence level of 92.5 percent the intervals representing the estimates of the values of $\mathbf{P}_X(E)$ and $\mathbf{P}_K(E)$ would be narrow enough that they would just fail to overlap. So the data from the first part of the experiment actually comes pretty close to providing good evidence for the causal hypothesis.

A Program for Evaluating Causal Hypotheses

Before evaluating the evidence from the second generation of the Canadian saccharin study we will develop a general program for evaluating causal hypotheses. This will be the last of the three programs developed in this text, and the second focusing on statistical studies.

Step 1. The Real World Population and the Causal Hypothesis. Identify the real world population that is the intended object of the study. Identify the cause variable and the effect variable and state the causal hypothesis to be evaluated.

Step 2. The Sample Data. Identify the real world sample and the particular data from that sample whose relevance for the causal hypothesis you wish to evaluate.

Step 3. The Design of the Experiment. Identify the model that best represents the design of the experiment. Explain briefly the reasons for your choice.

Step 4. Random Sampling. How well does a random sampling model represent the actual process by which the sample was selected from the population? Answer: a) very well; b) moderately well; c) somewhat well; or d) not very well. Explain the factors relevant to your answer.

Step 5. Evaluating the Hypothesis. Diagram the study as it applies to the real world population, the sample, and the data you have identified. Label all relevant parts of the diagram. Assuming a random sampling model is applicable, what can you conclude about the causal hypothesis? Evaluate the effectiveness of the causal factor.

Step 6. Summary. Review your evaluation, particularly your analyses in Steps 4 and 5. Give a summary statement of how well the data support the causal hypothesis.

The Second Generation Saccharin Experiment

Now let us evaluate the Canadian saccharin experiment focusing on the data from the second generation of rats in the experiment. These rats, recall, were conceived by parents who were members of the original experiment. Rats in the experimental group were thus exposed to saccharin from conception.

Step 1. The Real World Population and the Causal Hypothesis. The population of interest consists of rats bred for laboratory experiments. The cause variable is exposure to saccharin, and the effect variable is bladder cancer. The hypothesis at issue is that ingestion of saccharin is a positive causal factor for bladder cancer in the population of laboratory rats.

Step 2. The Sample Data. The overall data from Table 8.1, including both male and female offspring, show $^{14}/_{94}$ cases of bladder cancer in the experimental group exposed to saccharin, and $^{0}/_{89}$ cases of bladder cancer in the control group not exposed to saccharin.

Step 3. The Design of the Experiment. The experiment fits the model for randomized experimental design. The sample from the population is randomly divided into two groups, one of which is administered the causal factor by the experimenters.

Step 4. Random Sampling. The fact that this was a professional study reviewed by the OTA suggests that standard, scientific measures were taken to insure a random selection process. It should fit a random sampling model very well.

Step 5. Evaluating the Hypothesis. Figure 8.4 exhibits a diagram of the resulting data. The difference between $\mathbf{f}_X(E)$ and $\mathbf{f}_K(E)$ is reported as being statistically significant. And $\mathbf{f}_X(E)$ is obviously greater. So there is good evidence that exposure to saccharin from conception is a positive causal factor for bladder cancer in rats. Because of the low values of $\mathbf{f}(E)$, the margins of error are smaller than stated in our usual rules of thumb. Assuming **ME** = .05 for the k group and **ME** = .08 for the x group, the effectiveness of the causal factor would be estimated as between .02 and .23.

Step 6. Summary. This is clearly a careful study in line with random sampling models. The report indicates that one could raise the confidence level to 99.7 percent before the confidence intervals would become so large as to overlap, rendering the observed difference between $\mathbf{f}_X(E)$ and $\mathbf{f}_K(E)$ statistically insignificant at the .003 level. That is very strong evidence in favor of the causal hypothesis.

Of Rats and Humans

In all the furor over the proposed ban on saccharin, no responsible critics questioned whether the Canadian study proved that saccharin causes cancer *in laboratory rats.* Most of the controversy centered on the Delaney Clause. Many questioned whether it is wise to base public policy on such a cautious rule. Others questioned the scientific basis of the Delaney Clause, which is that animal studies are relevant to questions about human diseases.

Figure 8.4

A diagram of the data from the offspring generation in the rat-feeding experiment.

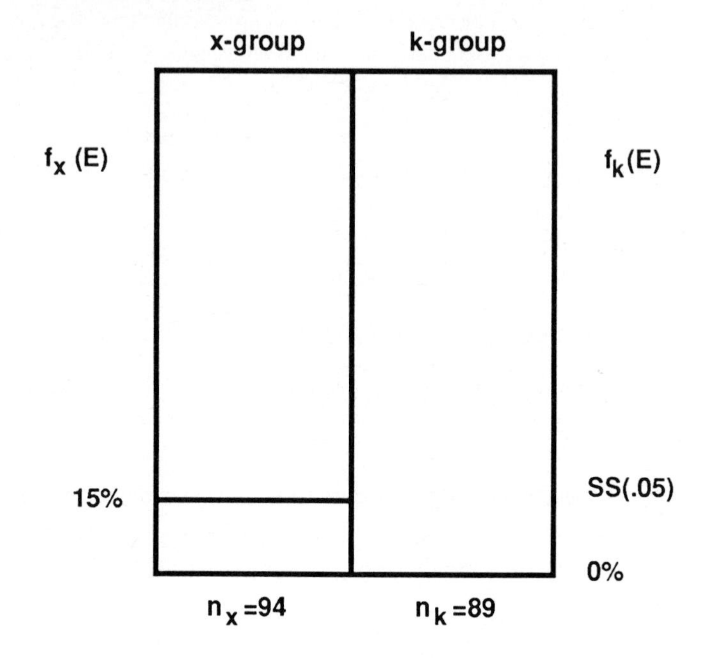

The relevance of animal studies was questioned for two reasons. First, humans are different from rats. Or, to put it in our language, the population studied was not the human population. Second, the amount of saccharin used (5 percent of the rats' diet) was quite large. Some critics calculated that 5 percent of a human diet corresponds to the amount of saccharin in 800 bottles of diet soda. Who drinks 800 bottles of diet soda a day? There is something to these criticisms, but not nearly as much as many critics thought.

Most environmental hazards are not highly effective. If they were, many more of us would be dead—or would never have been born. The typical cancer-producing agent strikes only one person in thousands, but that is a great number. In a population of 200 million people, one in 1000 works out to 200,000 people. Even only one in 50 thousand is 4000 people.

One would not expect saccharin, or most substances, in *normal* doses, to cause more than one cancer in a thousand cases. To detect so low a degree of effectiveness using normal doses, however, would require an experiment using many thousands of rats. To breed and maintain so many experimental animals over a period of several years would be almost impossible. So, instead, typical animal studies use many fewer animals and larger doses.

The assumption behind this strategy is that the number of cases of the effect is roughly proportional to the dose. Double the dose and roughly you double the number of cancers. Therefore, knowing what large doses do, one extrapolates back to what normal doses would do. Now, the assumption that the effectiveness is proportional to the dose can be legitimately questioned. In particular, there could be a *threshold* for the effect. That is, it might be that below a certain dose there is no effect at all in anyone. The number increases with increasing dose only after the threshold is reached.

The problem is that almost no biological agent has been shown, in fact, to have a threshold. Saccharin, of course, might be an exception, but there is no positive justification for believing that it is. Nor, as we shall see in Part Three, would it be a wise strategy to base decisions on the assumption that saccharin is an exception when there is no specific knowledge one way or the other.

It is often said by people who know nothing about medical research that any substance will cause cancer if given in big enough doses. That simply is not true. Many chemical substances have been tested on animals using large doses, and only a relatively few have been found to cause cancer. It may be true that large doses of anything will kill you, but few substances give you cancer in the process.

As for the statement that humans are not rats, that is obviously true. But of the approximately thirty agents known definitely to cause cancer in humans, all of them cause cancer in laboratory rats—in high doses. From this fact it does not necessarily follow that anything causing cancer in rats will also do so in humans. Again, it is difficult to justify basing practical decisions on the assumption that saccharin is an exception. And taking account of differences in dose and body weight, those fourteen cancers in ninety-four rats translate into about 1200 cases of bladder cancer in a population of 200 million people drinking less than one can of diet soda a day.

You can easily calculate that 1200 cases of bladder cancer in a population of 200 million means that any individual is facing a $6/1{,}000{,}000$ chance of getting bladder cancer. So it may be argued that the risk is small and people ought to be allowed to decide for themselves whether to take this risk. That, however, is an entirely different question than the scientific question of whether there is any risk at all, and if so, how much. The answer to the scientific question seems to be that there is some risk, with six in a million being a very rough estimate as to how great the risk might be.

8.3 SMOKING AND HEART DISEASE

Ever since Columbus took tobacco from the New World back to the Old, people have been saying that it is an evil weed that causes all manner of illness and disease. Others claimed it to be a blessing and cure for diverse ailments. It was not until after World War II that anyone undertook deliberate, well-planned studies of possible causal relationships between smoking and serious illnesses, such as cancer and heart disease.

You can easily appreciate the difficulties in studying these questions. Laboratory experiments of the type used to study the effects of saccharin (and many other drugs) are out of the question. It is practically impossible to subject enough animals (such as laboratory rats) to high-enough doses of tobacco smoke to produce statistically significant differences. At normal doses, you could not expect more than a few cases in a thousand of any disease over the lifetime of the subjects. At above normal doses of tobacco smoke, however, experimental animals tend to die of smoke inhalation.

Direct experiments on humans are, of course, out of the question. One would have to randomly select several thousand nonsmokers, say age 12, and randomly divide them into two groups. Everyone in the "experimental" group would be forced to smoke a pack of cigarettes a day, and everyone in the control group would be prevented from smoking at all. In twenty to forty years, there might be enough data to conclude either that smoking is a causal factor for some diseases or that its effectiveness in producing these diseases, if any, is very low.

Such experiments, though, cannot be performed on humans. It would be immoral, if not actually illegal. It is not even very practical either. The effort and expense of maintaining such a regime for thirty years would be too great.

The studies undertaken in the 1950s, therefore, were not experimental studies. They were *prospective* studies. As you read the following description of these studies, try for yourself to pick out the essential differences between prospective and randomized experimental designs. That will prepare you for the discussion of prospective models in the following section.

The Framingham Study

In 1950, the National Institutes of Health (NIH) sponsored a projected twenty-year study of the causes of several diseases, particularly coronary heart disease (CHD). The study was carried out in Framingham, a town of about 30,000 people, eighteen miles west of Boston. Among the disadvantages of Framingham for such a study was that its population was rather more homogeneous than the population of Americans in general, let alone the population of humans in general. For example, there were relatively few African-Americans in Framingham. A homogeneous population however has advantages too, and Framingham had other advantages as well. Among these were a cooperative and very competent group of local doctors, and a published list of the names, addresses, ages, and occupations of all residents over 18 years of age.

From these lists, 6507 individuals aged 30 through 59 (3074 men and 3433 women) were selected at random and invited to come into a clinic for a physical examination. All were informed of the long-term nature of the study, including physical examinations every two years for the next twenty years. With the help of the local physicians, 4469 of the initial sample agreed to participate, and, of these, 4398 were found upon examination to be free of any symptoms of CHD. In addition, another 734 disease-free volunteers were added to the sample for a final total of 5132 (2287 men and 2845 women). Figure 8.5 shows the statistical distribution of ages at entry in the study for both men and women.

The study was finally concluded twenty-four years later in 1974. Among the findings, one was that coronary heart disease appears to be predominantly a male disease. For example, the 24-year incidence of coronary heart disease for men and women in the middle age group (40–49) is shown in Figure 8.6. The difference between the percentage of men and the percentage of women who eventually exhibited CHD is clearly statistically significant.

Figure 8.5

The statistical distribution of ages at entry for both men and women in the Framingham study.

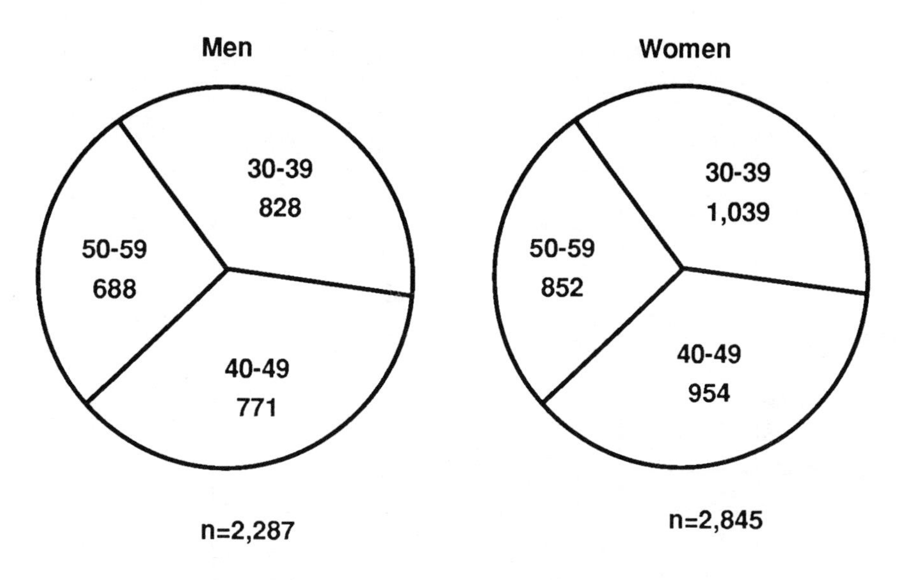

Figure 8.6

The 24-year incidence of coronary heart disease for men and women in the middle age group (40–49).

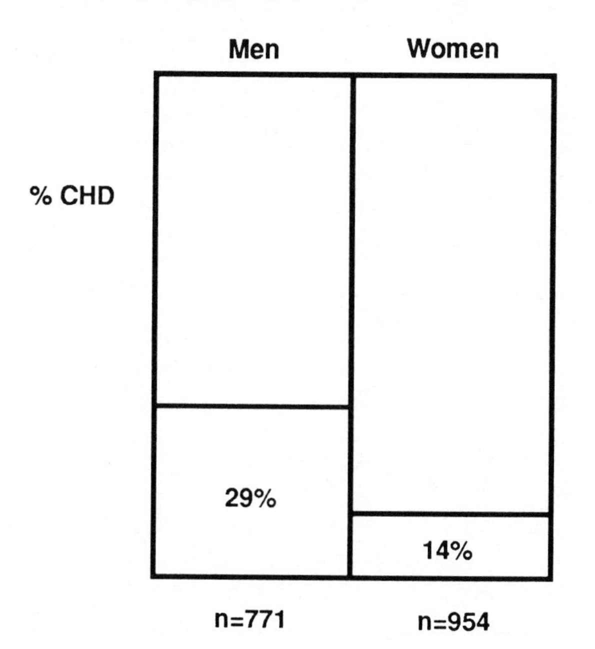

Figure 8.7

The data on coronary heart disease for smokers and nonsmokers among men and women.

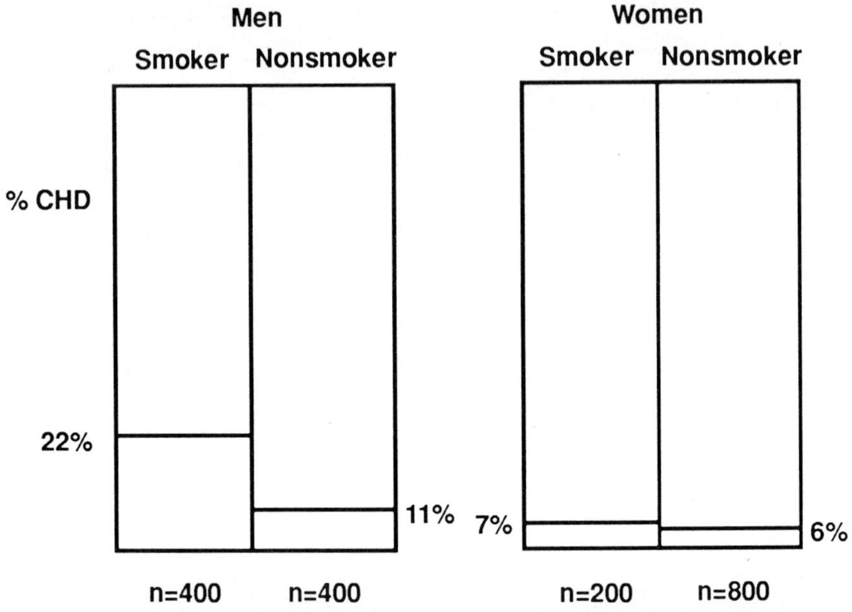

Men
Smoker Nonsmoker
% CHD
22%
11%
n=400 n=400

Women
Smoker Nonsmoker
7%
6%
n=200 n=800

Some of the things that make men more prone to coronary heart disease are documented in the Framingham study. One of them appears to be smoking. About half of the men in the study were smokers. In the youngest age group, at the start of the study there were, thus, roughly 400 each of smokers and nonsmokers. The data on coronary heart disease for these two groups are shown in Figure 8.7. This difference in incidence of CHD is statistically significant.

By contrast, only about twenty percent of women in the same age group were smokers. The data for women are also shown in Figure 8.7. The reported difference is clearly not statistically significant.

8.4 PROSPECTIVE DESIGNS

The Framingham study utilized a *prospective* design. The essence of such a design is that it begins with two groups of subjects, one exhibiting the suspected causal factor (such as smoking) that was absent in the other. Unlike experimental studies, however, the subjects are not assigned to the two groups by the investigators. The members of the population have already selected themselves into one of these two categories. The investigators merely sample from the two subpopulations (for example, smokers and nonsmokers) as they already exist. The prospective feature of these studies is that any occurrences of the *effect,* in either group, would be in the *future* relative to the time the sample is selected. At the beginning of the study, no subject in either group exhibited the effect under investigation. People who already exhibited symptoms of coronary heart disease were excluded from the Framingham study.

The overall strategy in prospective studies is to get two groups that are, on the average, similar in every feature *except* the expected causal factor. If there is a statistically significant difference in the frequency of the effect, then that provides evidence for the causal hypothesis. This type of design is pictured schematically in Figure 8.8. The sample of people having the suspected cause is still called the experimental group even though there is no actual experimentation involved. The investigators do nothing to the subjects. They merely collect information about them and then wait to see what happens.

Comparing Figure 8.8 with Figure 8.2 makes it clear why prospective studies do not provide as direct a test of a causal hypothesis as do randomized experimental studies. In a randomized *experimental* study, the two groups can be regarded as samples from the hypothetical populations in our model of a causal hypothesis. These samples are created through the process of random selection and direct experimental manipulation. *Prospective* studies, on the other hand, are more like tests of correlations in that they are based on samples from the actual population as it exists. Prospective studies, however, differ from tests of correlations in that you are not merely looking at coexisting properties. The two groups are selected from among members of the population who do not yet have the effect. The effect shows up later.

Figure 8.8

Prospective design.

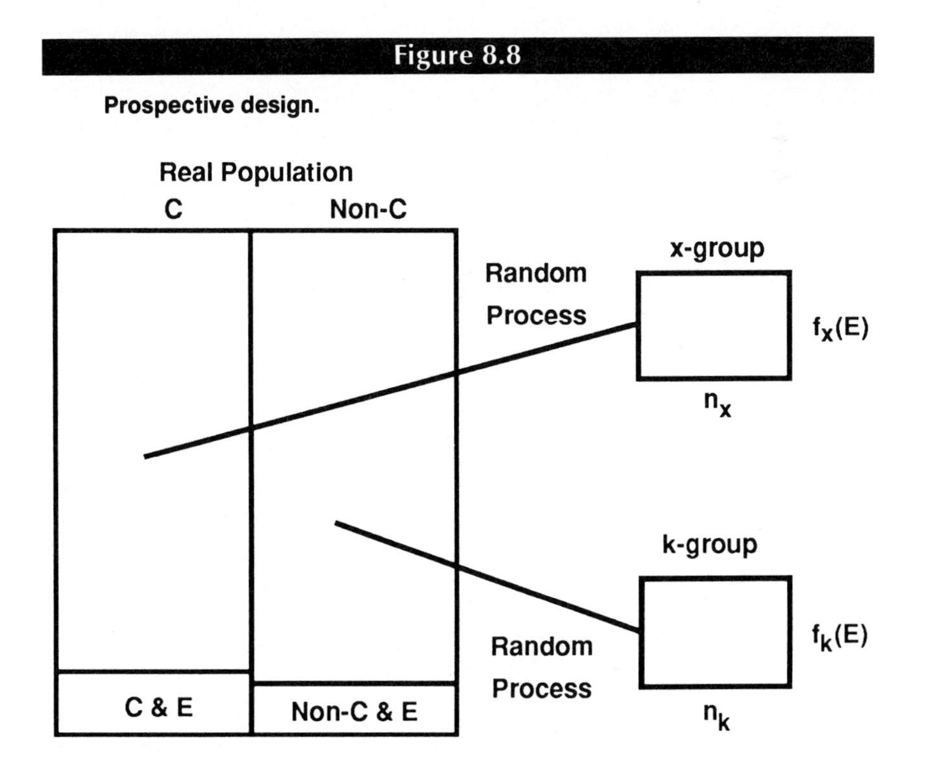

In reading reports based on statistical studies, you must sometimes read very carefully to discover whether the data are based on a one-shot survey or on a genuine prospective study in which subjects initially without the effect are followed through time to see how many in each group later develop the effect. In a one-shot survey, you can ask people whether they had some condition earlier. The reliability of the information you get about the prior condition, however, is likely to be strongly influenced by whether the effect occurred. For example, if you ask people with lung cancer about their previous smoking habits, you are likely to get an exaggerated report, though the exaggeration might go either way. The subjects might feel, "Why me? I didn't smoke that much," or they might feel, "That's what did it. I smoked too much." Either way, one is not getting reliable information. That is one of the main virtues of genuine prospective studies over mere surveys.

Controlling for Other Variables

Step 4 in our program for evaluating causal studies requires us to consider how well the design of the study fits a random sampling model. It is here that prospective studies exhibit a distinct disadvantage relative to randomized experimental studies. This difficulty is illustrated by some results from the Framingham study.

One of the causal hypotheses investigated in the Framingham study is that the consumption of *coffee* might be a positive causal factor for coronary heart disease. Thus, subjects enrolled in the study were originally questioned about their coffee-drinking habits. A first look at the data after twenty-four years suggested that there might indeed be a causal connection between coffee consumption and coronary heart disease. Dividing all the subjects into those who drink coffee and those who do not yields a statistically significant higher rate of coronary heart disease among coffee drinkers than among all others. Figure 8.9 provides a schematic picture of this data.

A more careful examination of all the data, however, reveals that the causal hypothesis is not warranted. The reason is that there is a *correlation* in the population between coffee consumption and cigarette smoking. In particular, smoking is positively correlated with drinking coffee. That is, the percentage of smokers among coffee drinkers is greater than the percentage of smokers among those who do not drink coffee. So the stereotype of the person holding a cup of coffee in one hand and a cigarette in the other has some basis in fact. This correlation is pictured in Figure 8.10.

What this means for the study is that a randomly selected sample of coffee drinkers is very likely to contain a higher percentage of smokers than a randomly selected sample of those who do not drink coffee. If smoking does cause coronary heart disease, then there may be a statistically significant difference between the percentage of cases of coronary heart disease in the two groups, even if drinking coffee is causally irrelevant to coronary heart disease. The extra cases of coronary heart disease among the coffee drinkers may be caused by smoking, not by drinking coffee.

The standard remedy for this difficulty is to *control* for the other variable, which in this case is smoking—coffee drinking and coronary heart disease

Figure 8.9

A schematic picture of the overall Framingham data on coffee consumption and coronary heart disease.

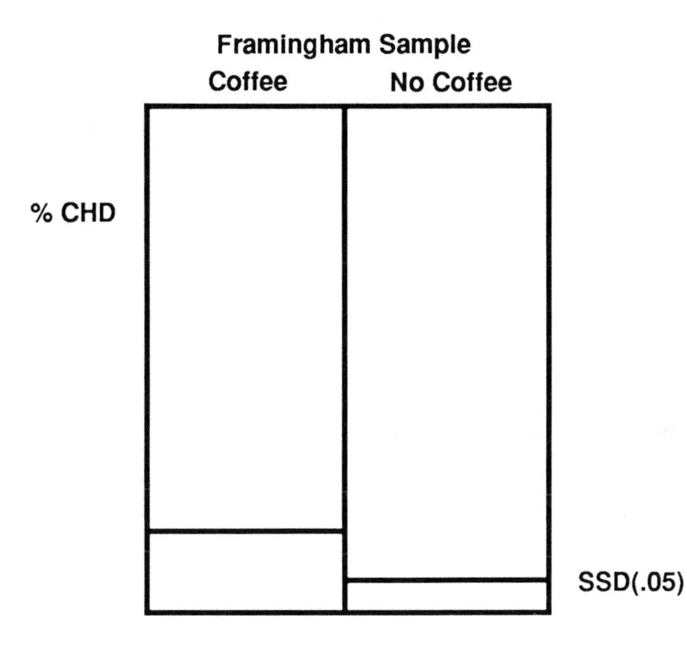

being the original two variables mentioned in the causal hypothesis we are considering. There are several ways to control for a third variable.

One way of controlling for a third variable is to restrict *both* groups to individuals *not* exhibiting the third factor. If there is still a statistically significant difference, it would have to be ascribed either to the originally suspected causal factor or to yet another factor correlated with it. If there is no statistically significant difference between the samples in which the suspected third factor has been eliminated, that is evidence that any effectiveness of the original causal factor in producing the effect by itself is not detectable by the experiment in question.

Thus, investigators in the Framingham study considered only the *nonsmokers* in their sample and then compared those nonsmokers who drink coffee with those who do not. There was no longer a statistically significant difference in the rate of coronary heart disease between these two groups. This data is pictured schematically in Figure 8.11.

An alternative way of controlling for smoking would be to consider *only smokers*. Once again a statistically significant difference in the rate of coronary heart disease would be evidence that coffee is a positive causal factor. Lack of a statistically significant difference would mean there is no evidence that coffee is a positive causal factor.

The disadvantage in this second way of controlling for smoking is that there might be some *interaction* between coffee and smoking that

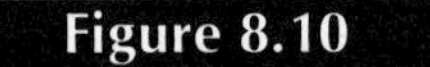

Figure 8.10

The correlation between coffee consumption and smoking in the general population of American adults.

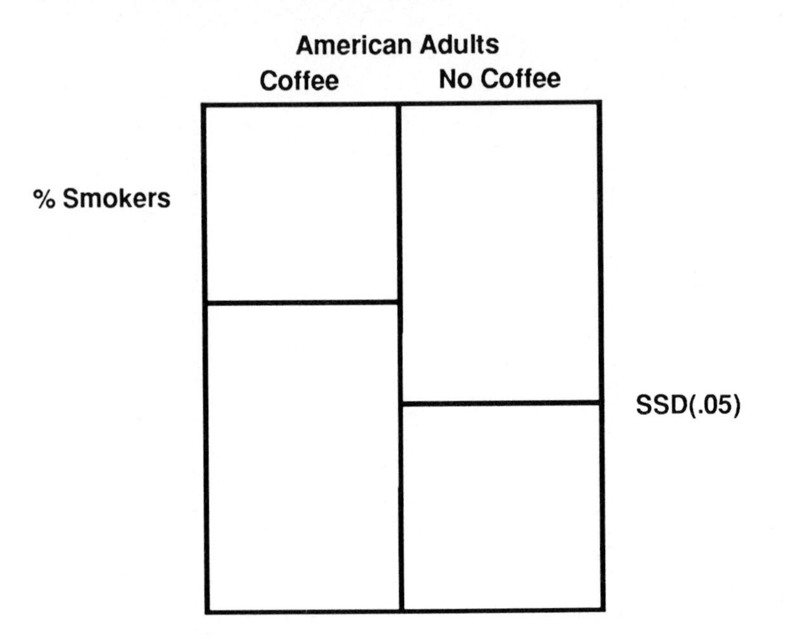

contributes to coronary heart disease even though neither by itself is a positive causal factor. That is, coffee drinking and smoking might be like so-called binary chemical weapons consisting of two relatively harmless chemicals that, when combined, form a highly toxic gas. It is to avoid this sort of possibility that scientists prefer to control by eliminating the suspected third factor altogether.

Controlling for other variables has a cost. It reduces the size of the samples. Thus, there were fewer subjects in the Framingham study who drank coffee but did not smoke than there were who drank coffee, period. A reduced sample size makes it less likely that there will be a statistically significant difference even if the original causal factor is somewhat effective.

To achieve adequate sample sizes, careful designers of prospective studies often proceed by *matching* the incidence of the suspected third factor in the control group with that in the experimental group. Thus, having obtained randomly selected samples of coffee drinkers and those not drinking coffee, the Framingham investigators could have examined both samples for smokers. Then, by randomly eliminating a few smokers from the sample of coffee drinkers, they could have insured that there would be the same number of smokers in both groups. Thus, the effect, if any, of smoking would be equalized in the two groups. Persistence of a statistically significant difference in the rate of coronary heart disease between the two matched groups would be evidence

Figure 8.11

A schematic picture of the Framingham data on coffee consumption and coronary heart disease when controlled for smoking.

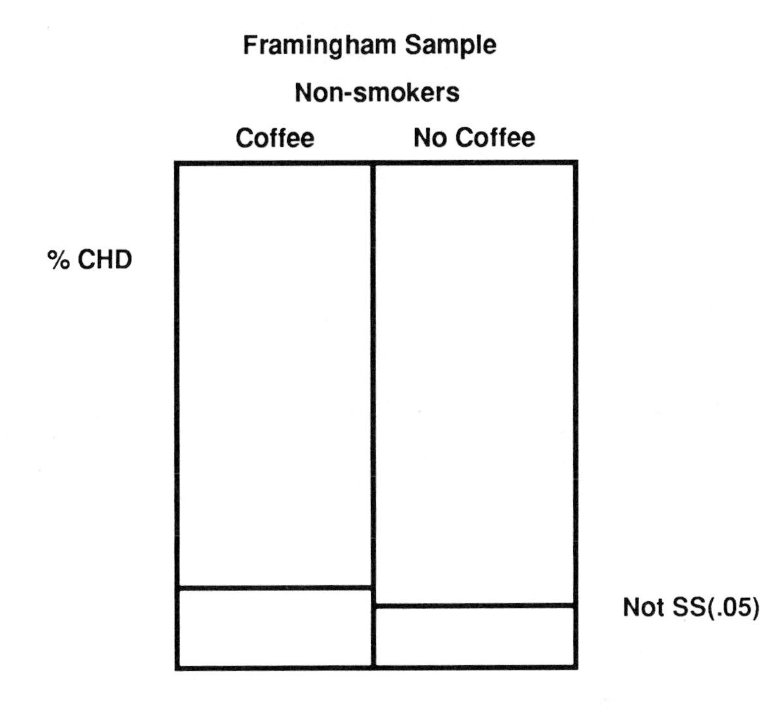

that coffee drinking was indeed a positive causal factor for coronary heart disease.

A drawback to matching samples for suspected third factors is that this strategy does not protect against interactive effects. If drinking coffee contributes to coronary heart disease only for smokers, and not at all for non-smokers, then the experimental group would tend to exhibit more cases of coronary heart disease than the control group with an equal number of smokers. Investigators could be misled into thinking that coffee by itself is a positive causal factor when it is a factor only in interaction with smoking.

Prospective Versus Randomized Experimental Designs

We can now see why randomized experimental designs provide much better evidence for a causal hypothesis than prospective designs. In a randomized experimental design, every characteristic possessed by members of the population is, on the average, matched in the experimental and control groups. This is true even for factors that are totally unknown to anyone. Thus, if, unknown to anyone, half of the population possessed a Q factor, roughly half of both the experimental group and the control group would possess Q.

Of course, because we are dealing with a random process, there is some small probability that the experimental group would end up with a very high percentage of Q's, and the control group would end up with a very low percentage of Q's. If there are many different potentially relevant Q factors, then it may even be quite probable that one of these might end up with very different percentages in the two groups. However, that this particular Q factor should turn out to be highly relevant to the effect would be an unlikely accident. So even randomized experimental designs are not perfect, but the chances of their being terribly misleading are small. Not so with prospective designs, as illustrated by the example of coffee drinking and smoking.

In evaluating a study based on a prospective design, one must always look for indications that other potentially relevant variables have been controlled. If there are no such indications, one must use whatever general knowledge one has to decide whether lack of control for specific, known factors, like smoking, could make the study very misleading. If so, one's conclusion regarding the quality of the evidence for the causal hypothesis must be tempered appropriately.

Evaluating the Framingham Study

Now let us apply our program for evaluating causal studies to the Framingham data on smoking and coronary heart disease. We will begin with the data for men.

Step 1. The Real World Population and the Causal Hypothesis. The population of interest is humans, more particularly, adult American men and women. The causal variable is smoking and the effect variable is coronary heart disease. The particular causal hypothesis to be evaluated is whether smoking is a positive causal factor for coronary heart disease.

Step 2. The Sample Data. The sample consisted of residents of Framingham, Massachusetts. Included in the sample were about 400 each of male smokers and male nonsmokers all aged 30 to 39. There were also about 200 female smokers and 800 female nonsmokers aged 30 to 39. Among male smokers at the end of the study (24 years later), 22 percent had experienced some form of coronary heart disease. Among male nonsmokers the percentage was 11 percent. Among female smokers at the end of the study, 7 percent had experienced coronary heart disease. Among female nonsmokers, the percentage was 6 percent. These data are diagramed in Figure 8.7.

Step 3. The Design of the Experiment. The design fits a prospective model. It is not experimental because there was no random division into experimental and control groups. The subjects selected themselves into the categories of smoker or nonsmoker. All subjects were originally free of the effect, coronary heart disease.

Step 4. Random Sampling. Framingham is not a random sample of all American adults. Within Framingham, however, the sampling was randomly done, except for the 734 volunteers. There was no mention of any other variables controlled for in the data presented.

Step 5. Evaluating the Hypothesis. The study for male smokers in the 30 to 39 age group is diagramed in Figure 8.12. In the observed range of sample frequencies, the margin of error for samples of 400 should be no greater than 5 percent. So the observed difference in the rate of coronary heart disease between smokers and nonsmokers is statistically significant, although not by a large margin. So there is good evidence that smoking is a positive causal factor for coronary heart disease. Its effectiveness in this group of subjects is estimated to be in the range from .01 to .21.

Step 6. Summary. This study is about as well done as any prospective study can be. The main worry is that there were no controls for other possible causal factors for the data presented. Assuming that coffee had been ruled out as an important causal factor, however, there are no other obvious candidates. So, unless there are other causal factors correlated with smoking, the evidence for the hypothesis that smoking is a positive causal factor for coronary heart disease is quite strong.

Figure 8.12

A diagram of the Framingham study as it applies to the hypothesis that smoking is a causal factor for coronary heart disease.

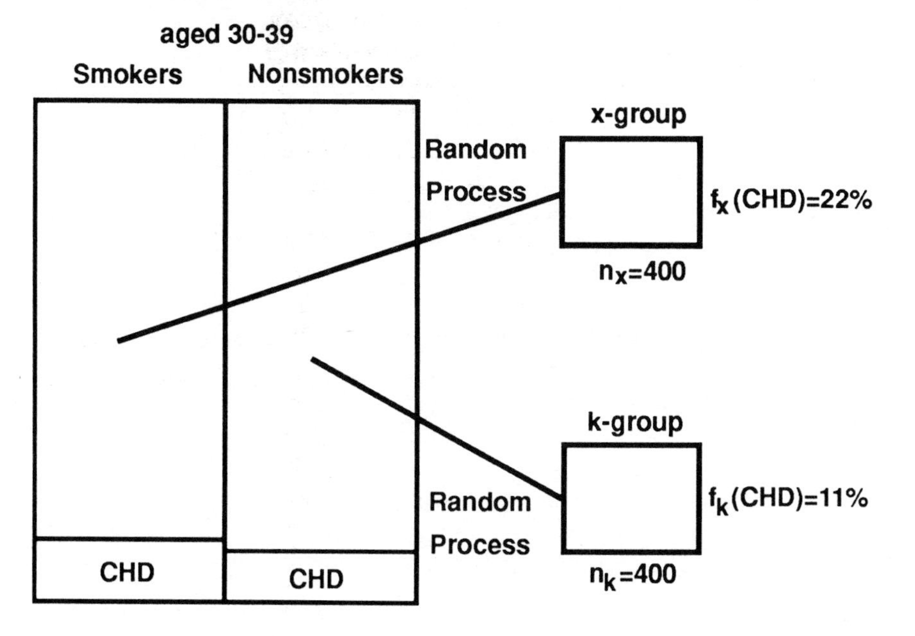

We need not go through the whole analysis for the corresponding data on women. In the observed range of sample frequencies, the margin of error for n = 200 would be around .05 and for n = 800 around .02. Given that the recorded frequency of coronary heart disease among smokers and non-smokers is 7 percent and 6 percent respectively, the interval estimates would clearly overlap. Effectiveness would be estimated in the range from −.06 to +.09, which is comfortably around zero. There is no good evidence that smoking is a positive causal factor for coronary heart disease in this population of women. The evidence is that the effectiveness of smoking in producing coronary heart disease in this population of women is at most quite low.

8.5 BLOOD CLOTS AND THE PILL

In the 1960s, doctors in Great Britain and the United States discovered a surprising number of fatal blood clots in otherwise-healthy, young women whose medical histories made such a condition extremely unlikely. Even though they had only a relative handful of such cases, this was many more than they had experienced in the past in such seemingly healthy, young women. Looking at the medical histories of these few women, they discovered that many of the victims had begun taking oral contraceptives less than a year before the fatal clot appeared. Because the use of such contraceptives had only recently become widespread, the doctors immediately suspected that the pill was responsible for the sudden increase in fatal clots among young women.

The connection between the pill and these few cases of fatal clots could have been a coincidence. So the question arose how to test the hypothesis that the pill really was a causal factor for fatal clots. Oral contraceptives had already been tested on animals, and this particular problem had not arisen. However, no one had been looking for clotting in particular. It would have been possible to begin new animal studies utilizing randomized experimental designs, but those would take several years to complete.

The situation was similar regarding prospective studies. Existing data included no information on blood clots. Given the relatively low incidence of fatal clots in the general population, it would take a year to set up a sufficiently extensive, prospective study and three to five more years for the data to build up. Was there not some faster way to test the hypothesis?

The desire for a faster method of testing the hypothesis is medically understandable because, if it is true, one wants to know as soon as possible in order to warn potential victims. Even if the effect in question had been less serious, there is still good reason for seeking a simpler test. Experimental and prospective studies are costly and time-consuming, and it is impossible to test for every possible effect. So one wants at least some reason for thinking that a particular causal hypothesis might really be true.

The type of study often employed in this sort of situation is based on what is called a *retrospective* design. As you read on, try to figure out why the investigation should be called "retrospective." That will provide you with the main clue to recognizing the essential differences between retrospective and other types of designs. We will explore these differences in greater detail in the following section.

To test the suggested causal hypothesis, doctors in Britain began a search of records in nineteen different hospitals for the previous two years (1964–1966). They looked for women who had been treated for nonfatal clots of the legs or lungs (clots in the heart or brain tend to be fatal). They needed nonfatal cases so that they could locate the women in question and find out if they had indeed taken birth control pills prior to developing the clot. Such information was not generally available in regular hospital records. Their search turned up fifty-eight cases of clots suffered by young, married women with no prior disposition for developing blood clots. Of these fifty-eight women, twenty-six (45 percent) were discovered to have been taking oral contraceptives the month preceding their admission to the hospital for treatment.

By now, you should be aware that the 45 percent figure by itself means nothing. If 45 percent of all British women in that age group had been taking oral contraceptives, this is exactly what one should expect. One needs a *control* group to provide a basis for *comparison.* In the British study, the control group consisted of 116 married women admitted to one of the participating hospitals for some serious surgical, or medical, condition other than blood clotting. Members of the control group were carefully chosen so as to match members of the experimental group in age, number of children, and several other possibly relevant factors. Of these 116 control subjects, only ten (9 percent) were found to have been taking oral contraceptives.

This difference in the relative frequency of pill users in the two groups was reported as being significant, meaning *statistically significant.* The publication of this study quite naturally produced some panic among women taking the pill. After a brief examination of the data, officials of the U.S. Food and Drug Administration issued statements saying, in effect, "Yes, oral contraceptives do seem to cause blood clots, but, no, you should not stop taking them on this account because the risk is very small."

Evaluating the risk and assessing its relevance for individual decisions whether to take the pill is a topic for Part Three of this text. For the moment let us concentrate on the prior question of whether there was indeed good evidence in favor of the suggested causal hypothesis.

8.6 RETROSPECTIVE DESIGNS

The British study of blood clotting and the pill utilized a *retrospective* design. The name comes from the fact that the study begins with a sample of subjects that already have the *effect* (a blood clot) and attempts to look back in time to discover the *cause* (oral contraceptives). The study is backward looking—unlike both experimental and prospective designs, which are forward looking.

Random sampling plays almost no role in retrospective studies. What is still called the *experimental* group is chosen from among subjects who already have the *effect* being studied. Women who have been treated in a hospital for a nonfatal clot obviously are not a random sample of women. Also, the *control* group is deliberately chosen so as to match the subjects in the experimental group for other variables that might be causally relevant.

You will note that the British study matched each experimental subject, fifty-eight in all, with two control subjects, 116 in all. The number of experimental subjects they could find was fairly small. They could not increase this number very easily. However, they could increase the number of control subjects, so they did. This increased their chances of obtaining a statistically significant difference, if there was indeed any difference in the population to be found. The only situation in which random sampling might be used in a retrospective study is if there are more potential experimental subjects or matched-control subjects than are needed. At that point, the actual groups might be randomly selected from among the available subjects.

Figure 8.13 pictures the essentials of a retrospective design. You should note in particular that the frequencies observed in the two groups are not frequencies of the effect but frequencies of the *cause.* Similarly, the subpopulations from which the samples come are not those that do or do not have the suspected cause, but those that do or do not have the *effect.* In retrospective studies, the roles of the cause and effect variables are, roughly speaking, reversed. Thus, you can almost always tell when a reported study is retrospective because the frequencies given will be the frequencies of subjects with the cause in groups of subjects that either all have or all do not have the effect.

Figure 8.13

Retrospective design.

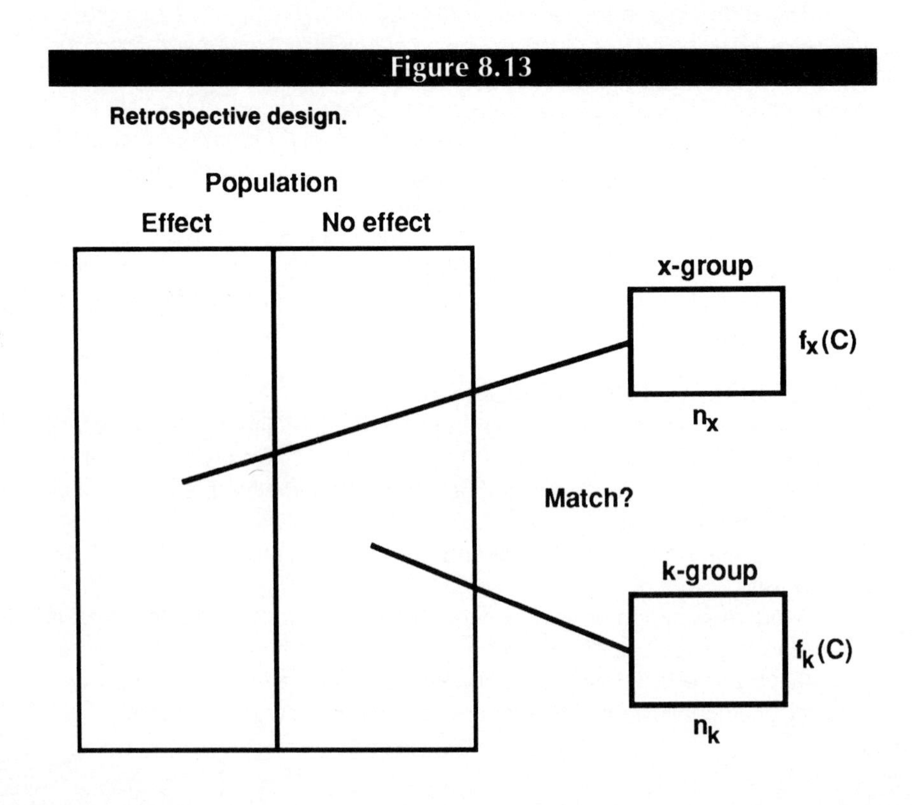

Reports of experimental and prospective studies always report the frequencies of the effect in the two different groups. Note that reference to all other controlled variables is included in this diagram.

This reversal of the roles of cause and effect in the design also shows up, of course, in the way one would diagram the observed frequencies. The data from the British study are pictured in Figure 8.14.

Controlling for Other Variables

Like prospective studies, retrospective studies suffer from the possibility that the *self selection* of subjects into subpopulations with the effect and without the effect may be biased in a way relevant to the percentage of subjects exhibiting the suspected causal factor. It is for this reason that an attempt is made to match the control subjects with the experimental subjects.

The subjects in the British study, for example, were matched according to whether they had any children. Women who have been pregnant have a higher risk of getting a blood clot. Also, in England in 1965, most of the women taking the pill had already had at least one child and were taking the pill so as not to have any more, or at least not for a while. So, if in looking at clot victims you automatically find more women who have had children, you would also find more women on the pill, whether or not the pill had anything to do with producing clots. Matching the two groups for the existence of previous pregnancies evens out the possible influence of this causal factor.

Figure 8.14

The data in the British pill study.

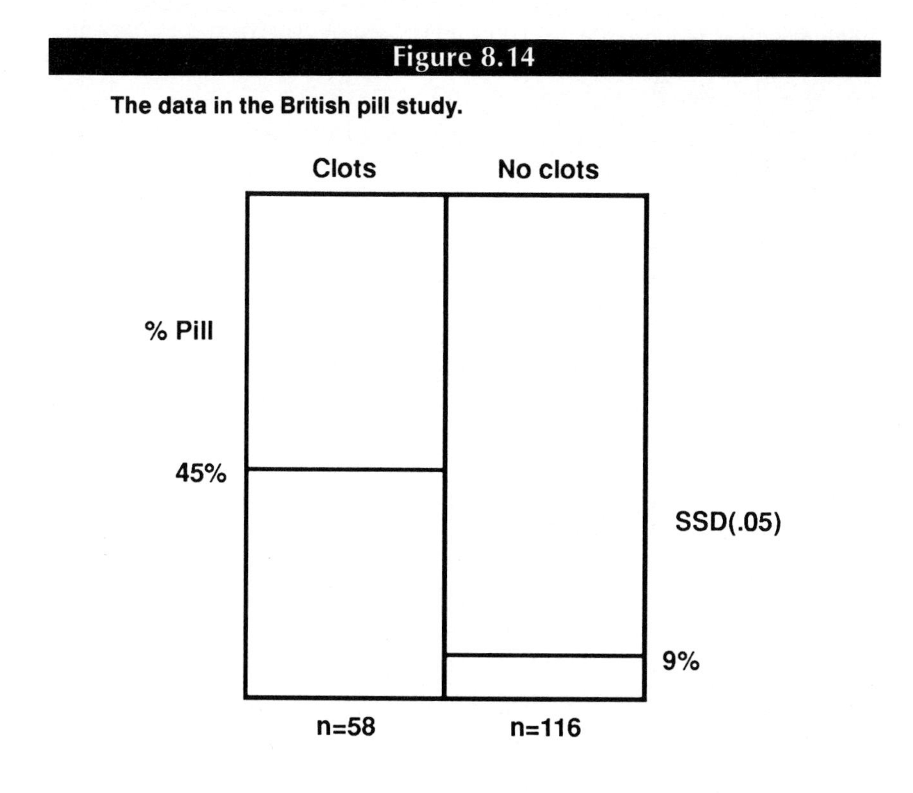

As in the case of prospective studies, it is impossible to control for every variable that might be correlated with taking the pill. The assumption that there are no relevant correlations among uncontrolled variables is always somewhat questionable. However, if the control group was matched with the experimental group for most variables that, on the basis of previous experience, might be relevant, that is the most one can do. If someone suggests a further variable that might be controlled, that can sometimes be done after the study is completed.

Nonrandom Selection of the Experimental Group

Retrospective studies suffer from a further possible failure to fit a random sampling model, a possibility that is not so serious in the case of prospective studies. This is that the experimental group itself could hardly be said to be a random selection from the population—even from the subpopulation of subjects with the effect. Subjects generally get into the experimental group because something special has happened to them—for example, they were hospitalized or saw a doctor. These special features may well be correlated with the suspected cause.

It is well known that people who have reason to believe they may develop some conditions are more likely to call a doctor when something suspicious occurs than people who have no reason to be suspicious in the first place. This fact is relevant because women taking oral contraceptives for the first time are generally warned that they may experience various side effects. Such women are, therefore, much more likely to call their doctor about various conditions than women who have not had this warning and are not wondering whether what they are experiencing might be due to the pill. This means that among all the women who may develop minor clots in their legs or lungs, those taking oral contraceptives are more likely to call a doctor, and therefore more likely to have the pain diagnosed as a clot and to be hospitalized for treatment. Maybe the higher frequency of pill users among the fifty-eight clot patients was simply an artifact of this "reporting factor."

Although it is easy to think of nonrandom selection processes that might produce misleading data, it is generally very difficult to control for such processes. The present example might be dealt with by considering only women whose initial symptoms were serious enough that they would have been hospitalized no matter how weak or strong their inclination to call a doctor for minor aches and pains. It is doubtful, however, that the initial study contained enough subjects in this category to produce statistically significant results. In any case, there are many other selection factors that might have been operative and would be more difficult to control.

Effectiveness and Retrospective Studies

There is an additional disadvantage to retrospective studies. The data they yield allow no estimate of the *effectiveness* of the causal factor. Effectiveness is defined in terms of the percentage of the population that would or would not get the *effect* depending on whether they all had the *cause*. The frequencies of the effect in two samples, as in a prospective study, may be used to estimate the effectiveness of the causal factor in the population. But retrospective studies give you the frequency of the *cause* in groups with and without the *effect*. There is no way to use these frequencies to estimate the effectiveness of the causal factor. For example, knowing that 45 percent of women who had been hospitalized for clots had been using the pill tells you nothing about the percentage of women using the pill that were hospitalized for clots. However, this is the percentage you need to estimate effectiveness.

This feature of retrospective studies is especially serious because the effectiveness of the causal factor is the most useful information when it comes to making decisions. Just knowing that something is a positive causal factor is of limited usefulness in decision making. But that is the very most you can get out of a retrospective study.

On the other hand, it is sometimes possible to combine information from a retrospective study with other information to reach an estimate of effectiveness. In the case of the pill studies, British scientists were able to use other information on the occurrence of blood clots and the use of oral contraceptives to determine a rough estimate of the probability of blood clots among women using the pill. This probability turned out to be roughly fifteen per million. That was about seven times the probability of clots in the population at large, but still quite small. So the FDA was right. It is a causal factor, but the risk is small.

Evaluating Retrospective Studies

The basic strategy for evaluating retrospective studies is the same as that for evaluating experimental and prospective studies. One looks for a statistically significant difference between the frequency of the cause in the experimental and control groups. If there is such a difference, that is evidence that there is a connection between the cause variable and the effect variable in the population. If there were no connection at all in the population, and all selections of subjects were random, any differences in frequency of the cause in the two groups could only be due to the chance effects of sampling. By definition, statistically significant differences are unlikely to be due to chance alone.

The trouble with retrospective studies is that selections are generally not random. Thus, the observed statistically significant difference between the two sample groups may all too easily be merely a reflection of biases in the selection process that produced the samples. In addition, there is the possibility of correlations existing in the population that, even with random sampling, may lead to a statistically significant difference in the sample groups even though there may be no causal connection between the variables being studied.

As a rule of thumb, it is probably best always to regard conclusions based on retrospective studies as more or less tentative. A causal hypothesis supported by retrospective data is not to be ignored and a positive result should be taken seriously by scientists and responsible agencies. Retrospective data may provide a good reason to undertake other studies, either experimental or prospective. Evidence for causal hypotheses based on retrospective data alone, however, cannot be regarded as being as good as evidence based on prospective or experimental studies.

Evaluating the British Study

As a summary of our discussion of retrospective studies, let us apply our standard program for evaluating causal studies to the British data.

Step 1. The Real World Population and the Causal Hypothesis. The population of interest consists of adult British women. There is a strong presumption that the results would apply also to other European and American women. The hypothesis is that taking birth control pills (of the type available in Britain in the middle 1960s) is a positive causal factor for serious blood clots.

Step 2. The Sample Data. Among fifty-eight women who had developed nonfatal blood clots, twenty-six (45 percent) had been taking oral contraceptives. Among 116 other women matched for age, number of children, and several other possibly relevant factors, only ten (9 percent) were found to have been taking oral contraceptives. These data are diagramed in Figure 8.14.

Step 3. The Design of the Experiment. The study was retrospective. This is made clear by the fact that the experimental group consisted entirely of subjects exhibiting the effect (blood clots), while the control group consisted entirely of subjects not exhibiting this effect.

Step 4. Random Sampling. There was almost no random sampling. The experimental group consisted of just about all the cases that could be found in a two-year period. Members of the control group, however, were carefully matched with members of the experimental group.

Step 5. Evaluating the Hypothesis. Figure 8.15 pictures the study. Assuming a random sampling model, the difference in percentages of pill use in the two groups is clearly statistically significant. The difference is 36 percentage points (45 − 9). The margin of error for the experimental group is about 15 points and the margin of error for the control group is about 10 points. The confidence intervals do not overlap. There is, therefore, evidence supporting the causal hypothesis.

Step 6. Summary. The careful matching of control subjects to experimental subjects somewhat compensates for the nonrandom selection of the experimental group. Also, the observed difference in rates of pill use between the two groups was quite large. The study must be taken as providing at least some evidence that pill use is a positive causal factor for serious blood clots.

It must be emphasized that the relevance of this particular study for the use of currently available contraceptive pills is limited. In the years since this British study, there has been much improvement in the quality of contraceptive pills, particularly in lowering the doses of active ingredients. The study is useful now mainly because it provides a dramatic example of an importantly successful retrospective study.

Figure 8.15

A diagram of the British pill study.

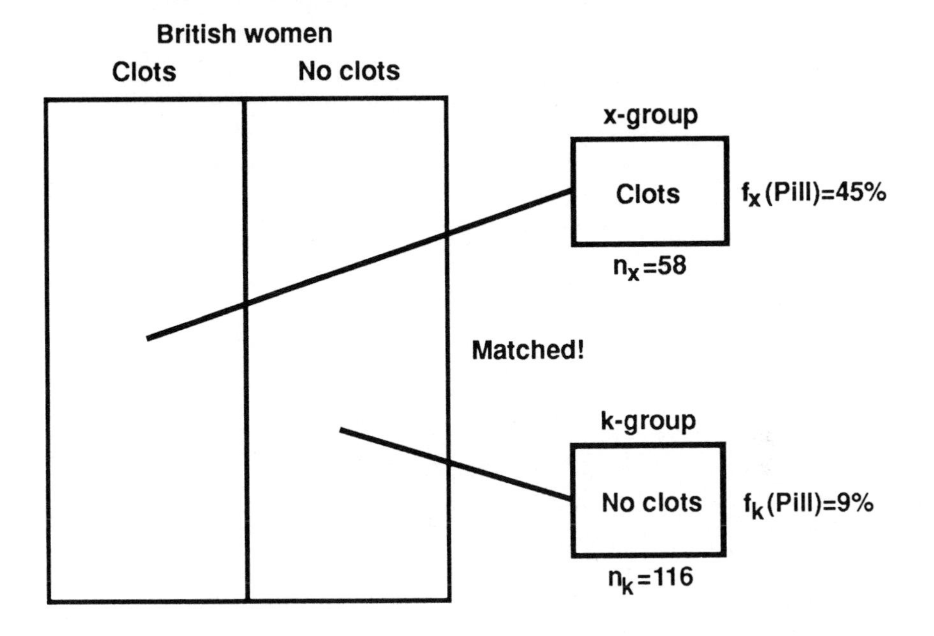

8.7 STATISTICAL EVIDENCE FOR CAUSAL HYPOTHESES

Although the use of statistical studies to investigate causal hypotheses is widespread, there are recurrent controversies about this practice. The objections cut both ways. The tobacco industry provides a classic example of one sort of objection. For twenty-five years, spokesmen for the industry have publicly disputed whether statistical studies can possibly prove the existence of a causal connection between smoking and any diseases. Often, this dispute centers on the crucial difference between experimental and prospective designs. There are, of course, no experimental studies of the effect of smoking on humans. Prospective studies, however, always leave open the possibility that there is an unknown factor correlated with smoking that is the real culprit, but smoking itself is innocent. Thus, according to this suggestion, people who choose to smoke are already different from those who do not so choose, and it is this unknown difference, not smoking, that causes heart disease, lung cancer, and other diseases.

On the other side, advocates of unorthodox treatments for new and deadly diseases, such as AIDS, often argue that retrospective data based on a few cases provide sufficient evidence to license new treatments. Experimental or prospective studies, that may take considerable time, are said to be unnecessary and to deny unjustifiably the new treatment to members of the control group and to the affected population at large.

To conclude this chapter, we will examine both objections. Both are dangerously misleading, but for different reasons.

Can Prospective Studies Prove a Causal Connection?

In prospective studies, scientists can control only for factors with which they are already familiar. Moreover, the number of variables that can be controlled in a single study typically is not very large. The more variables controlled, the more difficult it is to maintain a large enough sample for a sensitive and reliable test. This means that in practice a study cannot eliminate more than a moderate number of ways that the observed significant difference could have been produced by biased self-selection rather than by the proposed cause. In theory, things are even worse. In principle, there are innumerable ways that the self-selection process might be biased. So no matter how many variables might be matched, the theoretical possibility of biased self-selection still remains. Because of this possibility, no prospective study can ever provide as good evidence for a causal hypothesis as a randomized experimental study. A question mark always hangs over even the best prospective studies.

In some cases, however, the question mark has been reduced to the barest, theoretical possibility. In the mid-1960s, the National Cancer Institute sponsored a prospective study that enrolled well over 400,000 men between the ages of 40 and 80. Using a computer, the investigators matched almost 37,000 smokers with an equal number of nonsmokers for the following remarkably large number of other variables: age; race (white, black, Mexican, Indian, Oriental); height; nativity (native born or foreign born); residence (rural or urban); occupational exposure to dust, fumes, vapors, chemicals, radioactivity, etc.; religion (Protestant, Catholic, Jewish, or none); education; marital state; consumption of alcoholic beverages; amount of sleep per night; usual amount of exercise; nervous tension; use of tranquilizers; current state of health; history of cancer other than skin cancer; and history of heart disease, stroke, or high blood pressure. Every pair of men exhibited the same values of all these different variables, but in each case one was a smoker and one a nonsmoker.

After nearly three years and over 2000 deaths, the rate of death from all causes was twice as great for the smokers as for the nonsmokers. Nearly half of all the deaths was because of heart disease, with the rate for smokers being double that for nonsmokers. Even more dramatically, the death rate from lung cancer for smokers was nine times that for nonsmokers.

In addition, the larger study from which the matched pairs were selected provided evidence for correlations that are very difficult to reconcile with any factor other than smoking. Death rates for smokers increase with years smoked, amount smoked, and with amount of inhalation. For former smokers, death rates decrease with the length of time since quitting. It is difficult even to imagine what other factor, or factors, might produce these correlations. The tobacco industry continues to claim that smokers are "a different kind of people" who smoke because they are "different," and also get lung cancer, heart attacks, and so on, because they are different. But this is just a theoretical possibility for which there is no positive evidence, whatsoever.

So even though it is technically true that statistical data from a prospective study cannot prove the existence of a causal relationship, the overall evidence can be so strong that there is no basis for questioning the conclusion apart from the bare, theoretical possibility that there is an unknown factor correlated with smoking.

Double-Blind Studies

Although randomized experimental studies are more easily performed with nonhuman subjects, special care must always be taken to insure that the experiment itself does not introduce complicating factors that might affect the result. Thus, for example, in animal studies requiring surgical operations, the control subjects are typically given "sham" operations to insure that the effect is not produced by the overall surgical procedure itself rather than by the particular cause under investigation.

Experiments involving human subjects require even more subtle precautions. Consider, for example, the following experiment as described in Linus Pauling's famous book, *Vitamin C and the Common Cold*.

> A Swiss investigator, Dr. G. Ritzel, reported the results of a double-blind study. He studied 279 skiers, half of whom received 1 gram of ascorbic acid [vitamin C] per day and the other half an identical, inert placebo. He reported a reduction of 61 percent in the number of days of illness from upper respiratory infections in the vitamin C group as compared with the placebo group. These results have high statistical significance (probability in two samples of a uniform population only 0.1 percent).

To understand what double-blind means we must first consider what it means to do a study "blind." By definition, a study is *blind* if the subjects cannot tell whether they are in the experimental or control group. Because informed consent requires that the subjects know the overall purpose of the experiment, the control subjects must be given something that they cannot detect as not being vitamin C. Otherwise, all subjects would know immediately in which group they were. In the case Pauling cites, control subjects were given pills made of a known inactive substance that they could not distinguish from vitamin C. Any such inactive substance that is given to control subjects as a substitute for the suspected cause is known as a *placebo.*

The reason for going through all this trouble is to avoid a *placebo effect.* That is, it can happen that merely knowing that one is in the experimental group and believing that the suspected causal agent does work is enough to produce some effect. For example, it has been well documented that patients suffering from headaches or other minor pains often report a lessening of the pain when given pills that they believe to be pain killers but that in fact contain no pain-relieving drugs at all. So one can produce an effect with a placebo. In the present example, experimenters must worry that people might be able psychologically to suppress a mild, incipient cold, or that they might describe inaccurately how they feel, simply because they believe they are taking something that prevents colds. That could lead to a finding of fewer colds in the experimental group, even if there were in fact no difference between the two groups. There are corresponding worries about subjects in the control group.

A study is called *double-blind* if the experimenters making the diagnoses are also kept in the dark about which subjects are in which group. The reason for this is that an experimenter may diagnose borderline cases in the direction that favors personal bias if it is known to which group a particular subject belongs. For example, there may be a question whether to diagnose a particular slight cough as a cold symptom. If the experimenter wants the experiment to show that vitamin C does prevent colds and knows this particular subject is taking vitamin C, there may be a bias in favor of not calling it a cold. Such biases, even if completely unconscious, might produce an apparent difference when none really exists. Of course, someone in charge of the overall experiment must have a list of which subjects are in which group so that the data can be compiled at the end of the experiment. That person, however, cannot be performing diagnoses on subjects.

The clever use of placebos makes it possible sometimes to use the *same* subjects both as experimental and control subjects. For example, in one of the few experimental tests of the effects of marijuana on humans, nine marijuana-naive volunteers were tested for various psychomotor and psychological abilities, including hand-eye coordination and ability to concentrate. In this experiment, the same subjects served in both the experimental and control groups. That is, they were given either a measured amount of marijuana or a placebo in a series of sessions. Which subjects got the marijuana and which got the placebo in any particular session was determined randomly by an experimenter other than those actually conducting the tests. The test scores achieved by the subjects when they in fact had the marijuana were regarded as the experimental data. The scores achieved by the same subjects when they in fact had the placebo were regarded as the control data. In this experiment, the average score of the subjects was significantly lower when they received the drug than when they received the placebo.

The advantage of this sort of design is that the subjects in the experimental and control groups are automatically matched for every possible additional characteristic. They are, after all, the same people in both phases of the experiment.

A similar experiment was carried out with nine experienced marijuana users as subjects. These latter tests, however, could not be done blind because it proved impossible to devise a placebo that could fool the subjects into thinking that they were getting marijuana when they were not. The average scores achieved by these subjects were not significantly different when given the drug and when not. However, the small size of the sample made it unlikely that any but a quite strong effect would show up as a statistically significant difference.

The Ethics of Experimental Design

You should now have a clear idea why, of all statistical studies, randomized experimental designs provide the best evidence for the existence of a genuine causal factor. It should also be clear why, in some cases, insisting on a randomized experimental design can create severe ethical conflicts.

A clear example of such a conflict occurs in the testing of new treatments for life-threatening diseases such as AIDS. Use of a randomized experimental design requires that subjects run a fifty-fifty chance of ending up in the control group. If that happens, then, by necessity unknown to them, these subjects would not be receiving the new treatment, but another, possibly less effective, treatment. This, however, is only a possibility. Because it has not yet been tried as part of a randomized experimental study, there can be no really good evidence regarding the effectiveness of the new treatment. The whole point of the experiment is to determine if it is, indeed, more effective than existing treatments. Nevertheless, there must at least be some evidence that the new treatment might be better. Otherwise there is no basis for the experiment in the first place. And control subjects end up being deprived of the new treatment.

The situation is further complicated by the fact that the new treatment may not be approved for general use until the results of the experiment are known. Successful experiments typically are part of the approval process. So victims not part of the experiment have no legal means of getting even a chance to experience the new treatment. Is that right? Should not anyone have the right to any treatment they want, regardless of the evidence available?

Making approval of a new treatment depend on favorable results in a randomized experimental trial puts a premium on having available only treatments that are in fact very likely to be effective. Allowing approval with lesser evidence would mean that more people would end up using treatments that not only are not the most effective available, but also might in fact be worse than no treatment at all. That could, in the end, mean greater expenditures of public funds.

The best course of action cannot be determined by scientific experimentation. That is a problem for the political process. It is a matter of public policy, not science, under what conditions a new treatment should be available to the public. Science can only assure us that using randomized experimental trials is the most reliable way to determine the actual effectiveness of a new treatment. It provides no basis for insisting that the most reliable method must always be followed. There may be strong moral or political reasons to settle for less.

8.8 SUMMARY

Figure 8.16 provides a scheme for beginning an analysis of most studies appealing to statistical data. The first decision is whether the study is merely statistical or an investigation of a causal hypothesis. If just statistical, then you must determine whether you have data for a distribution or a correlation. A distribution has only one variable; a correlation has two. Here you apply the program for evaluating statistical hypotheses developed in Section 6.6. If the study is not just statistical, but causal, then you must determine the type of design employed.

Table 8.2 summarizes important characteristics of all three designs for investigating causal hypotheses. Familiarity with differences among the three types will help you in identifying the type of design in reports of studies as you encounter them. Retrospective studies are easily distinguished from the other two types because of the nature of the experimental and control groups and the fact that reported percentages are percentages of subjects who had the

Figure 8.16

A classification schema for studies using statistical data.

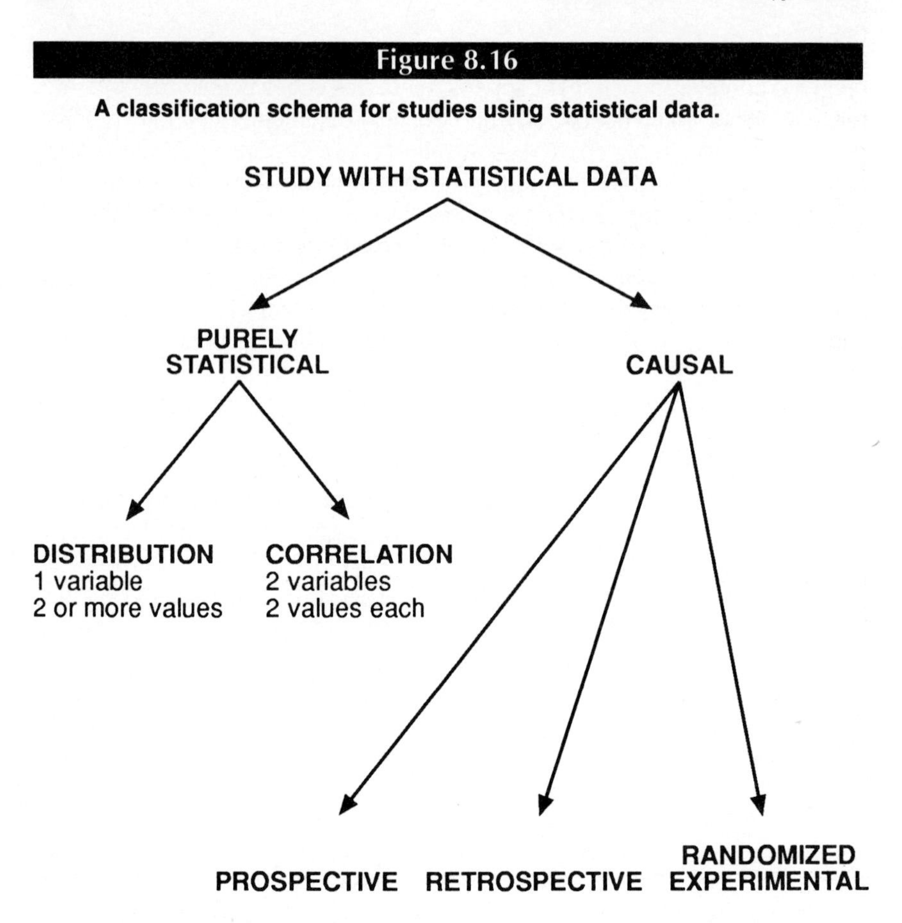

cause, not percentages of those who developed the effect. If, by this indicator, the study is clearly not retrospective, then your best strategy is to look for indications of random assignment, perhaps with use of a placebo. That would indicate a randomized experimental design rather than a prospective design. You must learn to make these determinations because your best indication of the reliability of the conclusion may be simply the type of design employed. Here you employ the program for evaluating causal hypotheses developed in Section 8.2.

EXERCISES

Analyze these reports following the six-point program for evaluating causal hypotheses developed in the text. Number and label your steps. Be as clear and concise as you can, keeping in mind that you must say enough to demonstrate that you do know what you are talking about. A simple "yes" or "no" is never a sufficient answer. Many of these reports are taken directly from recent magazine or newspaper articles and are presented here with only minor editing.

Table 8.2

Characteristics of Designs for Evaluating Causal Hypotheses

	Random Assignment	Random Sampling	Other Control Variables	Description of Groups	Description of Percentage Data	Estimate of Effectiveness
Randomized Experimental	Yes	Maybe	Not necessary	X- All C K- All Non-C	%E	Possible
Prospective	No	Maybe, sometimes not	Possible	X- All C K- All Non-C	%E	Possible
Retrospective	No	Maybe, usually not	Possible	X- All E K- All Non-E	%C	Not possible

Exercise 8.1

British Study Says Pill Increases Breast Cancer Risk

Young women who take birth control pills for more than four years run a significantly increased risk of breast cancer, according to a major British study. The study, published in the prestigious medical weekly The Lancet, *is the most comprehensive look at the pill and breast cancer undertaken in Britain and one of the largest in the world.*

Earlier studies on the relationship between the pill and breast cancer have reached conflicting results. A U.S. Food and Drug Administration committee of experts has said that recent research on possible links between the pill and breast cancer was inconclusive.

The new British study involved 755 British women under age 36 who were diagnosed as having breast cancer between the beginning of 1982 and the end of 1985, and an equal number of similar but cancer-free women for comparison. The researchers said 94 percent of the women with breast cancer and 86 percent of the comparison group had used the pill at some time.

The differences were similar for women who used the pill before and after their first full-term pregnancy, the researchers said.

The researchers calculated that among women younger than thirty-six, there is a 43 percent increase in the risk of breast cancer after four years of pill use and a 74 percent increase after eight years. They stressed, however, that the risks had to be kept in perspective. "Breast cancer is uncommon below age 36, the age group that was studied. Only one woman in 500 develops the disease before age 36, so even a 70 percent increase in risk would only put the chances of developing breast cancer by this age up to about one in 300," the researchers said.

Britain's Family Planning Association and National Association of Planning Doctors said they were advising pill users who are concerned about the findings to continue their course and discuss their cases with their doctors when convenient. They said the risks had to be balanced against the pill's benefits in protecting women against ovarian and other types of cancer.

The Committee on the Safety of Medicine said it was writing to all doctors saying that, in light of the uncertainties, there was no need for a change in prescribing practice. They called on doctors to tell women about the possible risks of breast cancer before they prescribe contraceptive pills, but they did not advocate avoiding oral contraceptives.

Clair Chilvers of the Institute of Cancer Research in London, one of the authors of the study, advised women to aim for the lowest-dose pill available and stay on it for the shortest possible time. "Women who are not having intercourse should stop taking it," she said.

Exercise 8.2

Vitamin C and the Common Cold: The Salisbury Study

The following report on a study done at the Common Cold Research Unit in Salisbury, England, is adapted from Linus Pauling's book, Vitamin C and the Common Cold.

These investigators reported observations on human volunteers and concluded that "there is no evidence that the administration of ascorbic acid has any value in the prevention, or treatment, of colds produced by five known viruses." Of the ninety-one human volunteers, forty-seven received 3 grams of ascorbic acid per day for nine days and forty-four received a placebo. They were all inoculated with cold viruses on the third day. In each of the two groups, eighteen developed colds. The incidence of colds observed in the subjects receiving ascorbic acid ($^{18}/_{47}$) was 6 percent less than that in the control group ($^{18}/_{44}$). This difference is not statistically significant.

The number of subjects, ninety-one in the two groups, was not great enough to permit a statistically significant test of a difference as large as 30 percent in the incidence of colds in the two groups to be made, although a difference of 40 percent, if it had been observed, would have been reported as statistically significant (probability of observation in a uniform population equal to 5 percent).

Exercise 8.3

Study Finds Decaf Coffee May Raise Cholesterol

Drinking decaffeinated coffee appears to raise levels of harmful cholesterol in the blood and might play a role in heart disease, according to a recent study.

Drinking decaffeinated coffee boosts low-density lipoprotein, or LDL—the so-called bad cholesterol—an average of 7 percent, the research found. That could translate into about a 12 percent increase in the risk of heart disease, said Dr. H. Robert Superko, director of the Lipid Research Clinic at Stanford University, who directed the research.

"Does coffee cause heart disease?" Superko said. "I don't know. We can only say that coffee causes cholesterol changes. It's not like going out and eating cheesecake, but the overall impact could be great" considering decaffeinated coffee's popularity.

In the latest study, presented at the American Heart Association's annual meeting, coffee-drinking volunteers were randomly assigned to one of three groups—regular coffee, brewed decaffeinated, or no coffee. Cholesterol levels remained steady in those who stayed on caffeinated coffee or gave up coffee entirely. But in the decaf drinkers, LDL levels rose 9 milligrams per deciliter, or 7 percent.

Superko said he doubts that the process of taking out caffeine changes the coffee so that it raises cholesterol. He said the difference may result from the kinds of beans that go into different types of coffee.

He said people who are concerned could experiment on themselves. He suggested that they get their cholesterol measured twice, then give up decaf for a couple of months and get it checked again to see if there is a difference.

Exercise 8.4

Sharp Cut in Serious Birth Defect Is Tied to Vitamins in Pregnancy

Women who take over-the-counter, multivitamin pills early in pregnancy sharply reduce the risk of having a baby with a type of serious neurological defect, a new study has found.

These neural-tube defects are among the most common and most devastating, causing effects varying from death to paralysis. They affect 4000 babies a year, or 1 to 2 in 1000.

In the study of about 23,000 pregnant women, about half said they had taken multivitamin pills containing folic acid in the first six weeks of pregnancy. The incidence of neural-tube defects in babies born to those who took the pills was about one-fourth that of women who did not take them. The study, the largest on the subject so far, is published in The Journal of the American Medical Association.

The chief author of the study, Dr. Aubrey Milunsky, director of Boston University's Center for Human Genetics, said he felt that all women should take over-the-counter, multivitamin pills containing folic acid while trying to become pregnant and during the first six weeks of pregnancy. "Our data are impeccable," Dr. Milunsky said. . . .

The idea that vitamin deficiencies might have something to do with neural tube defects originated after World War II, Dr. Milunsky said, when malnourished women in England, Holland, and Germany gave birth to unexpectedly large numbers of babies with neural-tube defects. Since then, researchers have been trying to see if vitamin supplements can prevent the birth defects. . . .

In the new study, Dr. Milunsky's group questioned about 23,000 women who were about sixteen weeks pregnant and were having prenatal tests for birth defects, including neural-tube defects. The women were questioned for about half an hour in an interview that produced twelve pages of information on all aspects of health practices. About half the women said they had taken multivitamin tablets containing folic acid in the first six weeks of pregnancy.

The researchers then determined which women had fetuses with neural-tube defects. They found that the incidence of neural-tube defects was 0.9 per 1000 in the women who took vitamins in the first six weeks of pregnancy but was 3.5 per 1000 in the women who did not.

Dr. Milunsky said he deliberately tried to avoid a methodological problem that had plagued previous studies. In those studies, including one by the Child Health Institute and one by the federal Centers for Disease Control that found that vitamins protected against neural-tube defects, women were asked after they had had their babies whether they had taken vitamins. Such studies can suffer from a recall bias because women can selectively remember, or forget, what they took if they have an abnormal baby.

But the new study does not meet what researchers consider the highest standard, a randomized controlled study. In such a study, women would take either a multivitamin pill or an inactive dummy pill in the first few weeks of pregnancy, and the incidence of neural-tube defects in the two groups would be compared. It remains possible that women who took vitamins on their own, and not because they were randomly assigned to take them, differed in some other way from women who did not. This difference, rather than the vitamins, could have accounted for the lowered incidence of neural-tube defects. . . .

Exercise 8.5

Heavy Smoking Costs Women Years of Life

A team of New York medical researchers said recently that a heavy upsurge of cigarette smoking among American women is causing a soaring increase in fatal heart attacks.

These warnings are based on a study of 194 Westchester County, New York, women who died between 1967 and 1971. The researchers autopsied the women and questioned relatives about the women's cigarette-smoking habits. The researchers said forty-one of the women suffered sudden, fatal heart attacks, and the other 153 women died from other causes.

Comparing the two groups, the researchers found 62 percent of the heart attack victims had been heavy cigarette smokers — twenty or more cigarettes daily. In contrast, only 28 percent of the women who died from other causes had been heavy smokers.

Exercise 8.6

Study: Cancer Patients in Support Groups Live Longer

In a new study of women with breast cancer that has spread, researchers were surprised to find that the women who belonged to a psychotherapy group lived almost twice as long on average as those who did not participate in a support group.

The research involved eighty-six women with metastatic breast cancer, as diagnosed by specialists.

They were randomly assigned either to a special therapy group or to a control group that received no psychotherapy. Women in both groups continued to get normal medical care.

After one year, the women in the therapy group reported less depression, anxiety, and pain than did the others.

More strikingly, when the researchers compared death rates after ten years, they found that those in the therapy group survived an average of thirty-seven months from the beginning of the study, while those in the other group lived an average of nineteen months.

Only three of the eighty-six women were still alive after ten years, and they were all in the therapy group.

"We were shocked when we saw the magnitude of the effect," said Dr. David Spiegel, a psychiatrist at Stanford University who conducted the study. "We expected no biological effect from the psychological one."

The study, published in The Lancet, *a British medical journal, does not mean that psychotherapy can cure cancer, Spiegel said. However, it does suggest that therapy can improve the quality of life for those dying of cancer, and it may extend their lives.*

But the scientists are at a loss to explain why the therapy has any effect on survival time and have urged continued research.

An editorial accompanying the study noted that because the therapy groups had such diverse elements, it was not clear precisely what might have accounted for the extended longevity.

The editorial said, however, that the methods of the study itself "are beyond criticism," and it encouraged other groups to pursue this method for the emotional management of cancer.

The research has also been well received by some cancer experts who have been skeptical of other psychological approaches to the disease, which are often well publicized but poorly researched.

In an issue of Science, Dr. Jimmie Holland, chief of psychiatry at Memorial Sloan-Kettering Cancer Center in New York, called it "the first study to show a change in survival from psychotherapy that is scientifically sound."

But Holland expressed a fear, shared by Spiegel, that some practitioners of alternative cancer treatments would cite the findings to encourage patients to abandon regular medical care in favor of a positive-thinking approach.

Spiegel said, "At no time did we take such an approach. Our focus was on living as fully as possible."

Exercise 8.7

The XYY Syndrome

After rumors that mass murderer Richard Speck had an XYY chromosome (normal males have XY), there was considerable speculation that this genetic abnormality was itself responsible for violent behavior. Investigations into the genetic makeup of men imprisoned for violent crimes, such as murder and robbery, yielded an XYY rate of roughly 1 percent for inmates of several large penitentiaries. The incidence of this abnormality among the general noncriminal population was determined to be roughly one-tenth of this (that is, .1 percent). Given the size of the samples involved, this difference is statistically significant at the .05 level.

Further investigation, however, has shown the XYY males tend to be taller and stronger, but less intelligent, than normal males. It is known that bigger men are more likely to commit violent crimes just because they are more powerful; and less intelligent men are more likely to be caught and imprisoned. This suggests controlling for the variables of size and intelligence—for example, looking at men over 6 feet tall with IQ's less than 100. When this is done, the percentage of XYY's in the criminal group remains about 1 percent, but that in the noncriminal group increases to roughly .8 percent. This difference is not statistically significant (.05).

Exercise 8.8

Study Shows Exercise Cuts Heart-Attack Risk

A study of nearly seventeen thousand Harvard alumni has found that there were fewer heart attacks among those who participated regularly in strenuous exercise than among those who were less active.

The study was begun in 1968 by determining the health and exercise patterns of men who had entered Harvard as freshmen between 1920 and 1950. Of these, 16,936 were found to be free of heart disease. Each man was classified as having a high, or low, exercise level depending on whether he expended more or less than 2000 calories a week on exercise. By 1976, 572 of the men had experienced a serious heart attack. However, the rate of heart attacks was 64 percent higher in the low-exercise group.

It was suggested that perhaps the high exercisers were simply those people who were more healthy and athletic all along. They exercised more because they were more healthy, not vice versa. This suggestion, however, was ruled out because a check of men who had participated in varsity sports as undergraduates showed the same pattern of heart attacks depending on their level of activity in later life.

Exercise 8.9

Drinking in Pregnancy Harmful, Study Says

Even moderate drinking by women in the first month or two of pregnancy, often before they realize they are pregnant, can impair the child's intellectual ability upon reaching school age, a new study indicates. . . .

The scientists interviewed 491 Seattle women in the fifth month of pregnancy and followed up with periodic assessments of their children, measuring intelligence, reaction time, and attentiveness.

The study also took into account factors such as parents' incomes and educations, which are known to affect a child's intelligence, and found that drinking had an effect apart from these influences.

The impairment was noticed even when the pregnant woman cut back her drinking in the first or second month of pregnancy. The most-published finding from the research involved 53 mothers who had on average three drinks or more a day in the first month or so of pregnancy.

Their children were found at age 4 to score substantially lower on intelligence tests than the children of mothers who did not drink at all. Specifically, the average score on IQ tests for these children was 105, 5 points below the average for the other children in the study.

Separately, the researchers have reported that children born to mothers who had as little as one to two drinks a day in the first months of pregnancy were found by their early school years to have a significantly slower reaction time and to have difficulty paying attention. These conclusions were also based on the 491 women and their children.

Although previous studies have indicated that very heavy drinking by a pregnant woman can cause mental retardation in her children, the Seattle research is the first to show effects on the intellectual capacities of children at school age whose mothers drank at moderate levels when pregnant.

It is also the first study to distinguish the effects of alcohol from those of other factors, such as smoking or caffeine use.

"We recommend that women who are trying to become pregnant, or who might become so, do not drink alcohol at all," said Dr. Ann Streissguth, a psychologist in the Psychiatry and Behavioral Sciences Department at the University of Washington School of Medicine, who directed the Seattle study. "The effects on children occur even at the social drinking level." The women in this study did not see themselves as having alcohol problems. . . . The Seattle researchers asked the 491 pregnant women in their study about drinking, smoking, and the use of marijuana, as well as the use of aspirin, caffeine, and other substances.

Exercise 8.10

Smoking, Heart Attacks, and the Pill

Since the use of birth control pills became widespread during the mid-1960s, many hospitals and clinics have followed the medical histories of their patients to determine whether women who choose to use the pill experience effects different from those who choose other contraceptives or none at all. In the early 1970s, it was discovered that women on the pill had a higher death rate from heart attacks than others. Typically, these studies found that the death rate from heart attacks for women using the pill was around 24 deaths per 100,000 women of age 40–44. The comparable rate for women not using the pill was around 7 per 100,000. Given the large samples, this difference is statistically significant.

Although the government issued official warnings based on this information, these studies were not controlled for smoking. This is hard to believe, given the then well-known connection between smoking and heart attacks in men. When researchers finally got around to controlling for smoking some ten years later, the death rates from heart attacks were found to be as follows: nonsmokers not on the pill, 7/100,000; nonsmokers on the pill, 7/100,000; smokers not on the pill, 14/100,000; and smokers on the pill, 59/100,000. Again, all reported differences are statistically significant.

Exercise 8.11

Scientists Study Aspirin, Heart Attacks

Aspirin, a drug that is taken for everything from arthritis to removing warts, may have still another use: preventing heart attacks. The most convincing evidence comes from two studies conducted by researchers at Boston University, who since 1966 have been compiling data relating drug exposure to the diagnosis of patients admitted to hospitals.

The first study involved 325 heart-attack patients and 3807 other patients. It revealed that only .9 percent of the heart-attack patients had been taking aspirin regularly before their attack (for arthritis, for example). Aspirin-taking was much higher—4.9 percent—among those patients who had not had heart attacks.

A second follow-up study of 451 heart-attack and 10,091 other patients at twenty-four Boston hospitals showed the same kind of results: 3.5 percent of the heart-attack patients said they had been taking aspirin regularly compared with 7 percent of those with other diagnoses.

Exercise 8.12

Findings on Aspirin Confirmed in Report

Healthy men over 50 years of age who take one aspirin every other day cut the risk of having a heart attack nearly in half, a nationwide study has found, confirming preliminary reports announced earlier.

The final report of the study's findings, which is published in The New England Journal of Medicine, *went beyond the preliminary one, however, by asserting that the benefit of aspirin was evident only for men over the age of 50. The study included men 40 years old and older. . . .*

"We have found a clear-cut, conclusive benefit from taking aspirin in preventing the first heart attack," said the study's leader, Dr. Charles H. Hennekens of Harvard Medical School and Brigham and Women's Hospital in Boston. . . .

In an editorial accompanying the study, doctors from Mount Sinai Medical Center in New York said the study indicated that it was reasonable to advocate aspirin for patients with a high risk of heart disease but that it should be used "cautiously, if at all," in others. . . .

The research, known as the Physicians' Health Study, examined 22,071 healthy, male doctors who were 40 to 84 years old at the time the study began in 1982. Of those, 11,037 took 325 milligrams of aspirin every other day and the remainder took placebos, harmless and ineffective substitutes.

The men who took aspirin had 44 percent fewer heart attacks than those taking placebos. Among those who received aspirin, there were ten fatal heart attacks and 129 non-fatal heart attacks. Among the nonaspirin group, twenty-six died of heart attacks and 213 had nonfatal ones. . . .

Exercise 8.13

AIDS Therapy Experiment Yields Encouraging Results

A new AIDS therapy that involves taking a drug, withdrawing some of the patient's blood, shining ultraviolet light on it, and then reinfusing it, has shown promising preliminary results in a study of five patients, researchers say.

In a report published in The Annals of Internal Medicine, *a leading medical journal, the researchers said the treatment appeared to stimulate the immune system in ways that were not clear.*

Until the findings are confirmed in larger studies, the treatment should be limited to research, but an editor of the journal wrote in an editorial that the findings were important enough to merit publication now so that other researchers could consider joining the research.

The therapy, called photopheresis (foto-fer-EE-sis), was tested on five patients with AIDS-related complex, a precursor to AIDS that produces swollen lymph nodes, night sweating, and other symptoms, but not the wasting, opportunistic infections and cancers of the disease itself.

In all five, the monthly therapy shrank swollen lymph nodes, stimulated the immune system, and stopped fevers and night sweating.

None of the patients in the study received any other treatment for AIDS. . . .

One patient who had difficulty climbing a flight of stairs before starting the treatment can now jog five miles a day and lift weights, the study's authors said in an interview. . . .

They said, however, that one woman, an intravenous drug user, lost seven pounds and did not regain the same degree of energy shown by the others, although laboratory tests showed that her immune system had improved.

A fifth patient who showed improvement dropped out of the study after nine months. The researchers said he disliked the hospital environment.

The authors and editors of the journal stressed that the study was preliminary, involved a small number of patients, and was not scientifically controlled. They said further research was needed.

The study was begun in 1988 to determine the safety of the treatment, which was adapted from treatments used for psoriasis and a rare form of cancer. In each treatment session, the patients took 8-methoxypsoralen, one of a class of compounds called psoralens that are derived from Ammi majur, *a weed native to the banks of the Nile. Two hours later, a pint of blood was removed and spun in a machine outside the body to separate red blood cells from the white cells of the immune system. The red cells were returned to the body.*

The white cells, which had absorbed the drug, were exposed to specific wavelengths of ultraviolet light in a device outside the body. The light activates the drug and the irradiated white cells are reinfused into the patient in a procedure that lasts four hours. A second treatment was given the next morning. The cycle is repeated monthly.

The authors speculated that the therapy worked by stimulating an immune system that had been weakened by the AIDS virus, but they said that they did not know the precise mechanism. They likened the weakened immune system to a partly dead battery and the photopheresis therapy to a charger that delivered a jump start.

The researchers had no control group, in part because it would have been unethical to give some AIDS-infected individuals a sham therapy that could have produced such dangerous side effects as shock.

They said there was no way to determine whether the treatment would improve chances of survival, or alter the course of AIDS. But they said the five patients did better than expected in the course of the fourteen-month study. For example, they said they would have expected a drop in an important measure of the immune system, the number of CD4 white cells in the blood. But the volunteers did not show such a drop, and none developed opportunistic infections.

Dr. Elias Abrutyn, associate editor of the journal, said it had published the study despite a number of serious, unanswered questions because "AIDS is a calamitous illness for which we need additional therapies, and positive responses were reported for a treatment unique for an infectious disease."

Dr. Abrutyn said "a hazard of such publication is the premature use of findings and treatments that are not yet fully evaluated and substantiated."

The Food and Drug Administration has given the Morristown researchers approval to expand the AIDS photopheresis study to twenty patients. The researchers said the study was fully enrolled and under way. . . .

Exercise 8.14

Benzpyrene and Lung Cancer

It has been known for some time that prolonged contact with the chemical benzpyrene can cause skin cancer. It has also been discovered that small amounts of this chemical are contained in cigarette smoke and in polluted, urban air. This suggests that benzpyrene may be the agent that causes lung cancer in smokers and in inhabitants of large cities.

To test this hypothesis, a specialist in environmental medicine studied the health of 5900 roofing workers. These workers breathe large amounts of benzpyrene, which is an ingredient in coal tar pitch and asphalt—both common roofing materials. The amount of exposure to benzpyrene for these workers was estimated to be equivalent to that from smoking thirty-five packs of cigarettes a day. At the beginning of the study, all of the workers had been members of the Roofers Union for at least nine years. Six years later, the number of lung cancer deaths recorded among these workers (43) was reported to be not significantly different from the number expected in the general population.

Exercise 8.15

Saccharin and Cancer: Human Link Is Found

The case against saccharin has been strengthened by a new Canadian study linking the artificial sweetener for the first time to human bladder cancer. An earlier study, also Canadian, had found a similar link in laboratory animals.

Researchers for the National Cancer Institute of Canada and three Canadian universities compared 480 men and 152 women who had developed bladder cancer with equal numbers of both sexes who had not. The scientists found that the frequency of use of saccharin among men with bladder cancer was 60 percent higher than that of the other group of men. Although this difference was regarded as statistically significant, the difference in saccharin use in the two groups of women was not.

Exercise 8.16

Artificial Sweeteners Do Not Increase Risk of Bladder Cancer

A recently published study concludes that the ingestion of artificial sweeteners, at least at moderate dietary levels, is not associated with an increased risk of bladder cancer. Earlier studies have shown that high doses of saccharin can cause such cancers in rats.

This study involved 519 humans, all patients with bladder cancer, at nineteen Baltimore-area hospitals between 1972 and 1975. These patients were matched with a like number of patients of the same age and sex who did not have bladder cancer.

The researchers questioned the cancer patients and the matched controls about their consumption of artificial sweeteners in carbonated and noncarbonated soft drinks, iced tea, and other liquids, and in salad dressings, candy, ice cream, pastry, gelatin, chewing gum, and other foods. All subjects were also questioned about smoking habits, occupational history, diabetes, and other factors that might have been involved.

Exercise 8.17

Study Casts Doubt on Link between Media, Teen Suicide

The notion that teenagers who read or hear about teen suicides will be more likely to kill themselves appears to be false, according to a study by the U.S. Centers for Disease Control. No evidence exists that media reports of suicide prompt others to take their own lives, said the study, published in the Journal of the American Medical Association.

The study examined two clusters of teen suicides in Texas that occurred from February 1983 to October 1984. Researchers based their conclusions on interviews with the parents of eight teenagers from one cluster and six from the second cluster.

Teenagers who committed suicide were found to have had lives disrupted by frequent changes in schools, residences, and parental figures. The youngsters also had been prone to emotional illness or substance abuse requiring hospitalization, the study found.

Such youngsters also had been more likely to injure themselves or others and to have threatened or attempted suicide. They were more likely to have lost either a girlfriend or boyfriend shortly before their deaths, the study found.

Information gathered from the victims' families was compared to that from forty-two living teenagers in a control group matched by school districts, grade, race, and sex. Those who committed suicide had been less likely than the control subjects to have talked about death or suicide as well as less likely to have seen television shows about suicide, the study found.

Dr. David Clark, director of the Center of Suicide Research at Rush-Presbyterian-St. Luke's Medical Center in Chicago, disagreed with the study's findings and said it was conducted using research standards that have changed in the five years since it was done. He said that current standards require interviewing not only the parents but also brothers and sisters of the suicide victim and friends because friends often know things about the victim the parent does not.

Exercise 8.18

High-Fat Diets in Pregnancy May Lead to Future Tumors for Fetus

Women who eat high-fat diets when pregnant may be disposing their fetuses to cancers of the reproductive system later in life, a researcher reported. Such a connection would be particularly worrisome because physicians routinely advise pregnant women to increase their consumption of red meat, whole milk, eggs, and cheese. These foods are loaded with the vitamins and minerals needed for fetal development, and with fat.

Prenatal exposure to fats dramatically increased tumors of the ovary, uterus, and pituitary gland in female mice, said Dr. Bruce Walker, a professor of anatomy at Michigan State University. He reported his findings at a meeting of the American Association for Cancer Research in San Francisco.

Studying 261 female mice over their two-to-three-year life spans, Walker found that among the daughters of mice who had received a high-fat diet during their pregnancy—equivalent to the fat consumed by the typical American—54 percent developed reproductive-system tumors. Among daughters of mice who consumed a low-fat diet—equivalent to the diet of a typical Japanese woman—only 21 percent developed such tumors.

Walker said there are enough critical similarities between mice and humans to justify concerns about cancers of the breast, ovary, and uterus in women and testicular and prostate cancer in men.

A link between exposure to fats in the womb and later reproductive-system cancers could clarify the continuing mystery over the role of dietary fat in the high rate of breast cancer in the United States, Walker said. Breast cancer occurs far more frequently among U.S. women, who obtain an average of 37 percent of daily calories in fat, than in Japanese women, who get about 20 percent from fat.

Researchers have not been able to confirm that dietary fat promotes breast cancer. Among Japanese-American women who eat typical U.S. diets, for example, some have breast-cancer rates approaching that of the United States, others approaching that of Japan.

Further examination of studies of these women demonstrates, Walker said, that Japanese-American women born in Japan have the lower breast-cancer rates, while those born in the United States have the higher rates. The difference can be explained, he said, by the fat contents of food these women's mothers consumed while pregnant.

Exercise 8.19

Limb Amputation and Heart Disease

A new study that says limb amputation can significantly increase the risk of death from heart disease may mean increased benefits for some military veterans. A National Academy of Sciences study, commissioned by the Veterans Administration, is the first hard evidence of a long-suspected link between amputation and heart disease.

Earlier studies on civilians have shown increased rates of heart disease for amputees as compared to the rest of the population. However, these studies are questionable since many amputations are performed because of vascular problems in the limbs, and such problems are common among people who already have a prior history of heart disease.

Similarly, a 1969 study of 5000 Finnish Army amputees of World War II showed higher rates of heart disease than among the civilian population. This study was also questionable because veterans, having originally been drafted partly because of good health, tend to live longer than civilians, and the incidence of heart disease increases with age.

In the new VA study, medical records of about 12000 wounded, World War II veterans were examined. Among these, roughly 3900 had lost one or more limbs and about 5000 others had suffered serious wounds not requiring amputation. Upon tracking the records down to the present time, the amputees were found to have a 150 percent higher death rate from cardiovascular disease than the other wounded veterans.

In addition, around 3000 veterans who had lost just a hand or a foot were found not to have a significantly higher rate of heart disease than wounded veterans without amputations.

The VA is planning to extend benefits for those who lost an arm or leg but not for those who lost hands or feet.

Exercise 8.20

Suspicious Link Seen between Coffee and Pancreatic Cancer

Harvard University scientists have found a suspicious and alarming association between coffee drinking and cancer of the pancreas, one of the fastest-killing of all cancers. Though the link is not yet proved and more studies are needed, the Harvard study's main author, epidemiologist Dr. Brian MacMahon, said Wednesday he gave up coffee a month ago. "I'm fairly convinced this is a real thing," he said, but "I would not give any advice to others yet. I'm in the business of collecting facts."

The data, reported in The New England Journal of Medicine, *shows that about 24,000 Americans a year get cancer of the pancreas, a large abdominal organ that secretes digestive enzymes and insulin. Only 1 percent or 2 percent will survive as long as five years. This dismal record means that pancreatic cancer ranks only after lung, colon, and breast cancers as a cause of death.*

MacMahon's Harvard School of Public Health team studied 369 patients with this cancer and compared them with a control group of 644 patients with other disorders. There were only twenty noncoffee drinkers among the pancreas cancer patients, but eighty-eight, or 13 percent, among the controls, enough to make a reliable statistical base. The inquiry showed drinking up to two cups of coffee daily had apparently increased the risk of pancreas cancer by 1.8 times or nearly two-fold, and drinking three or more cups increased it by 2.7 times or nearly three-fold.

If the coffee-pancreas cancer link is confirmed, just over 50 percent of such cancers may be caused, or partly caused, by coffee, the Harvard group believes. The chance of developing such cancer, like most cancer, is generally small in younger years and increases with age. Many kinds of cancer take twenty to thirty years to develop.

If coffee is indeed implicated in half of all pancreas cancers, said MacMahon, a noncoffee drinker aged 50 to 54 might have seven chances in 100,000 of getting pancreas cancer in any single year; one aged 60 to 64, nineteen chances in 100,000; one aged 65 to 69, twenty-seven chances—with coffee drinkers' chances about doubled or tripled.

The Harvard group also found a "weak association" between cigarette use and pancreas cancer, but none with cigars, pipes, alcohol or tea. The average cup of tea, one not too strong, has about half as much caffeine as the average cup of coffee but since they both contain caffeine, "we don't believe caffeine is the guilty substance," MacMahon said, "though it's not yet ruled out."

Coffee has been under investigation in recent years in connection with several disorders. There is no strong evidence that moderate coffee use increases the risk of heart disease, despite some allegations. But last year the Food and Drug Administration warned pregnant women they should stop or "minimize" coffee, tea, and cola drinking because the caffeine in all three may cause birth defects. The evidence here, too, is not yet conclusive, said FDA officials, but there is enough evidence to warn the "prudent."

Exercise 8.21

New Surgeries for Breast Cancer

Until very recently, the standard treatment for breast cancer was the Halsted radical mastectomy, a procedure developed in the 1890s by Dr. William S. Halsted. This operation removes the affected breast, the underlying pectoral muscles, all related lymph nodes, plus some additional skin and fat. The radical nature of this surgery has led many women to fear the treatment as much as the disease itself, often prolonging the time until diagnosis, and thus, tragically, increasing the need for the operation while decreasing its chances of success.

Recently, women's groups and some doctors have argued that, although this procedure may have been justified in Halsted's day, when most cancers were diagnosed only in later stages and other treatments did not exist, it is not justified today. Our understanding of cancer, the means for diagnosis, and methods of treatment are all much improved since that time. Some doctors have thus experimented with less radical procedures, including a simple mastectomy, which removes only the breast, and various forms of partial mastectomy, the most conservative of which removes only the tumor itself. All surgical treatments are now often combined with both radiation and chemical therapies.

From a patient's standpoint, the less radical procedures are clearly preferable, but are they effective? An early study in Finland compared 527 cases of radical mastectomy with 339 cases of partial mastectomy, all treated between 1948 and 1961. By 1971, there was no significant difference in the ten-year survival rates for the two groups. This study, however, was criticized on the grounds that the patients selected for the less radical treatment might have been those who, for other reasons, were more likely to survive, anyway. Similar criticisms were directed at a French study of 514 patients treated by partial mastectomy between 1960 and 1970. The latter group showed no significant difference in ten-year survival when compared to a similar group of 304 American patients treated by radical surgery.

In 1971, the National Cancer Institute began a study involving 1700 patients at thirty-four Canadian and American institutions. The women in the study agreed to be assigned at random to different treatment groups, including radical surgery and simple mastectomy. Six years after the last patient joined the study, there was no significant difference in the survival rates of the treatment groups.

Exercise 8.22

Zinc May Slow Blindness

Eye doctors are both excited and cautious about a new report suggesting that zinc may halt the progress of macular degeneration, a leading cause of visual impairment and blindness in older people. Macular degeneration, which is untreatable, damages the nerve and pigment cells of the macula, the area at the center of the retina that is responsible for sharp vision, such as for reading.

Dr. David A. Newsome of the Louisiana State University Eye Center in New Orleans studied 151 patients. Eighty of them took a 100-milligram zinc tablet twice a day; the others, a placebo. After a year to two years, the zinc-treated group had significantly less vision loss and fewer adverse retinal changes.

Because this was a small study and large amounts of zinc can be toxic, Newsome and other ophthalmologists are not recommending increased amounts of zinc for the general public. Still, preliminary findings "make us optimistic about further research," says Newsome. A larger study of 600 to 1000 patients is planned and should determine conclusively whether zinc helps.

Federal surveys reveal that most older people do not get enough zinc in their diets. Until further research is completed, perhaps the prudent thing to do is to eat good sources of zinc such as red meat, nuts, and oysters.

Exercise 8.23

Drug Is Found to Cure Some with a Deadly Liver Disease

For the first time, researchers have been able to cure patients with chronic hepatitis B, a devastating and often progressively debilitating liver disease that afflicts more than a million Americans and is the ninth leading cause of death worldwide.

Results from a clinical trial involving 126 patients at twelve medical centers throughout the country indicate that interferon alpha-2b, a synthetic version of a naturally occurring, immune-system hormone, cures 10 percent of patients and reduces the viral infection in 30 percent more. Even using the most exquisitely sensitive probes for the hepatitis virus, investigators could find no signs of the deadly microbe in the cured patients' cells. Chronic hepatitis B is a major cause of liver cancer and cirrhosis. Until now, there has been no way to treat this chronic condition. . . .

The new results, being published in The New England Journal of Medicine, *are "a major step forward," said Dr. Richard*

Aach, a hepatitis expert who is the director of medicine at Mount Sinai Medical Center in Cleveland and vice chairman of medicine at Case Western Reserve University School of Medicine.

"They are very important findings," said Dr. Jay H. Hoofnagle, who wrote an editorial accompanying the paper. He said interferon "is the first agent shown to be effective in this chronic viral infection."

The new study was sponsored by Shering-Plough, and led by Dr. Robert P. Perillo, the director of gastroenterology at the Veterans Affairs Medical Center in St. Louis and an associate professor of medicine at Washington University in St. Louis. Half the patients received a therapeutic dose of interferon for sixteen weeks and the rest received a much smaller dose or a dummy substance for comparison. . . .

Dr. Harvey Alter, a hepatitis expert at the Clinical Center of the National Institutes of Health, cautioned that the interferon treatment "is not easy." The drug was given in daily injections, and nearly everyone who took it suffered side effects, including fatigue, muscle aches, and irritability.

Dr. Hoofnagle said, "Although most patients can make it through the four months of treatment, all are quite anxious to get off." And for 60 percent of the patients, the treatment had no benefits.

"This is the first evidence we have that anything works," he said. "But the trouble is that for the remaining 60 percent, we don't know what to do. . . ."

Exercise 8.24

Margarine, Too, Is Found to Have the Fat That Adds to Heart Risk

Margarine, the spread millions use instead of butter in hopes of preventing heart disease, contains fatty acids formed during processing that actually increase coronary risk, according to a recent study.

These fats, called trans monounsaturated fatty acids, were shown, in a well-designed Dutch study, to raise blood levels of a harmful form of cholesterol and to lower levels of a form that protects against heart disease.

Consumption of margarine has been encouraged as a replacement for butter, which is high in saturated fat. Most saturated fats, which are also prominent in meats, eggs, poultry, high-fat dairy products, tropical oils, and other foods, raise the blood levels of a harmful form of cholesterol. Margarine consists largely of unsaturated fats, both polyunsaturates and monounsaturates, all forms of which had been thought to lower blood cholesterol.

The new findings, published in The New England Journal of Medicine, *"do not mean people should switch back from margarine to butter," because butter still has a more damaging effect on blood cholesterol than any margarine, said Dr. Scott M. Grundy, an expert on coronary risks of dietary fats who wrote an accompanying editorial in the journal. . . .*

The trans compounds are among the fatty acids formed when polyunsaturated vegetable oils are converted to margarines and vegetable shortenings that are solid or semi-solid at room temperature. The new study showed that trans fatty acids could raise blood levels of harmful LDL cholesterol and lower protective HDL cholesterol.

The amounts of LDL and HDL in the blood have been shown to significantly affect a person's risk of heart attack. LDL cholesterol appears to encourage the formation of fatty deposits that clog arteries, the precursor to heart disease, whereas HDL cholesterol washes excess cholesterol out of the body.

In the new study, Ronald P. Mensink and Martijn B. Katan, prominent nutrition researchers from the Agricultural University in Wageningen, the Netherlands, placed fifty-nine men and women on carefully controlled diets that differed only in the kinds of fatty acids they contained. Each of the diets was followed for three weeks by the participants, and the effects on their blood levels of cholesterol were measured.

The researchers showed that the manufactured trans monounsaturated fatty acids were as harmful to blood cholesterol levels as cholesterol-raising, saturated fatty acids, whereas the natural monounsaturated fatty acid was beneficial. . . .

Previous studies have yielded conflicting findings as to the possible role of trans compounds in heart disease. Dr. Grundy noted that the Dutch study was the first to isolate carefully the effects of individual fatty acids on serum cholesterol and thus, the study was able to produce an unequivocal indictment of the trans compounds.

The Dutch researchers noted that with only 2 to 4 percent of daily calories coming from trans fatty acids in the typical American diet, totally eliminating these compounds would not have a sharp effect on people's cholesterol levels. They conclude that "it would seem prudent" for patients at increased risk of atherosclerosis, or fatty deposits in the arteries, "to avoid a high intake of trans fatty acids."

Margarines sold in this country derive about 10 to 30 percent of their fats from trans compounds. Dr. Grundy insisted that even if as much as 30 percent of the fatty acids in margarine are of the trans type, "margarine is still better than butter" for a person's heart. . . .

Exercise 8.25

Project

Find a report of a causal study. You may find an example in a newspaper or newsmagazine. The Sunday supplement to your local newspaper is a good bet. When you have found something that you find interesting and substantial enough to work on, analyze the report following the standard six-point program for evaluating causal hypotheses.

PART THREE

Knowledge, Values, and Decisions

Chapter 9

Models of Decision Making

The results of scientific investigations often have implications for individual, or public policy decisions. This is particularly true of causal investigations, such as studies of possible causal connections between saccharin and bladder cancer or between smoking and heart attacks. To conclude this text, we will explore some relationships between scientific knowledge and decisions, whether private or public.

We will begin, as usual, by constructing some appropriate *models,* in this case models of decision making. In keeping with the tradition in which the study of probability and decision making focused on games of chance, we will take as our model of decision making a simple game. We will assume that the player is you, the reader. In the following chapter, we will apply these models to particular cases of real-life decision making.

9.1 OPTIONS

The game which will serve as our model for decision making involves the familiar jar of marbles, which this time may contain either red, green, or blue marbles. A single marble is to be selected at random. You are to bet on the color of the marble that is selected. Thus, you can bet on either red, green, or blue. In the language of decision theory, you have three options, it being assumed that you have already agreed to play the game.

According to our model, a decision is always a decision to do something. Making a decision is choosing a course of action. On this understanding, there is no decision to be made unless there is more than one thing that you might do. That is, you must be facing a choice between two or more possible courses of action. You have the option of choosing one from among a set of possibilities. These are your *options.*

Standard models of decision making require that the options in any decision be organized so that one, and only one, option can be chosen. In terms already familiar from our study of probability models, the options must be mutually exclusive and exhaustive.

Thus, one could not count betting on red and betting on red or green as two different options. These two options are not *mutually exclusive* because betting on the combination red or green is automatically to bet on red. The options must be specified in such a way that you could not choose more than one simultaneously.

On the other hand, you must be committed to choosing at least one option. If you are not, then the set of options is not *exhaustive.* The set of options in any decision must exhaust the actions regarded as possible in the circumstances. Thus, as we have described the particular decision in question, betting on yellow is not an option.

The options specified above (bet on red, bet on green, bet on blue) are both mutually exclusive and exhaustive.

9.2 STATES OF THE WORLD

In addition to the set of options, any model of a decision must include reference to the relevant, possible *states of the world.* Specifying this set is not always easy. For example, one might think that in our game there are two relevant, possible states of the world, winning and losing. Those two states, however, will not do for our model of decision making. For our model, the possible states of the world must be specifiable independently of the options. This is not so for winning or losing. Winning, for example, is a complex event consisting of an option and the result of the draw. Thus, one of the three possible ways of winning is betting on red when it turns out that the marble selected is red.

In our example, the way to specify the possible states of the world without involving any reference to the options is simply to list the three possible results of the draw. These are: red is drawn, green is drawn, blue is drawn.

As with the options, the possible states of the world must be specified in a way that makes them mutually exclusive and exhaustive. Thus, as far as any particular decision problem is concerned, the actual state of the world will end up being one, and only one, of the initially specified possibilities. If that turns out not to be so, the decision problem itself must be reformulated.

9.3 OUTCOMES

Figure 9.1 combines our options and the relevant states of the world into a *decision matrix.* The form of the matrix is standard in decision theory. The options are listed in rows off to the left, while the possible states appear across the top. The resulting boxes, or cells, define what decision theorists call *outcomes.* Thus the possible outcomes corresponding to a particular decision problem may be described as option-state pairs. The fact that each option-state pair determines an outcome means that the number of outcomes to consider will always be the product of the number of options and the number of states.

In this model, winning corresponds to a set of outcomes. Indeed, we see immediately that among the nine possible outcomes of the decision there are three outcomes associated with winning and six outcomes associated with losing. One of the outcomes corresponding to winning consists of betting on red when red turns out to be the color of the marble selected.

9.4 VALUES

Our model of your decision now lacks only one important component. How much is the bet? In particular, is the amount you stand to win, or lose, the

Figure 9.1

The decision matrix for a model decision problem.

	Red is drawn	Green is drawn	Blue is drawn
Bet on red			
Bet on green			
Bet on blue			

same for each option? In general terms, these are questions of *value*. Of what value to you is winning or losing? How much is it worth?

Your values attach not to options alone, nor to states alone, but to option-state pairs, that is, outcomes. It is time we looked at the role of values in decision making.

Once one has determined the available options and set out the relevant, possible states of the world, the next most important component in any decision is the relative values assigned to the possible outcomes. Determining the best option, in any model of decision making, is largely a matter of weighing the relative values of the various outcomes.

It is important to realize that assigning values to outcomes in a decision matrix does not require any kind of general or absolute value measure. It is enough just to be able to compare the relative value of any outcome with any other outcome in the particular matrix at issue.

Ranking Values

There are two fundamentally different ways of assigning values to outcomes. One way is simply to *rank* the outcomes in order of increasing value. Because all one needs is to be able to compare one outcome with another, how one labels the outcomes is quite arbitrary. For example, one could assign the least valued of all outcomes the value 1. The next most-valued outcome could get the value 2; the next, the value 3, and so on. In our example, the most-favored outcome would get the value 9, because there are, in all, nine possible outcomes. But one could equally well increase by twos, with the lowest-ranked outcome being assigned 2, the next-lowest assigned 4, and so on. This assignment of values would not change the relative ranking of the outcomes.

When one has only a ranking, there can be no presumption, for example, that the outcome ranked 3 would be one-third as valuable as that ranked 9. Because any numbers that maintain the same ordering will do for a value ranking, the concept of an outcome being three times as valuable as another simply does not make sense in this context. The most one can say is that the higher-ranked of any two outcomes is more valued. How much more is not defined.

Measuring Values

For a more fine-grained model, one would like an actual *measure* on the outcomes, so that one could say that one was worth twice as much as another. In our example, this is indeed possible. The standard way of assigning values to bets is in terms of money, which is generally regarded as a measure of value. Thus, an outcome that nets you $10 is twice as good as one that nets you only $5. There are, however, many real-life situations for which one might want to develop a decision model but for which one would be hard pressed to assign dollar amounts to outcomes—even though one could easily judge which were more valued than others.

Because all that matters is the relative measure of one outcome compared to other possible outcomes in a particular decision matrix, the scale for the measurement is arbitrary. A convenient, general strategy for measuring values is to begin by identifying the least-valued and most-valued outcomes. Assign the least-valued the measure 0 and the most-valued the measure 100. Completing the matrix is then a matter of fitting in all the other outcomes, according to their relative value on a scale of 0 to 100.

To complete our model decision problem, let us suppose for the moment that the game has the following simple value structure. If you win, by any

Figure 9.2

The decision matrix complete with value assignments for all outcomes.

	Red is drawn	Green is drawn	Blue is drawn
Bet on red	+10	-10	-10
Bet on green	-10	+10	-10
Bet on blue	-10	-10	+10

of the three possible ways of winning, you get $10. If you lose, by any of the six possible ways of losing, you must pay $10. With these value assignments, the completed decision matrix is that shown in Figure 9.2.

9.5 SCIENTIFIC KNOWLEDGE AND DECISION STRATEGIES

It takes a good deal of knowledge to construct a decision matrix. One has to know what are the available options, what are the relevant, possible states of the world, and the relative values of all the resulting outcomes. Yet, none of this is the kind of knowledge that results from typical scientific studies. What well-defined, scientific studies provide is evidence regarding the occurrence of one or another of the possible states of the world.

For purposes of decision making, there are three grades of knowledge one might have. Given the kinds of evidence that might actually result from a real-life, scientific study, each of these cases represents an idealized state of knowledge, but that is the way with all models. The important point is that for each of these idealized states of knowledge there are identifiable strategies for identifying the most desirable options. Understanding these strategies, and why they lead to desirable options, provides a basis for *evaluating* decisions, whether in the making or after a choice has already been made.

9.6 DECISION MAKING WITH CERTAINTY

One of the most important lessons of this text has been that we can never be absolutely certain of the truth of any scientific conclusion. No matter how good the evidence for any hypothesis, it is always possible that it is wide of the truth. So the idea that we might be absolutely certain which state of nature will occur is an idealization. What is called *decision making with certainty* is simply a situation in which the evidence for the occurrence of one particular state is so good that we ignore the small chance that we are wrong. If you are unwilling to ignore such small chances, then all your decision problems will involve what we will shortly identify as either "risk" or "uncertainty."

The Highest-Value Strategy

Decision making with certainty, then, includes any decision problem in which you know for sure which of the possible states of nature is, or will be, the real state. In our model, for example, you might be told by a completely reliable source that all the marbles in the jar are red. In that case, you know that the marble to be drawn must be red. Your course of action is clear. Bet on red. Let us now consider the situation a little more systematically.

Look back at Figure 9.2 and examine the outcomes included in the state that a red marble is drawn. These appear under the heading "red is drawn." Of these outcomes, the one corresponding to the option of betting on red has the value +$10. Those outcomes corresponding to the options of betting on green or on blue have the value −$10. So, of the three outcomes compatible with the state of red being drawn, the one associated with the option of betting on red has the highest value.

This analysis suggests the following general strategy for decision making when one has certain knowledge of the actual state of nature.

Highest-Value Strategy:

Given certain knowledge of the actual state of nature, choose the option associated with the highest-valued outcome compatible with that state.

If all decisions were made with certain knowledge of the state of the world, there would be little need ever to consider a full decision matrix. We did so in this example because we assumed we had the decision problem set up and then got the knowledge that made it a decision problem with certainty. This sort of thing might happen in real life, but usually one gets the information when considering the options. So one would not need to consider more than one state of the world. Apart from determining the available options, evaluating the decision problem is mainly a matter of ranking the outcomes. If you can manage just to rank the outcomes, it is obvious what option to take.

Unfortunately, there are many decisions for which this simple model does not apply. One often has to act in situations in which all the available information still leaves one undecided among several possible states of nature. We will now turn to these more interesting cases.

9.7 DECISION MAKING WITH COMPLETE UNCERTAINTY

In real life, one is hardly ever completely uncertain which of the states will occur. So any model of decision making with complete uncertainty, like that of certainty, is an idealization. It can be a useful idealization, however, in cases in which there really is very little relevant information and perhaps no clear way to utilize what little there is.

To evaluate a decision assuming complete uncertainty about the state of nature is really to consider the decision on the basis of the decision matrix all by itself. That does not seem to be an easy thing to do. Indeed, it is not. There are some systematic ways to go about it, however, and these are worth knowing.

Better and Worse Options

There is a way in which one option can be clearly better than another, quite apart from any information about the states. Figure 9.3 exhibits a set of values for our model decision that illustrates this possibility. Look at the values associated with the outcomes for betting on green and betting on blue. If red is drawn, the payoff for either option is –$5; if blue is drawn, the payoff is +$5. Thus, if either red or blue is drawn, having bet on green has the same payoff as having bet on blue. However, if green is drawn, having bet on green yields +$10, but having bet on blue yields only +$5. Thus, you are guaranteed to do as well betting green as betting blue, and there is at least one possible state

Figure 9.3

A value assignment illustrating the existence of one option that is better than another, but no option that is best overall.

	Red is drawn	Green is drawn	Blue is drawn
Bet on red	+10	-5	0
Bet on green	-5	+10	+5
Bet on blue	-5	+5	+5

for which having bet on green has a higher payoff. So, overall, green is clearly the better bet.

Let us generalize this idea by formulating it explicitly.

For any two options, A^1 and A^2, *option* A^1 *is better than option* A^2 if there is at least one state for which A^1 has a higher-valued outcome than A^2, and there is no state for which A^2 has a higher-valued outcome than A^1.

Note that this characterization of better and worse options can be used even if your value matrix consists only of a ranking of the outcomes.

If A^1 is better than A^2, we will say that A^2 *is worse than* A^1. Note that in the matrix of Figure 9.3 betting on red is neither better nor worse than betting on green. That is, on the above understanding of what makes one option better than another, neither of these options is better than the other.

Eliminating Worse Options

It would be silly to take a worse option if a better one is available. Thus, if a value matrix for any problem exhibits an option that is worse than some other option, you can simply forget about the worse of the two. Under no circumstances would you decide to take the worse option.

Here we have a general strategy for approaching a decision problem with complete uncertainty regarding which state obtains. Look to see whether there are any options that are worse than others and eliminate the worse ones from further consideration.

Best Options

If it is possible for one option to be better than another, then it is possible for one option to be the best option available. It merely needs to be better than all the others. More formally, if one option in a decision problem is better than every other option, that option is the *best option* available for the problem. Quite clearly, if your value matrix for a decision problem exhibits a best option you should take it.

We can state this simple idea as a general strategy for decision making with uncertainty.

Best Option Strategy:

Look for a best option. If one is found, take it.

In real life, best options are a rare commodity. Indeed, neither of the two matrices we have so far considered contains a best option.

Satisfactory Options

One might well wonder whether there is not some way to specify options that, although not best, are nevertheless good enough. What would it mean for an option to be "good enough"?

The general idea seems to be something like this. In any decision problem, the person making the decision has some idea of the lowest value that would be regarded as a satisfactory payoff of the problem. If there were an option that would guarantee at least this minimum value, and no option guaranteeing any higher value, one might well choose the option with the guaranteed, minimum-satisfactory value.

To make this general idea a little more precise, let us call the minimum value (or value rank) that the decision maker regards as a satisfactory result of the decision the *satisfaction level* of that decision maker for the given decision problem. A *satisfactory option* in a given decision problem would then be one for which every outcome associated with that option has a value at least as great as the decision maker's satisfaction level for that problem.

We can now formulate another strategy for decision making with complete uncertainty.

Satisfactory Option Strategy:

Look for a satisfactory option. If one is found, take it.

This formulation of the satisfactory option strategy leads to a preferred option only so long as there is just one option that is satisfactory. If there happens to be more than one, you have a further decision problem. You must apply some other strategy to the remaining set of satisfactory options. For the problem of Figure 9.3, betting on red is a satisfactory option if your satisfaction level extends as low as zero.

Playing It Safe

Figure 9.4 provides yet another version of our model decision problem. This version has no best option, as you should quickly verify for yourself. Indeed, no option is, by our earlier characterization, better than any other. Moreover, if you set your satisfaction level at any positive amount, then there is no satisfactory option either.

However, if you are willing to lower your satisfaction level to zero, then betting on blue would be a satisfactory option. This option guarantees you a payoff of no less than zero. No other option has so positive a guarantee. This is a completely general feature of decision matrices. By lowering your satisfaction level sufficiently, you are bound eventually to find an option with the highest possible satisfaction level.

There is another, more standard, way of characterizing the option with the highest attainable satisfaction level. For every option, there is a least-valuable outcome. In the context of decision theory, the value of this outcome is called the "security level" of the corresponding option. In other words, the *security level* of any option is the value (or rank) of the lowest-valued (or lowest-ranked) outcome associated with that option. Informally, the security level of an option is the value of the worst outcome you might get if you choose that course of action.

The option with the greatest security level is the same as the option with the highest attainable satisfaction level. Thus, there is a simple formula for

Figure 9.4

Another version of the model decision.

	Red is drawn	Green is drawn	Blue is drawn
Bet on red	+10	0	-5
Bet on green	+5	+10	-5
Bet on blue	0	+5	+5

finding this option. Just list the security levels for all options, and look for the option with the highest security level. This suggests the following play-it-safe strategy.

Play-It-Safe Strategy:

Choose the option with the greatest security level.

The reason I call this a play-it-safe strategy is that it seeks to minimize losses and pays no attention to possible gains. If you care more about potential gains than potential losses, there is another strategy you might consider.

Gambling

Every value matrix has at least one *highest valued* outcome. Most have only one. Let us assume only one for the moment. One way of characterizing a gambler is as a person who "goes for all the marbles" regardless of the risk. This means, in the present context, choosing the option that is associated with the highest valued outcome. If you do this, you may not get that outcome, of course, but you could not get it at all if you chose some other option. Formulating this idea as an explicit strategy:

Gambler's Strategy:

Choose the option associated with the highest valued outcome.

In case there is more than one outcome with the same highest value, the consistent strategy would be to look at the second-highest values, and take the option corresponding to the highest second-high value as well as the highest value overall. If there is more than one of these options, go to the highest of the third values, and so on. This strategy must in the end get you a unique decision. The overriding question, of course, is "Should you gamble?"

Gambling Versus Playing It Safe

Look again at Figure 9.4. The gambler's strategy recommends betting on green because it has the possibility of yielding payoffs of 5 and 10. The play-it-safe strategy recommends betting on blue because for this option the security level is 0, as opposed to −5 for the other two options.

In general, the gambler's strategy looks at the high values for each option and takes the option with the highest high value ("maximax"). The play-it-safe strategy looks at the low value for each option and takes the option with the highest low value ("maximin"). Is there some reason to think that one of these strategies is inherently better than the other in cases of decision making with complete uncertainty?

The accepted answer among philosophers and decision theorists alike is "no." Taking a best option does seem clearly to be the correct thing to do. Taking a satisfactory option seems, at least, a good strategy. If neither of those two strategies applies, however, there seems no obviously correct strategy to follow.

The main reason for taking the time to formulate the strategies for gambling and playing it safe is to have a clear characterization of these two attitudes toward a decision problem. If you want to gamble or to play it safe, you know clearly what to do. If you want to understand decisions already made, by others and even by yourself, one of these two strategies may correctly characterize what has been done. You may be able at least to understand what is happening, even if you cannot claim to know what is correct.

The underlying reason for this unhappy situation is easy to spot. We are dealing with decision problems in the absence of any specific information about the states of the world other than that they are possible. It should not be expected that there is always a clearly correct decision in the absence of specific scientific information. The situation is quite different if there is more information, as we shall now see.

9.8 DECISION MAKING WITH RISK

Making decisions in cases of nearly complete uncertainty about the states is obviously risky business. When decision theorists talk about risk, however, they have in mind *known,* or *controlled,* risk. The risk is known or controlled by *probabilities.* So decision making with risk is characterized as making a decision knowing the probabilities of all the possible states. Knowing the probabilities is a type of knowledge that is clearly intermediate between complete certainty and complete uncertainty.

Our study of interval estimation taught us that the most one can learn about probabilities from any sample is that the true probability lies within a specified *interval.* Even that can only be known with a given confidence, say 95 percent. For the moment, however, we will follow standard practice and imagine assigning specific probabilities to the possible states of nature. This amounts to introducing the idealization that the probabilities are known exactly and with complete certainty.

Up to now, all the decision strategies we have considered can be applied even if one's values are expressed merely in terms of a *ranking* of the possible outcomes. Strategies for decision making with risk are not so inclusive. They can be used only when the values are *measured* to the extent that ratios of differences in value are meaningful. This makes sense because probability can be thought of as a kind of measured knowledge. If you are to combine probabilities with values, the values have to be measured to roughly the same extent as the probabilities. If they were not, the combination would not be meaningful.

Expected Value

To combine probabilities and values we need a new concept, *expected value.* Before explicitly characterizing the expected value of an option, let us look at several ways of approaching it informally. Suppose that you are playing the game with values as given in Figure 9.4 and know that the jar contains 30 percent red marbles, 30 percent green, and 40 percent blue. You wish to bet on red.

One way to think of the expected value of an option is to imagine playing the game a large number of times. On the average, a bet on red would win three times in ten plays. So imagine just ten games. If things went by the average, you would win $10 each on three games, nothing on three games, and lose $5 on each of four games. That adds up to a net gain of $10, or $1 per game.

Another way of looking at the situation uses the probabilities more directly. If you play only once, betting on red, you have three chances in ten of gaining $10, three chances in ten of gaining nothing, and five chances in ten of losing $5. But these are exclusive alternatives, so you can add the results. Three-tenths of $10 is $3. Three-tenths of nothing is nothing, and four-tenths of $5 is $2. Three dollars minus $2 is $1. This makes it plausible to say that in betting on red your expectation is that you would win $1.

Now let us look at the explicit statement of the expected value of an option.

> The *expected value (EV)* of an option is the weighted sum of the values of its possible outcomes, the weights being the probabilities of the corresponding states.

One computes a weighted sum of values by first *multiplying* each value by its corresponding weight and then *adding* the resulting products. For example, the expected value of betting on red in the game represented by the matrix in Figure 9.4 is:

$$\mathbf{EV}(\text{bet on red}) = (.30 \times \$10) + (.30 \times \$0) + (.40 \times -\$5) = \$3 - \$2 = \$1.$$

This, of course, is just what we calculated earlier.

We can now explicitly state the *expected-value strategy* for decision making in cases of known risk.

Expected-Value Strategy:

> Choose the option with the greatest expected value.

You should verify for yourself that this strategy would lead you to bet on blue for the problem just considered. Betting on red would be the worst of the three options.

If there happens to be more than one option with the same maximum expected value, then you can simply eliminate all the other possible options and treat the remaining matrix as a case of decision making with uncertainty.

Having used your information about the probabilities to determine the expected values, you have no information left that could be used to decide which of several remaining options to choose. In practice, this does not happen very often, so you need not worry much about this eventuality.

Is the Expected-Value Strategy the Best Strategy?

It is worth pausing for a moment to consider whether the expected-value strategy is really the best possible strategy for cases of known risk. Consider a final version of our model decision problem as shown in Figure 9.5. Here betting on red has the lowest-expected value and betting on blue the highest.

What if someone confronted with this problem were to insist on ignoring the probabilities and playing it safe? That means betting on red, which has the highest security level. After all, this person reasons, if you play only once you cannot actually get the expected value of $7 when betting on blue. You are either going to win $15, win $5, or lose $5. There are no intermediate outcomes. Moreover, this person might say, losing $5 is not a satisfactory outcome, but losing nothing is. So betting on red is really the better option.

One response is that one should not think of the expected-value strategy merely as applying to one play of this particular game, but as a general policy to be followed in all sorts of situations. If one always ignored the probabilities and followed a play-it-safe strategy, one would not do as well on the average as one

Figure 9.5

A decision problem in which there is an intuitive conflict between the expected-value strategy and the play-it-safe strategy.

	P(R)=.20	P(G)=.30	P(B)=.50		
	Red is drawn	Green is drawn	Blue is drawn	Security Level	Expected Value
Bet on red	10	5	0	0	3.5
Bet on green	10	15	-5	-5	4
Bet on blue	-10	5	15	-10	7

Figure 9.6

A summary of decision strategies.

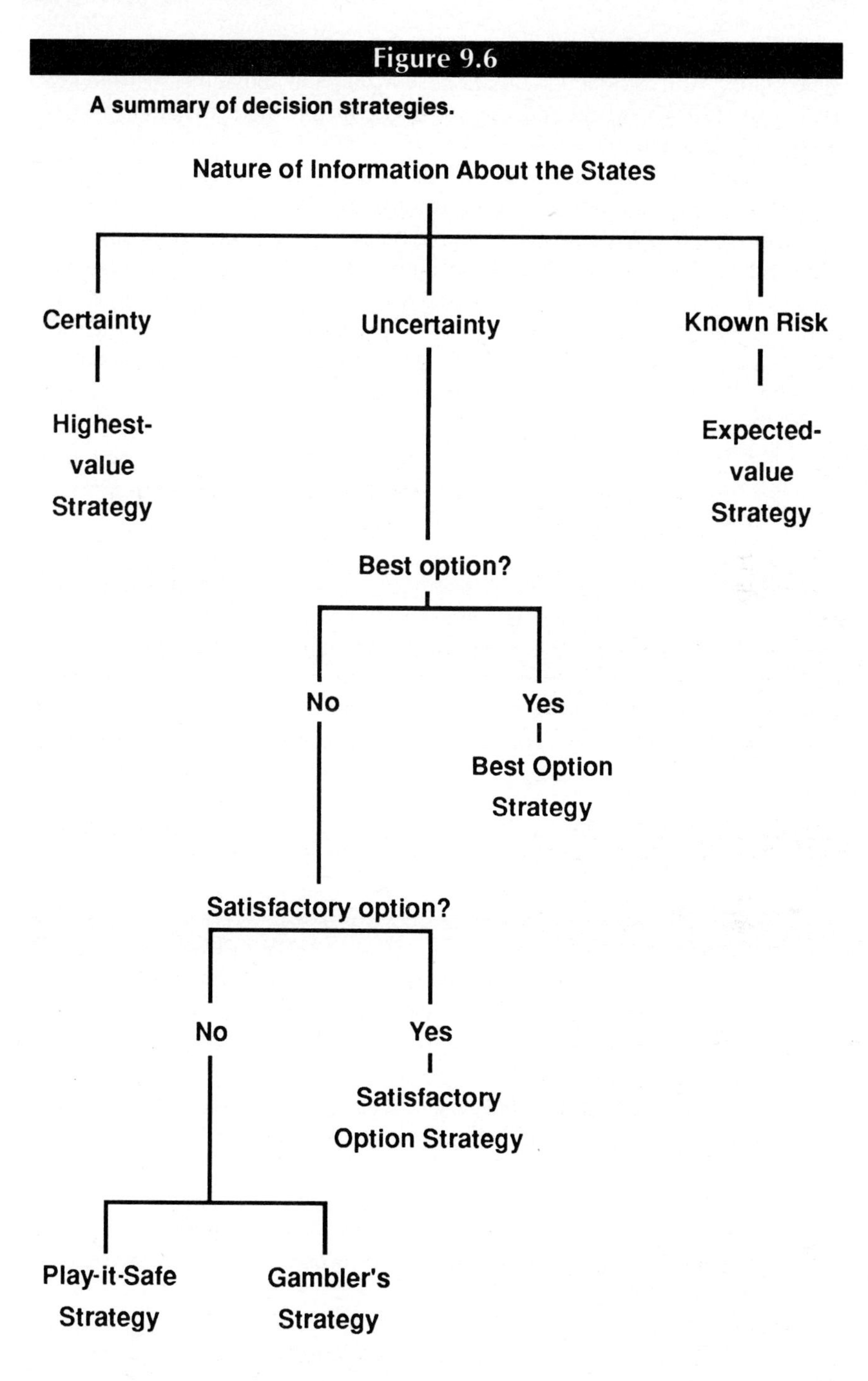

would by using the probabilities and following the expected-value strategy. When faced with a particular decision, however, it is sometimes hard to keep one's mind on the long run. After all, you might not be around to benefit from the expected, long-term gains.

Another problem is that the dollar amounts attached to the outcomes may not represent a person's real values regarding those outcomes. For example, if you had only $10 to your name, the prospect of losing it all by betting on blue, even if the chance of that happening is only 20 percent, would be quite serious. However, a fifty-fifty chance of winning $5 or $10 by betting on red, with no chance of actually losing anything, would look quite good. On the other hand, if $10 represented just a tiny fraction of your cash, then you might not worry so much about the 20 percent chance of losing $10 and care more about the expected gain of $7.

In sum, although there is much to be said for the expected-value strategy, it does not have the solid appeal of the highest-value strategy, the best-option strategy, or even the satisfactory-option strategy.

9.9 SUMMARY

All the decision strategies discussed in this chapter are summarized in the "tree" shown in Figure 9.6. Keeping this diagram in mind will permit you to categorize any decision problem and proceed to an appropriate decision strategy. The only complicated part of the tree is the branch for decision making with uncertainty. This branch has several subbranches, depending on whether the matrix exhibits a best option or a satisfactory option.

EXERCISES

Exercise 9.1

Suppose that the values shown in the decision matrix of Figure 9.7 represent a relative ranking of all the possible outcomes of the decision.

A. *Suppose it is known that state S_1 will occur no matter which option is chosen. Which option is preferable in this case? Explain briefly.*

B. *Suppose it is known that state S_2 will occur no matter which option is chosen. Which option is preferable in this case? Explain briefly.*

C. *Suppose it is known that state S_3 will occur no matter which option is chosen. Which option is preferable in this case? Explain briefly.*

Figure 9.7

	State 1	State 2	State 3
Option 1	1	9	4
Option 2	5	3	8
Option 3	6	7	2

Exercise 9.2

Suppose it is known with certainty that (1) performing the first option will lead to the occurrence of state S_3, (2) performing the second option will lead to the occurrence of state S_2, and (3) performing the third option will lead to the occurrence of state S_1. Which option is preferable in this case? Explain briefly.

Exercise 9.3

Now suppose that there is complete uncertainty regarding the outcomes of the decision problem shown in Figure 9.7.

A. Is any option better than any other? Explain briefly.

B. Is there a best option? Explain briefly.

Exercise 9.4

A. Suppose that a value ranking of at least 5 is regarded as a satisfactory payoff for the decision. Is there then a satisfactory option? Explain briefly.

B. Suppose that a value ranking of at least 3 is regarded as a satisfactory payoff for the decision. Is there then a satisfactory option? Explain briefly.

Exercise 9.5

A. Determine the security level for each of the options.

B. Which option is preferred if one wishes to play it safe? Explain briefly.

Exercise 9.6

A. Determine the highest-valued outcome for each of the options.

B. Which option is preferred if one wishes to gamble? Explain briefly.

Exercise 9.7

Now suppose that the values attached to the outcomes in Figure 9.7 provide not just a ranking, but also a measure of the relative value of the outcomes. Suppose further that the states have the following probabilities of occurring, regardless of which option might be chosen: $\mathbf{P}(S_1) = .2$; $\mathbf{P}(S_2) = .3$; $\mathbf{P}(S_3) = .5$.

A. Calculate the expected value for each of the options.

B. Which is the preferred option in this case? Explain briefly.

Exercise 9.8

Now suppose that the probability of the various states depends on which option is chosen. In particular:

$\mathbf{P}(S_1/O_1) = .2$; $\mathbf{P}(S_2/O_1) = .3$; $\mathbf{P}(S_3/O_1) = .5$
$\mathbf{P}(S_1/O_2) = .3$; $\mathbf{P}(S_2/O_2) = .5$; $\mathbf{P}(S_3/O_2) = .2$
$\mathbf{P}(S_1/O_3) = .5$; $\mathbf{P}(S_2/O_3) = .2$; $\mathbf{P}(S_3/O_3) = .3$

A. Calculate the expected value for each of the options.

B. Which is the preferred option in this case? Explain briefly.

CHAPTER 10

Evaluating Decisions

In this final chapter we will use our model of decision making to analyze and evaluate several real-life decisions. The examples represent types of decisions that must regularly be faced both by individuals and by governments or other groups.

10.1 DECISIONS INVOLVING LOW PROBABILITIES

A frequently encountered decision context requiring scientific information involves the discovery of a causal factor with very low effectiveness. This is the case with many environmental hazards. They are causal factors for various diseases, but their effectiveness at producing those diseases in the general population is very low. What is one to do? Our earlier discussion of blood clots and the pill provides a convenient example.

Deciding on the Pill: The Data

When data linking oral contraceptives with blood clots first became known (see Section 8.5), the headlines, of course, read "Pill Causes Blood Clots." Many people were, quite naturally, rather alarmed. Some women decided immediately to stop taking the pill. Meanwhile, those who read the fine print learned that the U.S. Food and Drug Administration (FDA) regarded the evidence, deriving solely from retrospective studies, as only preliminary. Moreover, they recommended that women not go off the pill in midcycle, and later only after consultation with a doctor about other methods of contraception. Were those who immediately went off the pill acting too hastily? Was the FDA giving good advice? These are questions we can now answer for ourselves and could have, at the time, using information that was readily available.

In the first place, we now know enough not to jump to conclusions merely because something has been shown to be a causal factor for some undesirable effect. What matters for decisions is how *effective* a causal factor it is. Once the effectiveness is known, one may find that the risk is outweighed by the benefits of the causal agent. We also know that effectiveness cannot be estimated from retrospective data alone. So even knowing the details of the relevant studies is not much help in making a decision. Finally, for reasons discussed earlier, the study in question focused on nonfatal clots. Most of the concern, however, focuses on fatal clots.

Nevertheless, it did not take demographers and epidemiologists very long to come up with fairly reliable estimates of the relevant probabilities. What they came up with, and what the FDA used in making its recommendations, was this: Among women of childbearing age, the frequency of fatal blood clots is about 2 per million. Among women in that age group who take oral contraceptives, the frequency is about 7.5 times greater, or 15 per million.

The newspapers, of course, latched onto the figure of 7.5 times greater. That is indeed a good indication that the pill is a causal factor for blood clots. The relevant information for making a decision about taking oral contraceptives, however, is the normal clot rate of 2 in a million. That is indeed a very small risk. Multiplying it by 7.5 still leaves a very small number. Fifteen per million is low compared with the chance of death when undergoing pregnancy and childbirth, which is about 200 per million. For comparison, the yearly chance of an American being killed in an auto accident is about 250 per million. Also, the yearly chances of being involved in any kind of auto accident are 1 in 20, which is equal to 50,000 in a million.

The Options

Contrary to what many people seem to think, including some doctors and public health officials, the choice is not between oral contraceptives and none at all. Formulating the options this way already biases the decision in favor of taking the pill. The realistic options are between oral contraceptives and other types of contraceptives.

The States of the World

Because the information at hand concerns fatal blood clots, the possible states of interest are whether one develops a fatal clot or not. Ultimately, this is much too narrow a set of relevant states on which to base the decision. The question we are considering at the moment, however, is whether it was wise to stop taking the pill on the basis of information about fatal clots. These are the right states to consider in answering that question.

Values

Values are a personal matter. However, we can learn much about the decision by considering what would be for most people a *minimal* positive value in favor of oral contraceptives. Other things being equal, most women prefer the pill to other available contraceptives. The question is whether other things are equal.

As suggested in Chapter 9, let us measure the values on a scale of 0 to 100. The least-valued outcome, using an alternative contraceptive and suffering a fatal clot, gets value 0. The most-valued outcome, using the pill and not suffering a fatal clot, gets value 100. A value difference of only one unit out of a hundred in favor of the pill over other contraceptives seems appropriately minimal. So using the pill and getting a fatal clot gets a value measure of one and using another contraceptive but not getting a fatal clot gets measure ninety-nine.

Note that we do not assign a lower value to using the pill because it is a causal factor for clots. Values apply only to the situation that characterizes the outcome. Any possible causal connections between options and states are accounted for by differences in the probabilities. In this case, the probability of the state of getting a clot is higher with the option of using the pill than with the option of using an alternative contraceptive. The difference in probabilities reflects the effectiveness of the pill as a causal factor for clots. The full decision matrix, complete with probabilities and expected values, is as shown in Figure 10.1.

Evaluating the Decision

If we treat this decision problem as one involving known risk, it is appropriate to apply the expected-value strategy. As you can see, the expected value of taking the pill is still greater than that of using other contraceptives, even though the positive value put on using the pill is minimal.

The reason for this should be intuitively clear to you by now. With a probability of only $^{15}/_{1,000,000}$ of experiencing a clot when taking the pill, the expected value for this option will be almost the same as the value associated with not getting a clot. In short, even though the probability of dying from clots is 7.5 times greater if you take the pill than if you do not, it is still much too small to offset even a very small positive value placed on using the pill rather than other contraceptives.

It seems fairly clear, then, that deciding not to use oral contraceptives because of the increased risk of fatal clots was not a good decision. In some cases, making this bad decision had tragic consequences. When this all happened, abortion was illegal in the United States. Some women who stopped using the pill got pregnant when they otherwise would not have, and some of these sought an illegal abortion. It is thus a plausible guess, though one hard to support with solid data, that among women who decided to go off the pill because of the increased risk of fatal clots, more ended up dying from the effects of a badly done, illegal abortion than would have died from blood clots.

In general, a decision whether to use oral contraceptives should not be based on a decision model that focuses on only one specific effect, such as blood clots. One wants a broader indicator of the health consequences of using oral contraceptives. Overall life expectancy would be a very good measure. Unfortunately, because oral contraceptives have been in general use only since

Figure 10.1

The completed decision matrix for deciding on the pill.

	Get fatal clot	Do not get fatal clot	Expected Value
Use the pill	1 15/1,000,000	100 999,985/1,000,000	99.99
Use other contraceptive	0 2/1,000,000	99 999,998/1,000,000	98.89

the early 1960s, reliable information on overall life expectancy for those who use oral contraceptives is not available.

Recent studies indicate some *decrease* in the chances of developing cancer of the uterine lining and the ovaries. This is offset, however, by an *increase* in the chances of developing heart disease and strokes, particularly for women who smoke. Moreover, several very recent studies have revived worries of an increased incidence of breast cancer among pill users. This data includes many women who took pills when the dose was much higher than it is now. The net result is that in 1990 there is greater uncertainty about the safety of oral contraceptives than there was in 1980. New information is actually making the decision more difficult. In any case, prolonged use of oral contraceptives at an early age should not be undertaken now without careful consultation with well-informed medical professionals.

10.2 DECISIONS INVOLVING MODERATE PROBABILITIES

Now let us consider an example of a decision for which the causal factor is of moderate effectiveness. Here the tradeoff between values and probabilities is more difficult to evaluate.

The Smoking Decision

We have seen (in Chapter 8) that smoking is a positive causal factor for coronary heart disease. It is tempting to regard this information by itself as sufficient to resolve the decision problem in favor of the choice not to smoke, but it is not. What matters for the decision problem is the effectiveness of smoking as a causal factor for coronary heart disease, which is to say, the relative probabilities of the disease for smokers and nonsmokers. In making the decision, a smoker (or potential smoker) must weigh these relative probabilities against the positive enjoyment of smoking (or the difficulty of giving it up).

It would be shortsighted, however, to base a decision on smoking on a single effect, such as coronary heart disease, or lung cancer. One wants a broader indicator of the health effects of smoking. This is provided by data on overall life expectancy that reflects the full range of risks to life.

Some Data

For purposes of illustration, we will focus on the data on overall life expectancy shown in Figure 10.2. For our evaluation, we will concentrate on two pieces of information: (1) 22 percent of twenty-five-year-old, male nonsmokers die before age 65, (2) 38 percent of twenty-five-year-old males who smoke a pack of cigarettes a day die before age 65. We can regard these as estimates from very large samples, so we can ignore the fact that each of these percentages should be a small interval rather than a single number. The data for people who do not fit the category of twenty-five-year-old males would be similar in predictable ways.

The Options

To keep the problem simple, we will restrict the options to smoking a pack a day (smoke), or not smoking at all (don't smoke). The resulting decision

Figure 10.2

The percentages of 25-year-old males expected to die before age 65 as a function of cigarettes smoked per day.

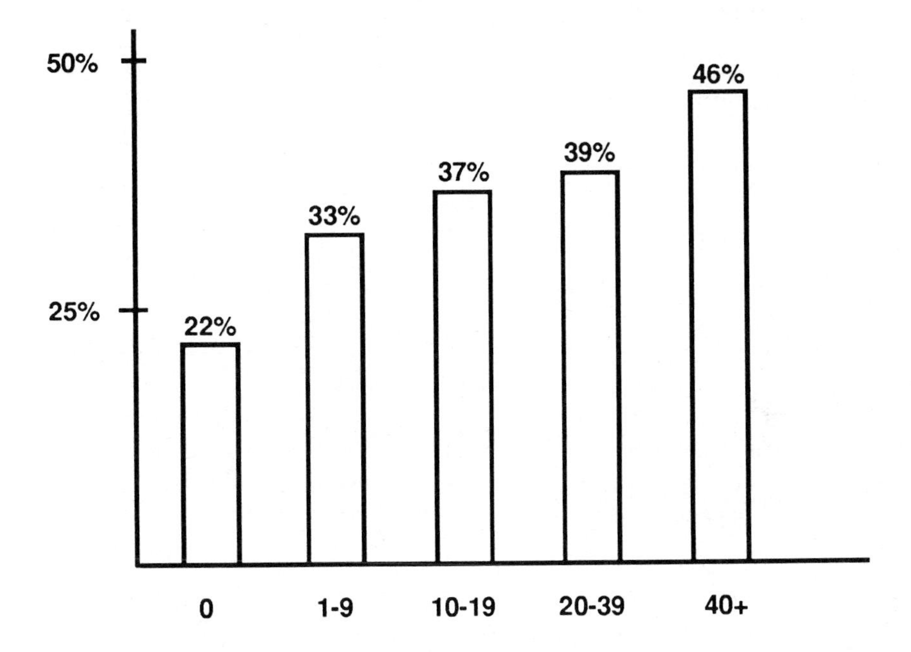

problem can be understood as applying either to a nonsmoker contemplating taking up smoking, or to a smoker contemplating quitting.

The States of the World

The information given restricts our attention to just two relevant states of the world: dying before age 65 and living past age 65. This gives us a matrix with only four *outcomes* to consider.

The Values

As before, let us measure the values on a scale from 0 to 100. The scale is fixed by stipulating that the least-preferred outcome has value 0 and that the most-preferred outcome has value 100. We will assume that, everything else being equal, the decision maker would prefer smoking to not smoking. We will also assume that, everything else being equal, the decision maker prefers living past 65 to dying before 65. This implies that the least-preferred outcome is choosing not to smoke and still dying before 65. The most-preferred outcome is smoking and still living past 65.

Evaluating the Decision

To evaluate the decision, let us first see what the decision matrix looks like if one places a very low value on smoking and a high value on living past 65.

Suppose that we assign the value 1 to the outcome of smoking and dying before 65. This is equivalent to supposing that our decision maker, caught in the state of dying before age 65, would pay only $1 out of a total fortune of $100 to get from the outcome associated with not smoking to the outcome associated with smoking. Similarly, we will assign the value 99 to the outcome of not smoking and living past 65. This is like supposing that a decision maker who paid $99 to get into the state of living past 65 would be willing to pay only $1 more to get into the outcome associated with smoking, rather than not smoking. In short, we are supposing that smoking is worth only $1 out of $100 to our decision maker. The completed matrix is shown in Figure 10.3.

Before considering the known probabilities, it is worth pausing to consider what the decision problem looks like with only the values and no probabilities. That is, suppose for the moment that this were a case of decision making in a state of total uncertainty. In these circumstances, what would be the preferred option? Try to answer this question before reading on.

Considered as a case of decision making under uncertainty the matrix in Figure 10.3 shows the option of smoking as being a *best option.* Check to see that this is indeed so. The fact that by ignoring the relevant evidence one gets a matrix favoring smoking may partly explain why smokers are so anxious to deny that there is any real evidence of a connection between smoking and health. So long as the connection is denied, smoking looks like a good choice. The trouble with this approach to the problem is that the evidence is undeniable. The connection exists and cannot be wished away. The real decision problem is one of *known risk,* not complete uncertainty. The recommended strategy is not the best-option strategy but the expected-value strategy.

To the right of the matrix is the expected value of each of the two possible options. You might want to check for yourself to see that these are indeed the correct values. Because the expected value of not smoking is clearly higher than that of smoking, not smoking is the recommended choice.

Figure 10.3

The completed decision matrix for the smoking decision with minimal preference for smoking over not smoking.

	Die before age 65	Live past age 65	Expected Value
Smoke	V = 1 P = .38	V = 100 P = .62	62.4
Don't smoke	V = 0 P = .22	V = 99 P = .78	77.2

What was the worst option in the absence of statistical information becomes the recommended choice once that information is taken into account.

In the matrix of Figure 10.3, the difference between a probability of .62 and a probability of .78 is more than enough to offset the one-point difference in values between the two corresponding outcomes. The only reply that a smoker might make is that the suggested value difference is just too small. A smoker in the state of living past 65 might well demand more than $1 out of $100 to move from the outcome involving continuing to smoke to the outcome involving quitting. Likewise, the same smoker caught in the state of getting lung cancer might well pay more than $1 to move up from the outcome involving quitting to the outcome involving continuing to smoke. How much more?

Let us turn the question around. By how much would we have to *decrease* the value difference between dying before 65 and living past 65 in order for the two options to have the same expected value? Equivalently, by how much would we have to *increase* the difference in value between outcomes involving smoking and not smoking in order to equalize the expected values of the two options? That value difference would mark the break-even point on the value scale.

This is a simple problem in high school algebra. We want to determine a particular value, call it V, such that the outcome of smoking and dying before 65 has value V. Likewise, the outcome of not smoking and living past 65 will get value (100 – V). Setting the expected value of the two options equal to each other,

$$(.38 \times V) + (.62 \times 100) = (.22 \times 0) + [.78 \times (100 - V)].$$

Solving for V, we get a value of about 14. So the matrix that makes the two options equally good is as shown in Figure 10.4. If you think smoking is a good decision, you must value living past 65 less than a difference of 86 points on a scale of 100. Similarly, you must value smoking over not smoking

Figure 10.4

The break-even value matrix for the smoking decision.

	Die before age 65	Live past age 65
Smoke	14	100
Don't smoke	0	86

more than 14 points on a scale of 100. That seems a high relative value to place on smoking over living into retirement. But we are not here in the business of criticizing a person's values. We are merely pointing out what relative values are represented by the decision to smoke.

Most people do not have much feeling for rating their values. Let us put these numbers in a form that is more easily understood. Suppose that you were giving grades to outcomes. At 100, the outcome of smoking and living past 65 gets an A+. But, rating only 86, living past 65 and not smoking gets only a B. If you smoke, you are saying that for you smoking makes the difference between an A+ life and a B life. Can smoking really be that important?

Sometimes it can be. The story is told of a famous scholar who tried hard to give up smoking. Once he stopped smoking, he found he could not write. He decided that being able to do his life's work was worth the risks of smoking. However, this is an extreme case. Few people can claim such an excuse.

Confronting Your Values

Decision making with risk is usually presented as a process of using known probabilities together with the values of outcomes to determine which option has the greatest expected value. This works well when the values are given in dollars and cents, as in the case of a lottery. For more qualitative values, however, such as the value of smoking to a smoker, this approach works less well. In such cases, it is sometimes helpful to put the decision problem in a form that does not require an explicit assessment of values. Rather, values are confronted *indirectly* through the contemplation of the choice to be made.

A simple way to do this is to construct a model of the decision in which each option is represented by a jar containing 100 marbles. You need one jar for each option. Each possible state of the world is assigned a color. So you need as many different-colored marbles as the problem has states. The probabilities of the states are represented by the relative number of marbles of the various colors. A state with a probability of .50 for a given option would be represented by having 50 of the 100 marbles carry the color assigned to that state.

Once all the jars are set up correctly, your decision to choose one of the options is equivalent to deciding which jar to pick. Having picked a jar, you must select one marble at random from that jar. The color of the selected marble tells you which state, and thus which outcome, you have achieved. Because you know the relative numbers of colored marbles in each jar, by picking a jar you are implicitly weighing your desire for the corresponding option against the probabilities of the various states. Imagining reaching in and picking a marble is a graphic way of confronting your true feelings about the decision.

Now we will apply this device to the smoking decision shown in Figure 10.3. Let a red marble represent the state of dying before the age of 65. A blue marble will represent the state of living past 65. The jar that represents the option of smoking will have 38 red marbles and 62 blue ones. The jar representing the option of not smoking has 22 red marbles and 78 blue ones. You have to select a single marble at random from one of these two jars. Selecting

Figure 10.5

A marble and jar representation of the smoking decision.

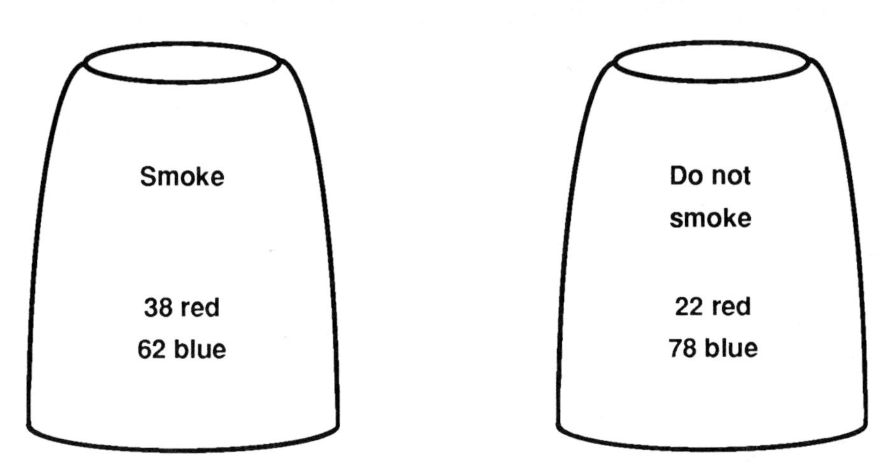

a red marble means that you will die before the age of 65. Your choice is pictured in Figure 10.5. Which jar do you choose?

This way of presenting the problem brings you face to face with your values because you have to ask yourself: Is smoking worth facing those additional 16 red marbles in the "smoking" jar? If you want to smoke, only you can answer that question.

10.3 POLICY DECISIONS WITH UNCERTAINTY

The previous examples have focused on decisions by individuals. The relevant values were those of the individual decision maker. We are now going to evaluate an example in which the decision is a collective, political decision. So the subject is really public policy rather than individual decision making.

Global Warming

In Chapter 2 we considered the problem of global warming (the greenhouse effect). Many scientists believe that the earth is experiencing significant warming because of the generation of carbon dioxide (and other greenhouse gases), mainly from burning fossil fuels, such as coal, oil, and gasoline. Even an increase of only a few degrees in the earth's average temperature could have severe effects on the ecology of the earth. It could, for example, result in faster melting of the polar ice caps, thus raising the levels of the oceans and flooding coastal cities, such as New York and Los Angeles. It could also turn current agricultural areas into vast deserts.

There have already been a number of international conferences aimed at reaching an agreement among the major polluting nations on steps that could be taken to lessen the threat of significant warming. Thus far, however, only minor agreements have been reached. Leaders of many governments, including the United States, have questioned whether enough is known to claim with sufficient confidence that global warming is really taking place, or that proposed measures could make a significant difference. Reluctance to take significant steps is understandable because any effective measures entail considerable costs. These include unemployment and higher prices for many goods.

The Options

In practice, the available options are many and complex. For our purposes, we will reduce the complexity to just two options: "do something," and "do nothing." It is to be understood that doing something means doing something on a scale that is likely to be effective. Doing nothing means proceeding as one would if the possibility of global warming were of no serious concern.

The States of the World

Here again, we will reduce the many possible states to just two: global warming takes place, "warming," and global warming does not take place, "no warming." Here, saying that global warming takes place means that it is sufficiently severe to have important consequences such as the flooding of major coastal cities or the destruction of farmlands. Otherwise, we would say that significant warming has not taken place. The resulting decision matrix is shown in Figure 10.6.

Values

Imposing an approximate value measure on the matrix of Figure 10.6 is not so difficult as it might seem. The worst outcome is making strenuous efforts to avoid warming and have it occur anyway. Give that outcome value 0. The best possible outcome would be to do nothing and have no significant warming. Give that outcome value 100.

Whether or not warming occurs, trying to do something to prevent it carries heavy costs. So in either state of nature the outcome of doing nothing

Figure 10.6

The basic decision matrix for a decision about global warming.

	Warming	No warming
Do something		
Do nothing		

carries a higher value than doing something. How much greater is not terribly important for evaluating the decision. A jump of ten points out of a hundred when moving from doing something to doing nothing is good enough for our purposes. The completed matrix is shown in Figure 10.7.

Evaluating the Decision

This decision problem exhibits a structure similar to that of the smoking decision examined earlier. The difference is that in the present case there is considerable controversy among scientists whether global warming is taking place, or would take place any time soon. So policy makers are inclined to treat the problem as one of decision making under uncertainty. In that case, paradoxically, doing nothing is a best option. That is because doing something is costly, whatever the eventual state of the world.

What is needed to make the decision problem look different to policy makers is strong evidence both that doing nothing will lead to warming and that doing something can prevent it. That would change the decision problem from one of uncertainty to one of certainty, or to one of known risk with high probabilities. Then, either a highest-value strategy or an expected-value strategy would apply.

Whose Decision Is It Anyway?

In decision making by individuals, it is the same person who benefits if the outcome is favorable, or suffers if the outcome is unfavorable. In policy decisions, those who would benefit if things go well may be different from those who would pay if things go badly. It becomes very important to determine who is making the decision and, thus, who is assigning values and choosing which decision strategy to follow. This is particularly important if the situation is one of uncertainty and there is no best option. Who decides what is a satisfactory outcome? If there is no satisfactory option, who decides whether to gamble

Figure 10.7

A completed decision matrix for the decision about global warming.

	Warming	No warming
Do something	0	90
Do nothing	10	100

or play it safe? These, however, are questions that have everything to do with politics and little to do with the use of scientific knowledge.

EXERCISES

Exercise 10.1

A Raffle

A friend stops you on the street saying that his fraternity is having a raffle and you now have the opportunity of buying the very last ticket for only $5. The prize, which has been donated by a rich alumnus, is worth $1000 cash. Being somewhat suspicious, you inquire as to the number of tickets being sold and are assured there are only 250. Indeed, the ticket in question has the number 250 on it. Having a spare $5, you pause to consider whether to buy the ticket.

A. Set up the decision matrix for your situation, including all the relevant information given above. Assume that your interest in buying a ticket is based solely on consideration of monetary gain or loss (and not on friendship, love of gambling, etc.).

B. What decision strategy applies to this problem?

C. What is the recommended choice? Exhibit any necessary calculations.

Exercise 10.2

Oral Contraceptives and Breast Cancer

Rework the example of Section 10.1 concerning oral contraceptives and blood clots using the information on oral contraceptives and breast cancer given in Exercise 8.1. The relevant probabilities from that study are that the chances of a woman contracting breast cancer before age 36 when not using oral contraceptives are about $1/500$. Using oral contraceptives for approximately eight years raises this probability to about $1/300$.

A. Using the same relative values as in the blood clot example, what is the recommended choice in the breast cancer case?

B. Using the information from Exercise 8.1 on oral contraceptives and breast cancer, represent the choice whether or not to use oral contraceptives as a choice between two jars from which one marble is to be selected at random.

Exercise 10.3

Generating Electricity: An Economic Analysis

Imagine that the officials of your state's electric company are trying to decide whether a needed new generator should be powered by coal or by a nuclear reactor. Suppose they have reached the point where the only unknown is the nature of new federal regulations on air pollution. If the regulations stay as they are now, a coal generator will involve a net cost of $50 million. On the other hand, if the regulations are tightened, equipment to "clean" the smoke will cost another $10 million. The officials are completely in the dark as to whether the regulations will be tightened. The cost of the nuclear plant is $55 million.

A. Set up the officials' decision problem as specified above. (Express the values in negative millions of dollars.)

B. Does the problem exhibit a best option? Explain briefly.

C. What decision would the officials make if they were to play it safe? Explain briefly.

D. What decision would the officials make if they were to gamble? Explain briefly.

Exercise 10.4

Generating Electricity: A Worst Case Analysis

Another way of looking at the decision between coal- and nuclear-powered, electrical generating plants is to focus on the possible environmental consequences of an accident. For a coal-fired plant, the worst possible accident could at most kill a small number of workers on the site and leave a city without electricity for some days. With a nuclear plant, the worst possible case is an accident such as the one that occurred in Chernobyl, USSR, or the one that almost occurred at Three Mile Island in Pennsylvania. Thousands of people may be exposed to damaging radiation. A meltdown could contaminate underground water supplies for several states, perhaps even the whole Mississippi River basin from St. Louis to New Orleans.

A. Set up the decision problem as specified above, assuming that a coal-fired plant with no accidents is better than a nuclear plant with no accidents. Rank the outcomes beginning with 1 for the worst outcome and ending with 4 for the best.

B. What would be the recommended option? Explain briefly.

C. Now set up the decision problem assuming that a nuclear plant with no accidents is better than a coal-fired plant with no accidents. Rank the outcomes beginning with 1 for the worst outcome and ending with 4 for the best.

D. Is there now a recommended option? Explain briefly.

Exercise 10.5

Vacation Plans

Vacation time is coming and you are trying to decide whether to go to Florida with your friends or to go home. Your main worry is the weather, about which there is total uncertainty. If the weather in Florida turns out to be good, it would be worth the money to go. If, however, the weather turns out to be bad, you would rather be home with your family. Besides, it is free. On the other hand, if you decide to go home, you prefer that the weather in Florida be bad so you will not feel that you are missing something. You do not reveal that preference to your friends.

A. Using the numbers 1, 2, 3, and 4 to rank the four outcomes, construct the decision matrix for this problem.

B. Does the matrix exhibit a best option? Explain briefly.

C. What option would you choose if you were to play it safe? Explain briefly.

D. What option would you choose if you were to gamble? Explain briefly.

Exercise 10.6

Flu Epidemic

Imagine that the Centers for Disease Control in Atlanta has issued a bulletin that there is an impending flu epidemic. The flu is serious but only rarely fatal—typically, leaving one incapacitated for thirty days. Fortunately, there is a very effective vaccine ready to be used, and it is being made available free to anyone who requests it. Unfortunately, this flu vaccine has the side effect that it may cause temporary paralysis, which, though again rarely fatal, also has the effect of leaving one incapacitated for thirty days. If one were so unlucky as to get both, one would be incapacitated for forty days. The officials are wary of making recommendations and are leaving the decision whether to take the vaccine up to individuals.

A. *Set up the decision problem giving the possible options, the relevant, possible states of the world, and values measured in negative days of incapacitation. Note that there will be four exclusive and exhaustive, possible states of the world.*

B. *If you have set up the problem correctly, there is no applicable decision strategy using only the information given above. Explain briefly why this is so.*

C. *In the whole population, the estimated risk of getting flu without a shot is 1/10; but with a shot only 1/10,000 (1000 times less). On the other hand, the probability of getting the paralysis without the shot is only 1/100,000; with the shot, it jumps to 1/100 (1000 times more). Determine the approximate expected values of the two options. Treat very small probabilities as 0 and very large probabilities as 1. Which option is preferable? Explain briefly.*

Exercise 10.7

The Prisoners' Dilemma

Suppose that you and a friend are engaged in some illegal activity and are "busted." The district attorney puts you in different cells and then comes to you with the following "bargain." The evidence against you both is quite good, but the case would be stronger if one of you would provide more information. So the DA promises that if you talk you will get off with at most one year in the state prison—provided your partner does not also talk. On the other hand, if you do not talk and your partner does, you are likely to get four years. Knowing a bit about decision theory, you realize that there are two other outcomes to consider. So you ask the DA what happens if neither of you talks, and if both of you do. Being an honest person, the DA admits that without some additional evidence, you will both get off with, at most, two years. If you both talk, on the other hand, you will both get three years.

A. *Set up the decision matrix for your own situation. Use negative years in prison as a measure of the value of the outcomes.*

B. If you treat your situation as being one of complete uncertainty as to whether your partner would talk, and follow recommended decision strategies, what should you do? Why?

C. The DA goes to your partner with the same deal. Suppose your partner also treats the situation as one of uncertainty regarding whether you would talk, and applies the standard rules. What is going to happen to each of you?

D. Do you think your overall situation could be improved if you and your partner were allowed to discuss the problem with one another? How might that change the decision problem?

Exercise 10.8

The Tragedy of the Commons

In the nineteenth century, it often happened that a number of families shared a small field in common. This field was used for raising supplementary animals. To keep the example simple, let us suppose that the families all raise cows, and that at the moment each family has nine cows. They all sell milk and butter to nearby townspeople, each earning roughly $95 annually. Things have been prosperous and each family has been adding a new cow every year or so for a number of years. The Commons, though, has about reached its limit. It cannot take much more grazing.

Suppose you are one of these people, and you are wondering whether to add a new cow this year. You realize that if everyone adds a new cow, the yield per cow will go down because of the lack of good grass, so everyone's net gain will drop to $90 a year. On the other hand, if you add a new cow and no one else does, that will not hurt the field so much and you can expect your income to go up to $100. The other side of this coin, however, is that if everyone else adds a cow and you do not, the yield per cow will go down and you will net only about $85.

A. Set up the decision matrix for your situation using the values given.

B. If you treat your situation as one of uncertainty regarding the actions of your neighbors, what decision rule applies? What should you do if you follow this rule?

C. What happens if each of the other families analyzes the situation in the same way?

Exercise 10.9

Preventative Attack

Imagine two countries, A and B, that have a disputed piece of territory between them. Tension is mounting. The ministers of country A meet in emergency session to plot their course of action. They conclude that there are really only two options open to them. One is to offer to negotiate the dispute with country B, and the other is to attack swiftly in an attempt to take the territory by force. The main uncertainty is over what country B will do. Will country B negotiate in good faith, or will it try to attack first?

The ministers of A agree that the best thing would be to attack while B is preparing to negotiate. That way they can get most of the territory with minimal losses. The worst possibility would be to have B attack as they are trying to negotiate. In that case, B would get most of the territory. However, if both countries attack simultaneously, being fairly equally matched, they will probably each end up with about half the territory, after fighting a short war. If B would negotiate, they could probably end up agreeing to split the territory without a war.

A. *Set up the decision problem from the standpoint of country A. Use the numbers 1, 2, 3, and 4 to rank the outcomes.*

B. *What decision strategy applies? If it follows this strategy, what will country A do?*

C. *Country B is in exactly the same position as country A. If it applies the same decision strategy, what will happen?*

Acknowledgments *(continued from copyright page)*

Science, September 16, 1988, Vol. 241, p. 1431. © AAAS. Reprinted by permission; Exercise 2.7, "Quartz Discovery Supports Theory That Meteor Caused Dinosaur Extinction," appeared in *Sunday Herald Times,* May 17, 1987, p. A10. Reprinted by permission of The Associated Press; Exercise 2.10, "Scientists Put a New Twist on Creation of the Universe," by Jim Dawson, *Minneapolis Star Tribune,* November 18, 1989, p. 1A. Reprinted by permission of the Minneapolis Star Tribune. Exercise 2.12, "Comet Source: Close to Neptune," by Richard A. Kerr, *Science,* March, 18, 1988, Vol. 239, p. 1372–3. © AAAS. Reprinted by permission.

Chapter 3: Exercise 3.11, "Clues to the Drift of Continents and the Divergence of Species," by Walter Sullivan, *New York Times,* May 30, 1982. © 1982 by The New York Times Company. Reprinted by permission; Exercise 3.12, "Continental Drift: Latest Fossil Finds Link Ancient Africa to the Carolinas," by Walter Sullivan, *New York Times,* October 19, 1982. © 1982 by The New York Times Company. Reprinted by permission.

Chapter 4: Exercise 4.1, "UFO Update," by Sherry Baker, *Omni Magazine,* September 1987, p. 91; Exercise 4.6, "The Shocking True Story: Tabloid Predictions Fail," appeared in *San Francisco Chronicle,* December 30, 1989, p. C10. Reprinted by permission of Reuters News Service; Exercise 4.8, "Debunking the Shroud of Turin," by Richard N. Ostling, *Time,* October 24, 1988, p. 81. © Copyright 1988 by Time, Inc., All rights reserved. Reprinted by permission; Exercise 4.9, "Thousands Plan Life Below, After Doomsday," by Timothy Egan, *New York Times,* March 15, 1990, p. 1. © 1990 by The New York Times Company. Reprinted by permission.

Chapter 6: Exercise 6.1, "Study: People Tell Lies to Have Sex," appeared in *Minneapolis Star Tribune,* August 15, 1988, p. 1A. Reprinted by permission of The Associated Press; Exercise 6.2, "Study Disputes View of Dyslexia, Finding Girls as Afflicted as Boys," by Gina Kolata, *New York Times,* August 22, 1990, p. 1. © 1990 by The New York Times Company. Reprinted by permission; Exercise 6.3, "Poll Shows Men Sinking in the Opinion of Women," appeared in *Minneapolis Star Tribune,* April 26, 1990, p. 1A. Reprinted by permission of The Associated Press; Section 6.4 "Bars to Equality of Sexes Seen as Eroding, Slowly," by Lisa Belkin, *New York Times,* August 20, 1989, including Tables 6.4, 6.5. © 1989 by The New York Times Company. Reprinted by permission; Exercise 6.7, "Homicide is Top Cause of Death From On-Job Injury for Women," appeared in *New York Times,* August 18, 1990, p. 8Y. Reprinted by permission of The Associated Press; Exercise 6.9, "Aggression in Men: Hormone Levels Are a Key," by Daniel Goleman, *New York Times,* July 17, 1990, p. B5Y. © 1990 by The New York Times Company. Reprinted by permission; Exercise 6.11, "AIDS Plague Hits Teen Heterosexuals," by Kim Painter, *USA Today,* July 20, 1989. p. 1A. Copyright 1989, USA Today. Reprinted by permission.

Chapter 8: Material in Section 8.7 and Exercise 8.2 reprinted by permission from *Vitamin C and the Common Cold,* by Linus Pauling. Copyright © 1970 by W. H. Freeman and Company, Bantam edition. Exercise 8.1, "British Study Says Pill Increases Breast Cancer Risk," appeared in *Minneapolis Star Tribune,* May 5, 1989, p. 4A. Reprinted by permission of The Associated Press; Exercise 8.3, "Study Finds Decaf Coffee May Raise Cholesterol," appeared in *Minneapolis Star Tribune,* November 4, 1989, p. 11A. Reprinted by permission of The Associated Press; Exercise 8.4, "Sharp Cut In Serious Birth Defect Is Tied To Vitamins In Pregnancy," by Gina Kolata, *New York Times,* November, 24, 1989. p. 1. © 1989 by The New York Times Company. Reprinted by permission; Exercise 8.6, "Study: Cancer Patients in Support Groups Live Longer," by Gina Kolata, appeared in *Minneapolis Star Tribune,* November 23, 1989. © 1989 by The New York Times Company. Reprinted by permission; Exercise 8.9, "Drinking in Pregnancy Harmful, Study Says," appeared in *St. Paul Pioneer Post Dispatch,* February 16, 1989, p. 19A. © 1989 by The New York Times Company. Reprinted by permission; Exercise 8.12, "Findings on Aspirin Confirmed in Report," *New York Times,* July 20, 1989, p. 18. © 1989 by The New York Times Company. Reprinted by permission; Exercise 8.13, "AIDS Therapy Experiment Yields Encouraging Results," by Lawrence K. Altman, *New York Times,* August 15, 1990, p. A12. © 1990 by The New York Times Company. Reprinted by permission; Exercise 8.17, "Study Casts Doubt on Link Between Media, Teen Suicide," appeared in *Minneapolis Star Tribune,* November 23, 1989, p. 5. Reprinted by permission of The Associated Press; Exercise 8.18, "High-fat Diets in Pregnancy May Lead to Future Tumors for Fetus," appeared in *Minneapolis Star Tribune,* May 25, 1989, p. 10A. Reprinted by permission from Scripps Howard News Service. 8.22, "Zinc May Slow Blindness," by Gina Kolata, appeared in *Readers Digest,* July, 1988, p. 139. © 1988 by The New York Times Company. Reprinted by permission; Exercise 8.23, "Drug is Found to Cure Some With a Deadly Liver Disease," by Gina Kolata, *New York Times,* August 2, 1990, p. 1. © 1990 by The New York Times Company. Reprinted by permission; Exercise 8.24, "Margarine, Too, Is Found to Have the Fat That Adds to Heart Risk," by Jane Brody, *New York Times,* August 16, 1990, p. A12. © 1990 by The New York Times Company. Reprinted by permission.

Index

* Asterisks following page references indicate material in exercises.